Planetary Gear Trains

Planetary Gear Trains

Kiril Arnaudov and
Dimitar Petkov Karaivanov

CRC Press
Taylor & Francis Group
Boca Raton London New York

CRC Press is an imprint of the
Taylor & Francis Group, an **informa** business

CRC Press
Taylor & Francis Group
6000 Broken Sound Parkway NW, Suite 300
Boca Raton, FL 33487-2742

Printed on acid-free paper

International Standard Book Number-13: 978-1-138-31185-5 (Hardback)

Library of Congress Cataloging-in-Publication Data

Names: Arnaudov, Kiril Borisov, author. | Karaivanov, Dimitar Petkov, author.
Title: Planetary gear trains / Kiril Arnaudov and Dimitar Petkov Karaivanov.
Other titles: Planetni zæubni predavki. English
Description: Boca Raton : Taylor & Francis, a CRC title, part of the Taylor & Francis imprint, a member of the Taylor & Francis Group, the academic division of T&F Informa, plc, [2019] | Includes bibliographical references and index. |
Identifiers: LCCN 2018055498 (print) | LCCN 2018058849 (ebook) | ISBN 9780429857997 (Adobe PDF) | ISBN 9780429857973 (Mobipocket) | ISBN 9780429857980 (ePub) | ISBN 9781138311855 (hardback) | ISBN 9780429458521 (ebook)
Subjects: LCSH: Gearing, Planetary. | Gearing, Planetary—Design and construction.
Classification: LCC TJ204 (ebook) | LCC TJ204 .A76126 2019 (print) | DDC 621.8/33—dc23
LC record available at https://lccn.loc.gov/2018055498

Visit the Taylor & Francis Web site at
http://www.taylorandfrancis.com

and the CRC Press Web site at
http://www.crcpress.com

Contents

Part IV Compound Planetary Gear Trains

From the Preface to the Bulgarian Edition

Planetary gear trains make up an extremely large technical field. The theory behind them is complex, with many unexpected challenges, while the ways they function are not obvious and easy to understand.

Thanks to their advantages and capabilities, planetary gear trains usage in machinery has expanded over the past years.

The presented book will discuss the following:

- Theory
- Calculations and design
- Load capacity and durability
- Manufacturing and quality control
- Applications of different types of elementary (simple, single-carrier) and complex compound (multi-carrier) planetary gear trains

Because planetary gear trains have a reputation for being complex and hard to understand—for some, they are borderline mystical—the authors have set themselves the following objectives for this book:

- *The structure* is logical, starting from elementary planetary gear trains to complex, compound gear trains. All the information about a certain type of gear is given in one place to help the gear train designer.
- *The information* treats all problems of various types of planetary gear trains so that a designer can receive complete clarity on them and creatively understand them instead of mechanically calculating and designing these gear trains.
- *The explanations* are succinct, systematic, easy to understand, and reader-friendly.
- *The visualization* is as complete as possible because it is of paramount importance for the engineer.

Although the book is addressed primarily to engineers who design, it can be used by professors, students (future engineers!), as well as by PhD students who can draw ideas and inspirations for their own research. In this aspect, the extensive cited literature will be of help.

The book reflects not only the teaching but also the practical experience of the authors. It was developed under the motto *"From practice to practice."*

Kiril Arnaudov and Dimitar Petkov Karaivanov
May, 2017

Preface to the American Edition

The American edition of this book is a translation of the Bulgarian original after being thoroughly revised. Small additions and improvements have been made in some places and appropriate omissions—at others.

The area of gearing and planetary gear trains in particular is very specific and more often than not the different countries have different terms and idioms. The authors and the translator tried to

- Preserve as much as possible not only the structure but also the style of the Bulgarian edition
- Use the terms that are agreed in the international standards [84,112–114]
- Comply with the North American vocabulary used in engineering [5,58,102,123], trying to find the best parallels with the European one [120,208]

This was not easy to do in some cases, and at times, synonyms were used in sequential sentences by referring to the sources (with regard to the planetary gear trains, we adhered to the American edition of Herbert W. Müller's book [166]). We truly hope that this will not impede the easy reading (perception) of the text and the American edition will preserve the typical characteristics of the Bulgarian one for which the authors strived—a succinct, systematic, easy-to-understand, and reader-friendly presentation.

Kiril Arnaudov—author
Dimitar Petkov Karaivanov—author and translator
Galina Koteva—translator

Acknowledgment

The authors would like to express their deepest gratitude to

- The CRC Press team that prepared this book for publication—Jonathan Plant, Sarah Head, Glenon C. Butler Jr., Kritheka, and Esther Saks
- Dr. Eng. Alexander Kapelevich for encouraging them to become involved in the American edition and for the invaluable help with terminology
- The translator Galina Koteva who helped to render the Bulgarian word order into English
- Eng. Georgi Stoimenov and Podemcrane AD, who sponsored the publication of the Bulgarian edition as well as its translation into English
- Prof. Dr. Marin Guenov and Eng. Vladimir Tzanov for advising about the terminology used in their scientific and engineering field
- The gear experts and colleagues from Balkan Association of Power Transmissions, with whom it was a pleasure to exchange experience and ideas at numerous scientific conferences
- The reviewers and readers of the Bulgarian edition, whose positive feedback gave them hope that the publication of the book in the United States is justified
- Last but not least, to their families and partners for the patience, understanding, and support without which not only this book but also other things would not be possible.

Originally published as

АРНАУДОВ, К., Д. КАРАИВАНОВ. *Планетни зъбни предавки*. София, Издателство на БАН „Проф. Марин Дринов", 2017, 368 с. ISBN 978-954-322–885-0.

ARNAUDOV, K. and D. KARAIVANOV. *Planetary gear trains*. Sofia: Bulgarian Academy of Sciences Publ. "Prof. Marin Drinov", 2017, 368 p., ISBN 978-954-322–885-0. (in Bulgarian)

Translated from Bulgarian:

Galina Koteva and Dimitar Petkov Karaivanov

Authors

Dr. Kiril Arnaudov holds a Master's Degree in Mechanical Engineering from the Technical University in Sofia (Bulgaria) and a PhD from the Technical University of Munich (Germany). His doctoral thesis (in planetary gear trains) has been cited over 20 times and translated in Russian. Having started as a designer and technologist, long years he served as a professor at TU-Sofia and as a research professor at the Institute of Mechanics of Bulgarian Academy of Sciences (BAS).

Dr. Arnaudov is the author of over 120 publications, including 4 monographs, 33 scientific publications on planetary gear trains, conference papers in 18 countries of America, Asia, Australia, and Europe, 4 patents on planetary gear trains—3 BG and a German one, 2 textbooks on Machine elements (in Bulgarian). Over 100 citations are noticed.

Dr. Arnaudov is:

- National expert of Bulgarian Scientific-Technical Union of Mechanical Engineering in Gear Trains and Power Transmissions.
- Initiator, founder and first President of the Balkan Association of Power Transmissions and Honor-President now.
- Former representative of Bulgaria in IFToMM and ISO Techn. Comm. 60 "Gearing".
- Member of Scientific/Program Committees of numerous conferences (incl. of the Board of Gear Excellences of the International Conference of Gears, Munich, Germany).
- Consultant of gear producers in Bulgaria, Russia, Switzerland.

In 2003, he was nominated for Corresponding member of BAS.

Dr. Dimitar Petkov Karaivanov holds a Master's Degree in Mechanical Engineering from the Technical University in Sofia (Bulgaria). He served as a designer in electric forklifts plant. His pedagogic work started in 1985 at the University of Chemical Technology and Metallurgy (UCTM)—Sofia, Bulgaria where he has been working until now. At UCTM he had his Dr. (PhD) degree (in 2000) and associate professor (Docent) rank (in 2002) in "Machinery and Machine Elements". Since 2015, he has been lecturing in the Faculty of Fire Safety and Civil Protection at the Academy of Ministry of Interior of Bulgaria. He has been a guest-lecturer in several Kazakhstan universities and co-mentor of 6 graduated PhD students (from Croatia and Kazakhstan). Dr. Karaivanov is the author of more than 90 publications, including 2 monographs and 2 BG patents. Over 60 citations are noticed.

Also Dr. Karaivanov is:

- Expert in Bulgarian Institute of Standardization (BIS).
- National expert of Bulgarian Scientific-Technical Union of Mechanical Engineering in Gear Trains and Power Transmissions.
- President of the Balkan Association of Power Transmissions since Oct. 2016.
- Reviewer of several international scientific journals and member of the Editorial Board of five of them.

Part I

General

1

Introduction

Owing to their positive features and various applications, the planetary gear trains (PGTs) are nowadays increasingly widely used in the different fields of engineering and in particular in the mechanical engineering sector:

- In both the largest and the smallest powers, such as powerful ship reducers or multipliers in wind turbines, as well as in miniature clock mechanisms.
- Also for the largest and smallest speed ratios.
- As gears with the highest efficiency and as gears with the required lowest efficiency, or self-locking.

1.1 Features and Capabilities of PGTs

PGTs have a number of advantages and applications [30, 41, 46, 54, 99, 102, 104, 121, 147, 149, 155, 158, 161, 166, 167, 179, 194, 200, 224, 236, 239, 240, 259].

They are distinguished for being very compact; i.e., they have small dimensions and low weight—two to three (5) times lower than the common non-planetary gear trains thanks to the adoption of the *multi-flow principle*; i.e., several planets are used to split the power flow. The diminished dimensions (Figure 1.1) have a number of beneficial consequences:

- Reduced material consumption and a light construction, respectively.
- A small mass moment of inertia, important for fast-paced drives.
- The diminished dimensions of the gears allow for both heat treatment and achieving higher accuracy in their production, which combined with the lower pitch line velocity leads to lower internal dynamic loads and to a quieter operation of the gear train, which is particularly important nowadays.
- Due to their compactness, the required smaller gear train bearer is important in some cases. Especially for lifting equipment, such as bridge cranes, this accounts for a substantial lightening, not only for the trolley but also for the entire construction, and hence reduction in the price.
- Another substantial advantage is the very high efficiency of some PGTs and vice versa—the possibility of self-locking when efficiency is low.
- The coaxiality of the input and output shaft also has advantages in some cases (e.g., vehicles, wind turbines, and aircraft engines).

The PGTs allow for the rational solution of a number of problems in the area of machine drives. They offer new layout possibilities that do not exist with the other types of non-planetary gear trains. They are used as follows:

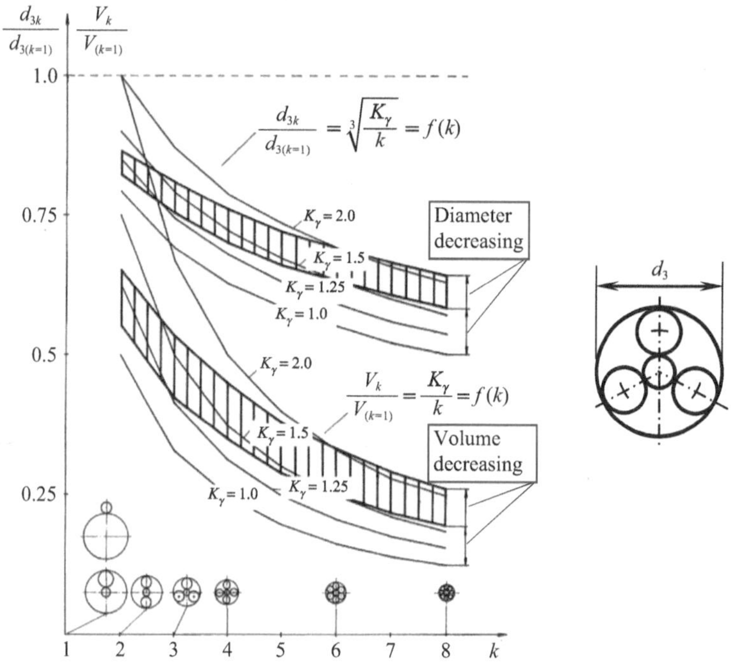

FIGURE 1.1
Constructive volume V and diameter d_3 of the planetary gears depending on the number of the planets k and mesh load factor K_γ (the uneven distribution of the load between planets is taken into account; see Chapter 8.)

- Gear trains, both with $F = 1$ and with $F = 2$ degrees of freedom
- Reducers or multipliers
- Differentials, i.e., power division or power summing gear trains
- A part of systems to make a stepless change of the angular velocity
- Change-gears (gearboxes) in vehicles—cars, buses, tugboats, tractors, tanks, etc.
- Reversing gears in ships, locomotives, etc.

1.2 Difficulties in Using PGTs

The application of the PGTs despite the advantages has, on the other hand, some *shortcomings* and *difficulties.*

Their theory is more complex than the one of the non-planetary gear trains. These are the processes that run inside the train, and hence problems such as differentiating the types of internal power—absolute, coupling, and relative (rolling) power (see Chapter 10), internal division and internal circulation of power, and load sharing between planets. These processes, especially in the complex compound planetary gears, are not so clear and easy to understand and have contributed to the reputation of PGTs as something complicated and difficult to understand. Some unsuccessful technical solutions and failures, apropos, have also contributed to this reputation. This leads to the difficulties with the accurate determination of the loads as a prerequisite for the proper calculation of the gear train elements, and the difficulties with the correct determination of efficiency, which is crucial for some cases. All in all, the theory and practice of planetary gears have quite a lot of "pitfalls."

The great compactness of PGTs, which is itself a considerable advantage, otherwise may mean a reduced cooling surface, which in some cases leads to difficulties in heat removal and complicated and costly arrangement due to forced lubrication and cooling.

Kilogram price of PGTs is higher than that of non-planetary ones. However, due to their lower weight, with a successful design, the cost of PGTs may get lower eventually.

It should also be noted that planetary gears require a higher precision of manufacturing.

There is also the danger of complete destruction of planetary gears when a single tooth is broken, which unlike the non-planetary gears cannot be discarded into a safe place.

All this means that the design and production of PGTs must be executed with extreme responsibility.

It should be taken into consideration that the number of the different types of PGTs is relatively large and this fact alone makes it hard for the designer to select a suitable gear train type. In addition, tooth geometry causes some difficulties as well.

1.3 Possibilities for Application of PGTs

PGTs have been known for a long time, mainly as clock mechanisms, but their industrial application began only at the end of the 18th century. This application increased during the 19th century and developed rapidly in the 20th century, and up to the present times. Interestingly, at the end of the 18th century, James Watt used a planetary gear train to drive his steam engine (Figure 1.2), because the crankshaft was then patented and he could not make use of it [155].

The chariot used in China 26 centuries BC [167] is also interesting, always pointing at the same geographic direction (to the south) regardless of the winding road it moved along

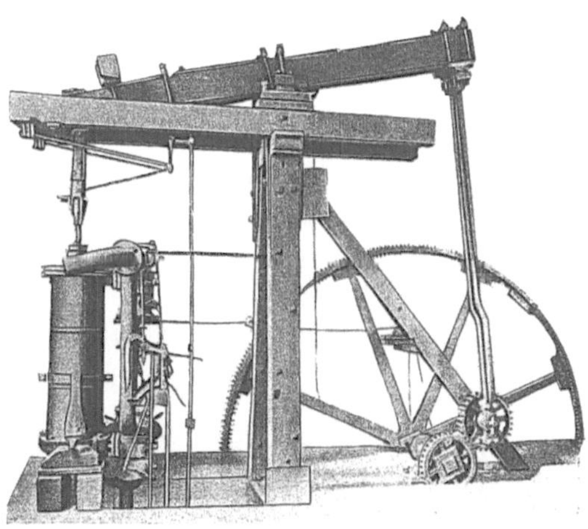

FIGURE 1.2
James Watt steam engine from the end of the 18th century with a planetary mechanism to convert the translational motion into rotational one.

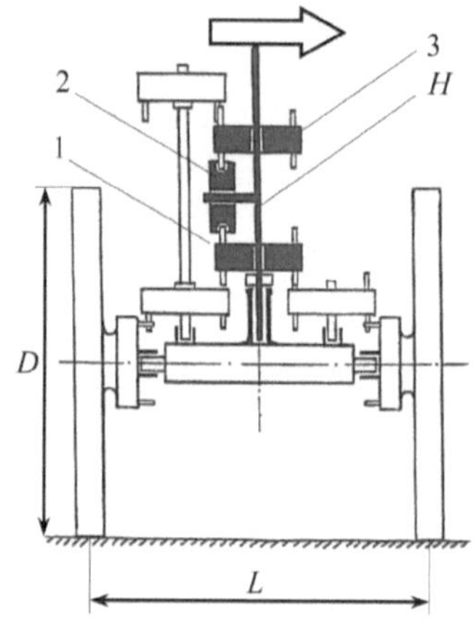

FIGURE 1.3
Chinese south-pointing chariot of the 26th century BC, indicating the geographical direction (always pointing south) when in motion (reconstruction by description of the 11th century.)

(Figure 1.3). It included a differential gear train and wheels in the most primitive form, which may be called gears conditionally. They had pins instead of teeth. The differential gear train shown in Figure 1.3 consists of "gear wheels" 1, 2, and 3 and the carrier H, onto which the south-pointing figure (arrow in the Figure 1.3) is fixed. Figure 1.3 shows the construction described in the 11th century AD, where the running wheels diameter is equal to the distance between them ($D = L$).

Even in a small country like Bulgaria, the PGTs find application in

- Hoists (Figure 1.4)
- Forklifts

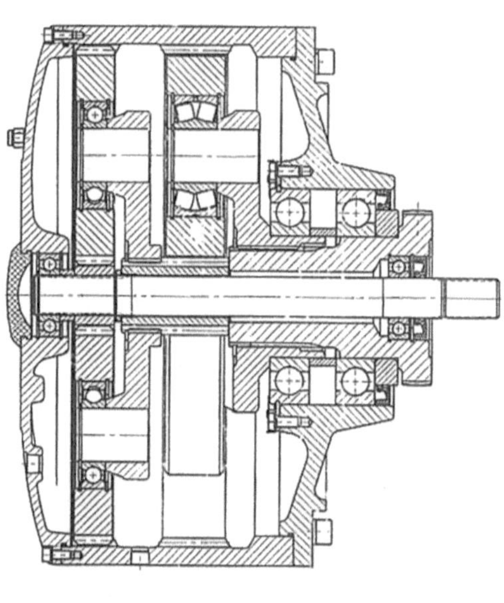

FIGURE 1.4
Planetary reducer for Bulgarian hoists. (Courtesy of Podemcrane AD, Gabrovo, Bulgaria.)

- Shipbuilding, in a variety of deck mechanisms
- Metal cutting machines
- Heavy machine building industry, etc.

Figure 1.5 shows a PGT with 20 planets, which is used as a reducer in aircraft construction, and Figure 1.6—a helicopter reducer.

PGTs for very large power are used in the cement industry (Figure 1.7).

Figure 1.8 shows a PGT, which is used as a gearbox in forklift construction.

Nowadays, the automatic gearboxes which involve PGTs are widely used in automotive industry.

FIGURE 1.5
Aircraft planetary reducer.

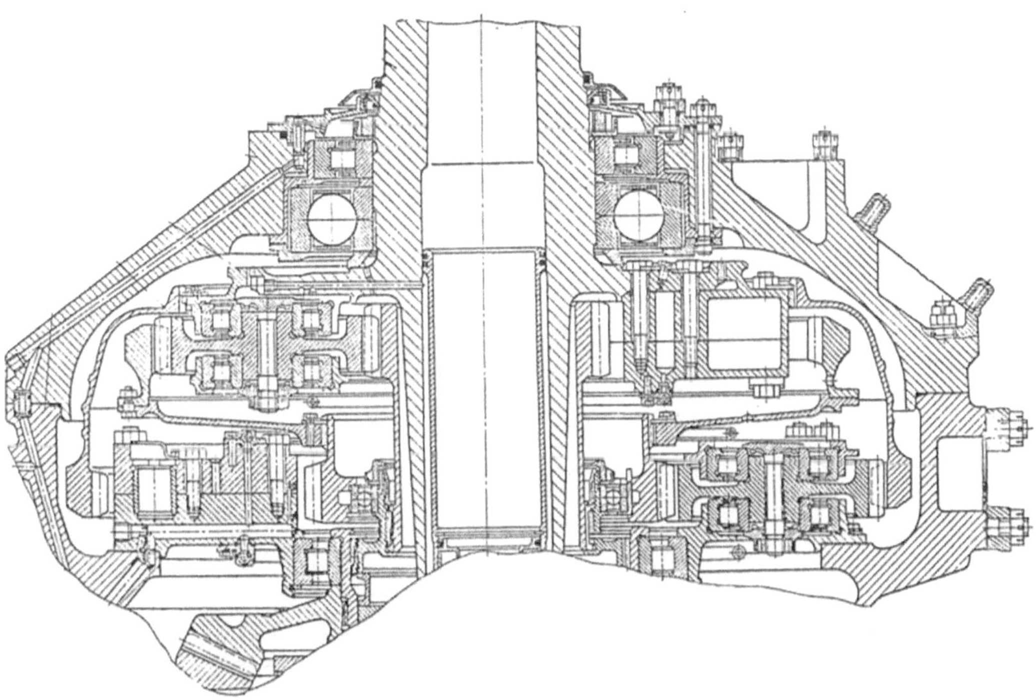

FIGURE 1.6
PGT as main reducer in helicopter.

FIGURE 1.7
PGT from the drive of a cement mill.

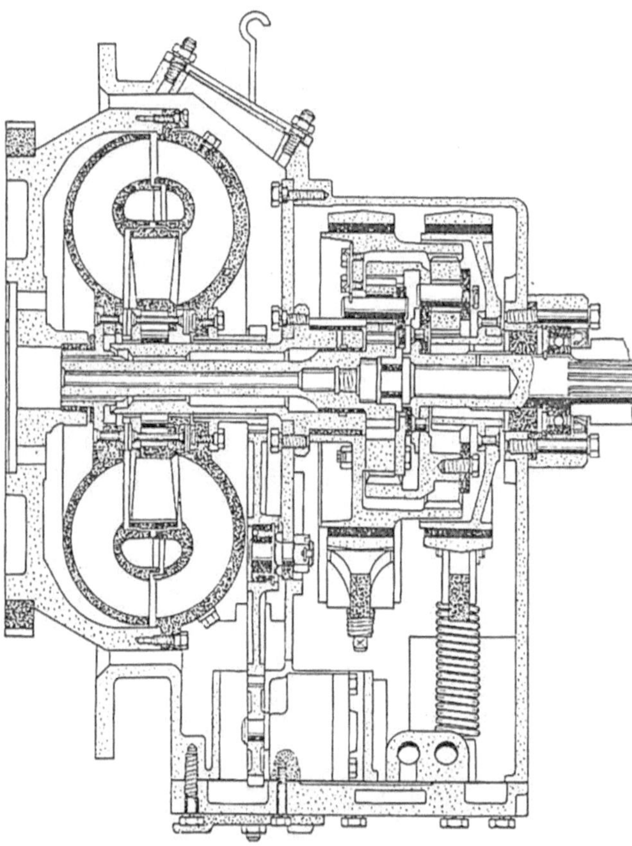

FIGURE 1.8
Planetary change-gear of a diesel forklift. (Courtesy of Georgy Stoilov.)

The aim of the authors is to provide the reader with a concise, compact, and orderly layout of the material, and above all, of the most commonly used types of simple PGTs, as well as some other simple and complex compound PGTs. This book particularly emphasizes the clear presentation of difficult and intricate questions and especially their visualization, which is the dominant requirement for the way an engineer should think.

2

Nature, Purpose, and Types of Planetary Gear Trains

2.1 Nature and Designation of Planetary Gear Trains

Planetary gear trains (PGTs) are most often coaxial gear trains and extremely rarely non-coaxial gear trains (see Chapter 27). They are designed in such a way that they possess at least one gear wheel, most often more than one, and these are called *planets*, mounted on the so-called *carrier*, which perform a double rotation—around their geometrical axis and together with the same axis *around the main (the central) geometrical gear train axis*. Thus, the complicated motion of the planets comprises a vector summation of *both rotations*—the *revolution* performed by the carrier H and the *relative* one (spin), performed by each planet 2 with respect to the carrier:

$$\overline{\omega_2} = \overline{\omega_H} + \overline{\omega_{2\,rel}}. \tag{2.1}$$

Figure 2.1 visualizes the most common case of a cylindrical PGT.

This typical planet motion explains the name of these gear trains, whose motion resembles the motion of the planets around the Sun, and this is *their typical distinctive kinematics characteristic*. Planets are meshing with one or two central gear wheels with external or internal teeth.

Unlike the non-PGTs, the PGTs, which are cylindrical, bevel, worm, and crossed-helical, encompass a comparatively extensive technical field. Only the most commonly used cylindrical PGTs are presented in this book.

About 20 types of trains can be identified as *simple (elementary) single-carrier PGTs* [167]. These trains of combinatorial principle generally build more complex *compound* or *higher PGTs* that are multi-carrier—two-, three-, four-carrier, etc. Their number is enormous—several thousand.

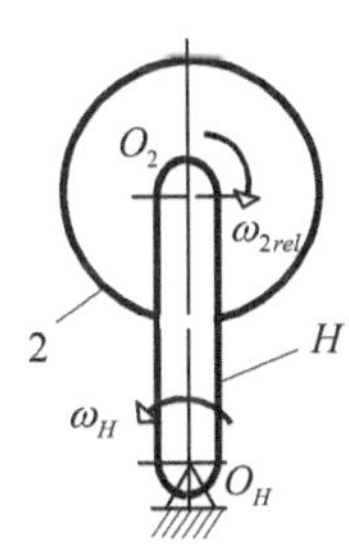

FIGURE 2.1
Revolution and relative rotation (spin) of the planets.

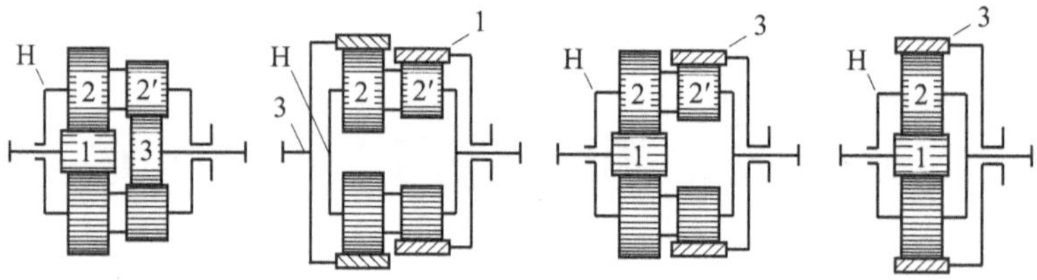

FIGURE 2.2
Examples of some simple (elementary) PGTs: 1—small central gear wheel; 3—big central gear wheel; 2 and 2′—planet (with one or two rims); H—carrier.

As *simple (elementary) single-carrier PGTs* are meant the gear trains, possessing the following *mandatory characteristics*:

1. The existence of only *one carrier* H (Figure 2.2), which generally rotates and must have an outlet; i.e., an external torque acts upon it. At fixed carrier ($\omega_H = 0$), the *term pseudo*-PGT is used.
2. The existence of one or more gear wheels with movable axes of rotation, called planets 2, that rest in the carrier H and perform the complicated rotation motions in question.
3. The existence of one or two (not more!) *central gear wheels* 1 and 3.
4. The existence of three shafts, coming out of the gear train and loaded with torques, which means the existence of a *three-shaft gear train*. It does not matter if some of the three shafts are immovable (fixed), or the three of them are movable. In the first case, the train works with $F = 1$ degree of freedom and in the second one—with $F = 2$ degrees of freedom.

Different ways exist to denote various types of PGTs, but unfortunately, there is not a uniform and internationally accepted classification. In this book, such characteristic features of the gear trains are used, namely the type of meshing (proposed by Prof. Tkachenko [236]). The following indications (letters) are used for the PGTs, as shown in Figures 2.3 and 2.4:

A—for external meshing

I—for internal meshing

At PGTs, using one-rim planets, a line is placed above the corresponding indications, whereas the indication of the gear train with two-rim planets is without a line.

2.2 Purpose of PGTs

Unlike the non-PGTs, the PGTs in general have considerably more capabilities and various applications. They are designed to

- Perform not only constant speed ratio $i = \text{const}$, like the non-planetary gear trains but also step or stepless change of speed ratio, i.e., $i \neq \text{const}$, in the gearboxes.

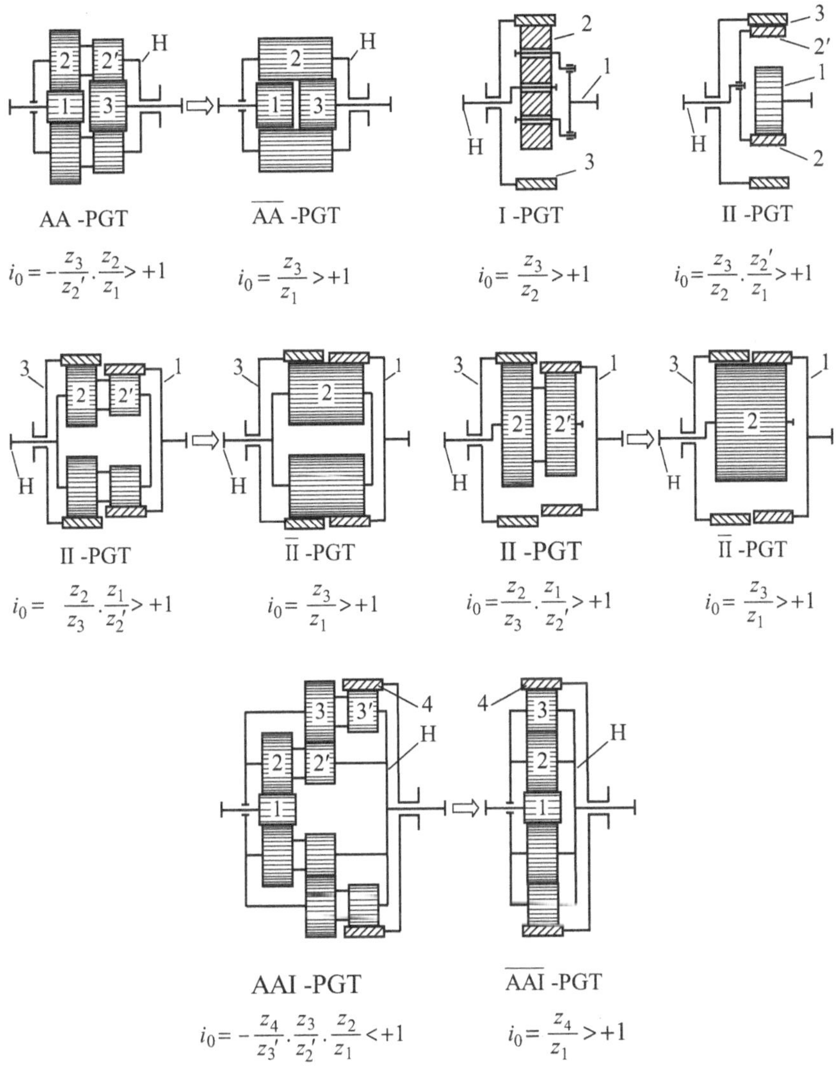

FIGURE 2.3
Positive-ratio simple cylindrical PGTs ($i_0 > +1$) (about i_0 of II- and $\overline{\text{II}}$-gear train; see Chapter 23).

- Perform summation or division of motion.
- Change the direction of rotation, i.e., reversing.

2.3 Types of Simple PGTs According to the Basic Speed Ratio

Figures 2.3 and 2.4 show only the most frequently used cylindrical PGTs from the big number of simple meaning single-carrier PGTs. The planets of the cylindrical PGTs can be either with one or with two rims, as it can be seen from the figures. The second one is less

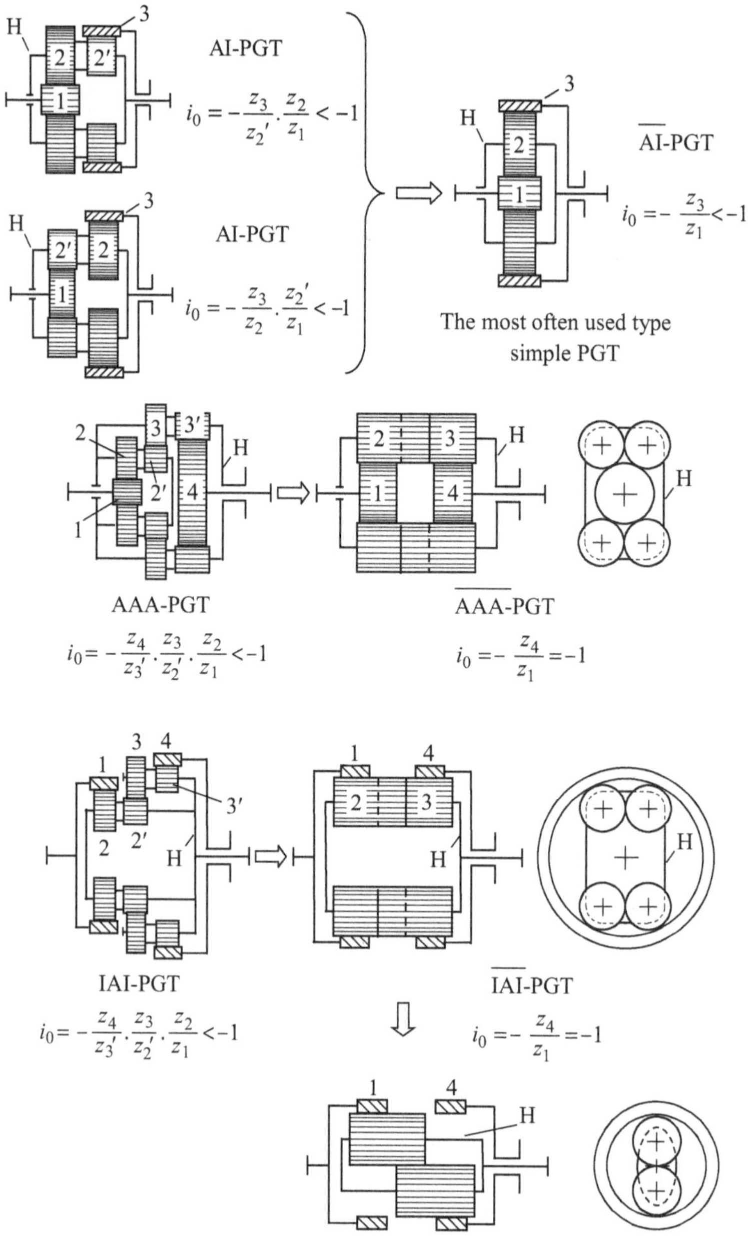

FIGURE 2.4
Negative-ratio simple cylindrical PGTs ($i_0 < -1$).

favorable from a technological point of view, because it requires the exact mutual disposal of both rims of all the planets of a given PGT and there are requirements for their assembly as well (see Chapters 21–23). Besides, it influences the uneven load distribution among the planets. These two variants are presented in Figures 2.3 and 2.4 for series of trains [162]. As it is also shown in Figures 2.3 and 2.4, the simple PGTs form two groups according to their *basic speed ratio* i_0, which two groups have very substantial differences in their kinematics behavior. The basic speed ratio i_0 represents the speed ratio of both central gear wheels 1 and 3 or 4 (meshed with the planets 2) in their rotation with relative angular velocities ω_{1rel} and ω_{3rel} or ω_{4rel}, with respect to the carrier H. Specifically, for the gear trains in Figure 2.2, the relation is

$$i_0 = \frac{\omega_{1\,rel}}{\omega_{3\,rel}} = \frac{\omega_1 - \omega_H}{\omega_3 - \omega_H} = \left\{ \begin{array}{l} \frac{z_3}{z_2'} \cdot \frac{z_2}{z_1} \\ \text{or} \\ \frac{z_3}{z_1} \end{array} \right\} = \left\{ \begin{array}{l} > 0 \\ \text{or} \\ < 0 \end{array} \right. . \tag{2.2}$$

In accordance with this speed ratio i_0, some of the simple PGTs where the speed ratio is positive, that means $i_0 > 0$, are denoted as *positive-ratio gear trains* (Figure 2.3), and the other group of simple PGTs, where $i_0 < 0$, are denoted as *negative-ratio gear trains* (Figure 2.4). The most often used types of simple PGTs according to this characteristic are therefore negative-ratio gear trains shown in Figure 2.4, and only they will be considered in this book. The other types of PGTs shown in Figure 2.3, only with external or only with internal meshing, are positive-ratio gear trains such as AAI-PGT. At the positive-ratio PGTs, central gear wheels 1, 3, or 4 have the same directions of rotation with respect to the carrier H, as with the negative-ratio PGTs—opposite directions of relative rotation.

It is necessary to emphasize explicitly that in this case, the terms "positive" and "negative" gear trains have nothing to do with the profile shift modifications, where the terms positive and negative gears are also used [175].

2.4 Possible Ways of Working of PGTs

As mentioned in Section 2.1, each simple PGT, whether positive-ratio or negative-ratio one, possesses three shafts, coming out of the gear that means it is a three-shaft gear train. Two PGTs possessing one and the same basic speed ratio i_0, regardless of their different kinematics scheme, can be looked over like black boxes (Figure 2.5), which are kinematics equivalent. It should be mentioned that the three external shaft torques of all the simple

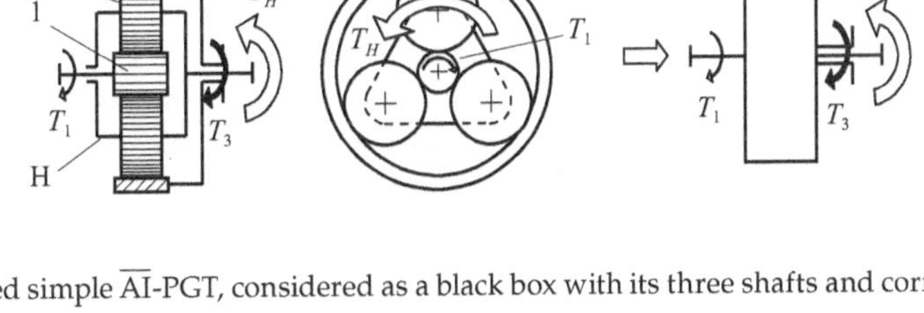

FIGURE 2.5
The most often used simple $\overline{\text{AI}}$-PGT, considered as a black box with its three shafts and corresponding external torques.

PGTs are in equilibrium. Taking into account the torques of the most often used type of PGT ($\overline{\text{AI}}$-PGT from Figures 2.4 and 2.5), this condition is

$$\boxed{\sum T_i = T_1 + T_3 + T_H = 0} \tag{2.3}$$

This condition is in power regardless of the PGT's way of work:

- Regardless of whether it works with $F = 1$ or $F = 2$ degrees of freedom
- Which one of the gear train elements is fixed at $F = 1$ degree of freedom
- Regardless of the direction of power transfer, i.e., if the PGT works as a speed reducer or multiplier at $F = 1$, or as a differential at $F = 2$ degrees of freedom (as a summation or division gear train);
- Regardless if the PGT works separately, or as a part of a compound, multi-carrier PGT.

As is clearly shown in Figure 2.5, two of the torques T_1 and T_3 are unidirectional and due to this are marked with single lines, but with different thickness with respect to the size of the torques, and the third torque T_H is opposite. It is the biggest one, equal to the absolute value of the sum of the other two torques, and is marked by a double line [254]. This can be expressed by the following inequality:

$$T_1 < T_3 < |T_H|. \tag{2.4}$$

At the other types of PGTs, the arrangement of the torques in the inequality is different (see Chapters 22–24). It is very important to know that simple single-carrier PGTs, as well as the compound, multi-carrier PGTs, can work in accordance with the degrees of freedom $F = 1$ and $F = 2$ in four different ways:

At $F = 1$

- As a speed reducer
- As a speed multiplier

At $F = 2$ (as a differential)

- As a summation PGT (e.g., at a twin-motor drive)
- As a division PGT

This means that the terms *reducer, multiplier,* and *differential* do not in any way mean different gears designs. They are not a design (arrangement) but an *operating characteristic* of a gear train; i.e., they characterize how it works.

Here, we should mention another common misconception that the term "epicyclic gear train" is a general term, and the term "PGT" is a specific term. As it is shown in Figure 2.6, it is exactly the opposite situation [21, 36]. Depending on the movement of the planets, the epicyclic, hypocyclic, and pericyclic gear trains shown on the figure are undoubtedly specific cases of the more general term "PGTs," which is a concept of higher rank, because PGTs always have the common characteristic kinematic feature—the planetary motion of the planets—around the main (the central) axis of the gear train.

In the figures of this book, P_A means input power and P_B—output power.

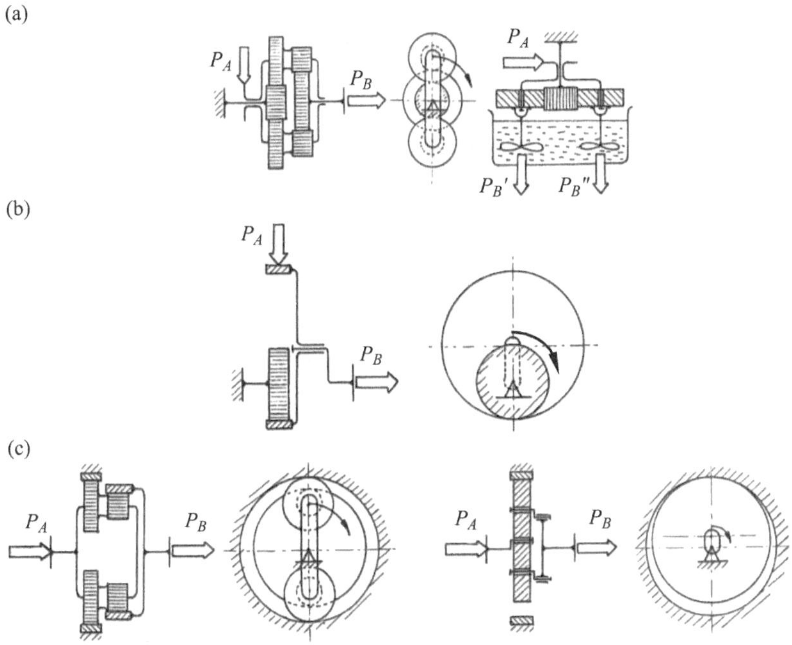

FIGURE 2.6
Types of PGTs according to planets movement: (a) Epicyclic; (b) pericyclic (Wankel engine); (c) hypocyclic.

2.5 Types of PGTs According to Different Features

PGTs differ in the following features:

1. According to the carrier number
   - Single-carrier, i.e., simple (elementary) PGTs (Figures 2.3, 2.4, and 2.6)
   - Multi-carrier, i.e., compound PGTs (Figures 2.7–2.11)
2. According to the component PGTs number
   - Two-carrier compound (coupled) PGTs or lower compound PGTs (Figure 2.7)
   - Multi-carrier compound PGTs or higher compound PGTs—three-carrier, four-carrier, etc. (Figures 2.10 and 2.11)
   - Reduced PGTs (Figure 2.8), where the two carriers merge into one
3. According to the gear wheels type
   - Cylindrical
   - Bevel
   - Worm
4. According to the gears tooting and meshing
   - With external meshing
   - With internal meshing
   - With both external and internal meshing

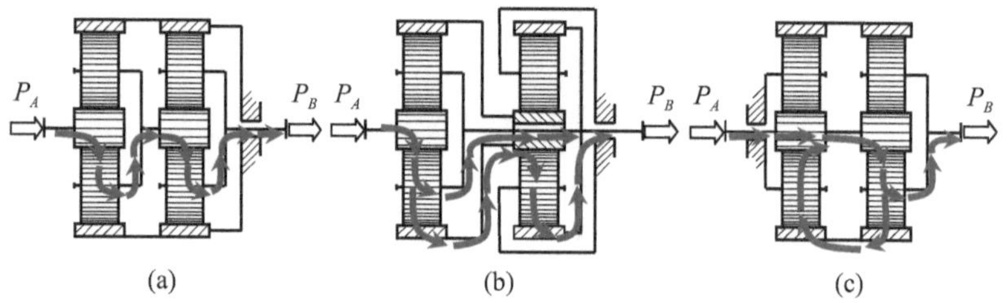

FIGURE 2.7
Compound two-carrier three-shaft PGTs: (a) Sequentially (in series) coupled component PGTs (see Figure 29.4 and Figures 35.1–35.9); (b) compound PGTs with internal division of power (A very clear distinction between the *power division between the shafts* and the *power sharing between the planets* has to be made. However, power sharing (*branching* according to [112, 166]) does not affect the operating characteristics of PGTs.) (see Figures 29.6 and 35.17); (c) compound PGTs with internal circulation of power (see Figures 29.8 and 29.9).

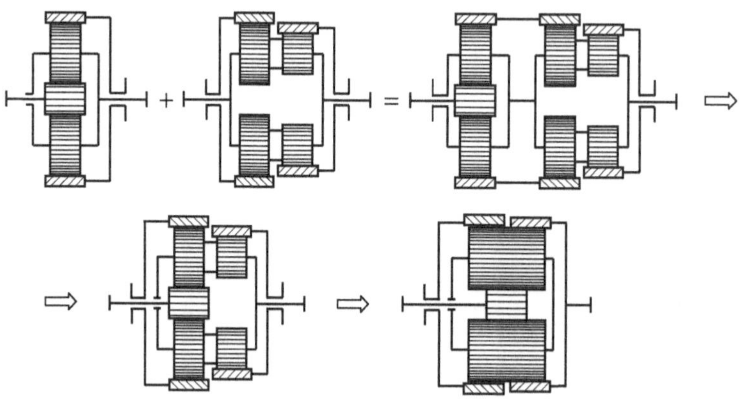

FIGURE 2.8
Reduced compound three-shaft PGT—Wolfrom gear train [255] (3*K*-PGT according to Prof. Kudryavtsev's classification [147–149]).

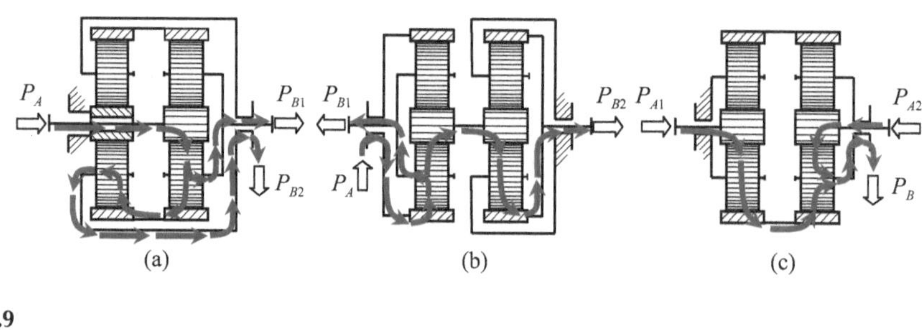

FIGURE 2.9
Compound two-carrier four-shaft PGTs (see Chapter 31): (a) and (b) Division PGT $(P_A = P_{B1} + P_{B2})$; (c) summation PGT $(P_{A1} + P_{A2} = P_B)$.

5. According to the meshing type
   - Involute
   - Cycloidal (cyclo-PGT)
   - With Novikov meshing

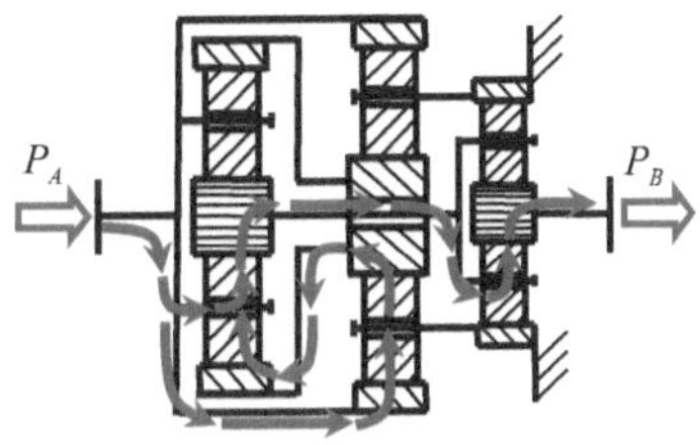

FIGURE 2.10
Compound three-carrier three-shaft PGT as a multiplier of wind turbine (see Figure 34.11).

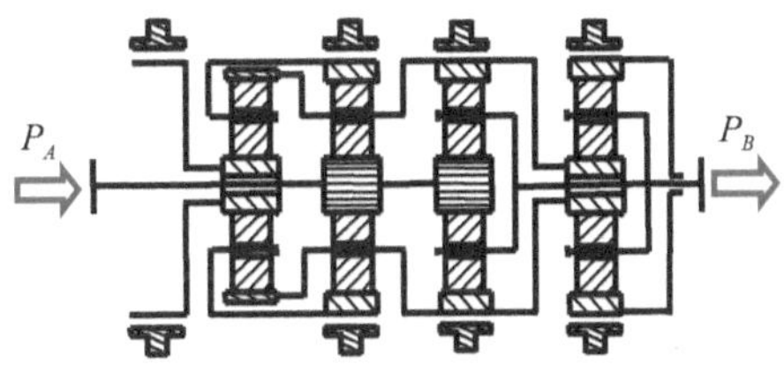

FIGURE 2.11
Compound four-carrier sixth-shaft planetary change-gear train (see Figure 34.12).

6. According to the basic speed ratio i_0
   - Positive-ratio PGTs (Figure 2.3)
   - Negative-ratio PGTs (Figure 2.4)
7. According to the external shafts number
   - Three-shaft PGTs, such as the single-carrier PGTs (Figures 2.3 and 2.4) and some of the compound PGTs (Figures 2.7, 2.8, and 2.10)
   - Multi-shaft PGTs, such as some of the compound ones (Figure 2.9—four-shaft PGTs, Figure 2.11—six-shaft PGTs)
8. According to the external shafts coaxiality
   - Coaxial PGTs—the most common (Figures 2.3 and 2.4)
   - Uncoaxial—very rare (Chapter 27)

Part II

AI-Planetary Gear Train (2K-H Gear Train)

3

Arrangement and Possible Ways of Work of $\overline{AI}$-Planetary Gear Train

3.1 Arrangement of $\overline{AI}$-Planetary Gear Train

This type of most often used negative-ratio ($i_0 < 0$, Section 2.3 and Figure 2.4) planetary gear train (PGT) (Figure 3.1) is signified in different ways. In many countries, the Kudryavtsev's designation 2K-H is popular. It means two central gears (2K) and a carrier (H) [147–149]. In this book, Prof. Tkachenko's designation is used [236] as more detailed one (A for external and I for internal meshing; see Section 2.1).

As shown in Figure 3.1, this gear train has two central gear wheels—a *sun gear* 1 with external teeth and a *ring gear* 3 with internal teeth. These two gears 1 and 3 mesh with one-rim *planets* 2 which are housed in *carrier* H. Their number most often is $k = 3$, more rare $k = 2$ or 4, but in special cases, there are PGTs with $k = 20$ planets (Figure 1.5). However, planets number k does not affect the gear train's kinematics. Increasing the planets number ($k > 1$), because of the torque sharing, leads to a few effects:

- Decreasing the overall dimensions and mass of the gear train
- Decreasing the mesh load
- Unloading the central element bearings—sun gear 1 and carrier H
- Decreasing the noise level because of the lower peripheral velocities and higher accuracy of the smaller gears

However, increasing the planets number k leads to some problems discussed next (see Chapter 8).

Central elements of the gear train (sun gear 1, ring gear 3, and carrier H) rotate around an axis—the so-called *main (central) geometrical axis of the gear train.* Typical of this simple PGT is that there are three shafts that go out of the train (external shafts). In Figure 3.1 the

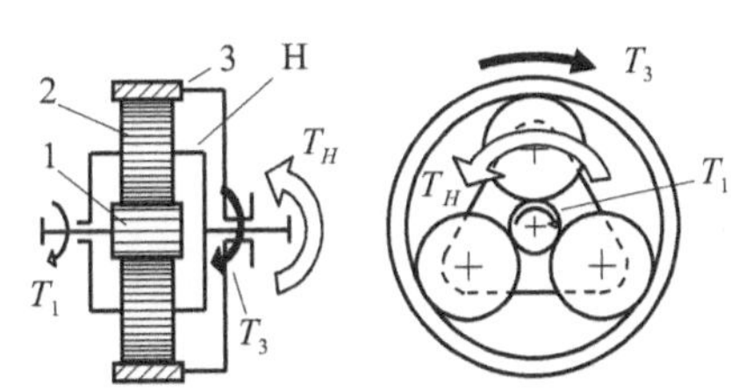

FIGURE 3.1
$\overline{AI}$- PGT (2K-H gear train).

corresponding external torques T_1, T_3, and T_H also are shown. Two of them T_1 and T_3 are unidirectional, and the third torque T_H is with opposite direction. Since the train is simple, of course, it has only one carrier, i.e., it is a single-carrier PGT.

3.2 Possible Ways of Working of $\overline{\text{AI}}$-PGT

This train, like other PGTs, can operate with both $F = 1$ and $F = 2$ degrees of freedom. With $F = 1$ degree of freedom, any one of three shafts (of the sun gear 1, ring gear 3, or carrier H) can be fixed. In Figure 3.2a and b, the six possible working modes in this case are shown—three modes as a reducer and three modes as a multiplier. Input power is denoted with P_A and output with P_B. With a fixed carrier ($\omega_H = 0$), the PGT works as pseudo-planetary. At $F = 2$ degrees of freedom (working as a differential), six working modes are possible, too (Figure 3.2c and 3.2d)—three as a summation PGT and three as a division PGT.

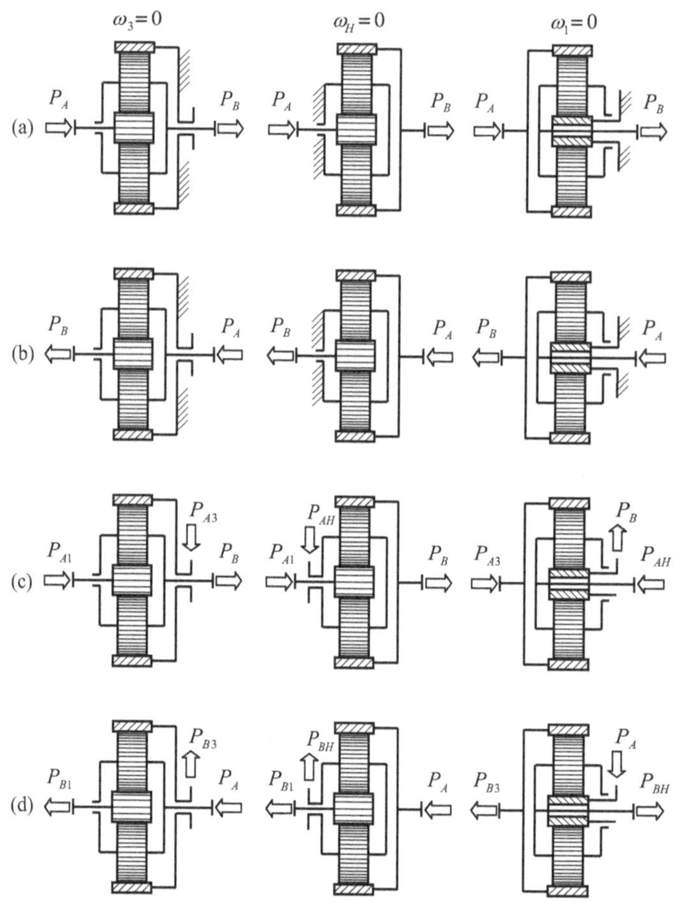

FIGURE 3.2
Working modes of $\overline{\text{AI}}$-PGTs: **With $F = 1$** degree of freedom: (a) As a reducer (reduce the speed); (b) as a multiplier (multiply the speed). **With $F = 2$** degrees of freedom (i.e., as a differential): (c) as a summation PGT; (d) as a division PGT.

All mentioned in Section 2.4, the possible modes of working of PGTs, in general, fully pertain also to the $\overline{AI}$-PGT.

From all mentioned so far, one can see what varied possibilities $\overline{AI}$-PGTs offer, which can be much more. The various examples given in Chapter 20 reveal other valuable possibilities of this train.

In this book, the $\overline{AI}$-PGT is considered relatively in detail, because it is mostly produced in the world of all other types of PGTs.

4

Specific Conditions of $\overline{AI}$-Planetary Gear Train

As for the other simple planetary gear trains (PGTs) and for $\overline{AI}$-PGT, it is necessary to observe certain conditions [5, 120, 149], namely,

- Mounting (assembly) condition
- Coaxiality condition
- Adjacent condition

4.1 Mounting (Assembly) Condition

If the PGT is with multiple planets (in principle, all PGTs are at least with $k = 2$, usually with $k = 3$ planets), the number of teeth of central gear wheels (sun gear 1 and ring gear 3) cannot be selected at random. The reason is that when a planet is mounted with an inadequate number of teeth z_1 and z_3, other planets cannot be mounted because the teeth of one wheel will stand against the teeth of the other wheel [107, 149].

For equally spaced planets (the central angle ψ between the planet centers is the same $\psi = 2\pi/k = \text{const}$), the mounting (assembly) condition can be defined as

$$\boxed{\frac{z_1 + z_3}{k} = \text{an integer.}} \tag{4.1}$$

One can get this condition as follows:

With fixed ring gear 3 ($\omega_3 = 0$) and motion transmission from sun gear 1 to the carrier H (the most frequent case of PGT operating as reducer), the speed ratio $i_{1H(3)}$ is (see Chapter 7 and formula 7.3)

$$i_{1H(3)} = 1 + \frac{z_3}{z_1}.$$

The relation between the sun gear rotation angle φ_1 and the carrier rotation angle φ_H depends on this speed ratio, which is given by

$$\varphi_H = \frac{\varphi_1}{i_{1H(3)}}, \quad \text{i.e.,} \quad \varphi_1 = i_{1H(3)} \cdot \varphi_H. \tag{4.2}$$

If sun gear 1 rotates at an angle pitch $\Delta\varphi_1$

$$\Delta\varphi_1 = \frac{2\pi}{z_1},$$

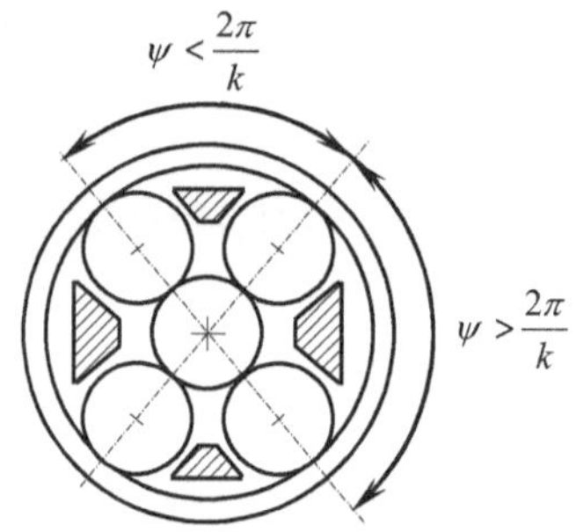

FIGURE 4.1
Asymmetric positioning of even number ($k = 4$) of planets.

the carrier rotates at angle

$$\Delta\varphi_H = \frac{\Delta\varphi_1}{i_{1H(3)}}. \tag{4.3}$$

If the carrier H rotates at angle $\varphi_H = \psi - 2\pi/k$, equal to the central angle ψ between planet centers (most often $\psi = 120°$ when $k = 3$), the sun gear 1 rotates at a few angle pitches $\Delta\varphi_1$, respectively, at angle given by

$$\varphi_1 = i_{1H(3)} \cdot \varphi_H = \left(1 + \frac{z_3}{z_1}\right)\frac{2\pi}{k} = \frac{z_1 + z_3}{z_1} \cdot \frac{2\pi}{k}. \tag{4.4}$$

In this rotation, a planet 2 moves at angle ψ. To be able to mount a new planet in the released space, it is necessary that the rotation angle of sun gear φ_1 should be equal to an integer (C) number of its angular pitches $\Delta\varphi_1$.

$$\varphi_1 = C \cdot \Delta\varphi_1 = C\frac{2\pi}{z_1}. \tag{4.5}$$

After replacing formula (4.4) in formula (4.5), the mounting (assembly) condition formula (4.1) is obtained.

It is possible to avoid this condition formula (4.1) with asymmetric positioning of planets (nonuniform angle ψ), as shown in Figure 4.1. This solution occurs in case of a load capacity increase of an already produced gear train by increasing its planet number from $k = 3$ to $k = 4$, e.g., or if the aim is to obtain a more rigid carrier by enhancing the cross section of the beams (arms) between its two sides.

4.2 Coaxiality Condition

This condition requires that the operating center distances a_{w12} and a_{w23} of both pairs of mesh gears—the sun gear 1 and the planet 2 as well as the planet 2 and the ring gear 3—should be the same in order to obtain the gear train coaxiality. For gears with standard proportions (x = 0), i.e., they roll without sliding on their reference circles (which are thus also operating circles), the equation for reference center distances a_{12} and a_{23} is

$$a_{12} = m\frac{z_1 + z_2}{2} = a_{23} = m\frac{z_3 + z_2}{2}. \tag{4.6}$$

Hence for the number of teeth is obtained

$$z_1 + z_2 = z_3 - z_2, \text{i.e., } z_1 + 2z_2 = z_3$$

or finally for the required number of teeth of the planets 2 is obtained

$$z_2 = \frac{z_3 - z_1}{2}. \tag{4.7}$$

In fact, the same result can be obtained from the sum of the reference diameters: $d_1 + 2d_2 = d_3$ or $z_1 + 2z_2 = z_3$, respectively.

However, this condition can easily be avoided by using the profile shift (see Chapter 5). Then, a smaller number of teeth is selected:

$$\boxed{z_2 < \frac{z_3 - z_1}{2}, \text{respestively, } z_2 = \frac{z_3 - z_1}{2} - \Delta z_2,} \tag{4.8}$$

where $\Delta z_2 = 0.5$; 1 or 1.5. When an integer is obtained by the formula (4.7), we choose $\Delta z_2 = 1$ and, in the case of a non-integer, choose $\Delta z_2 = 0.5$ or $\Delta z_2 = 1.5$.

When profile shifts are used, equality of the operating center distances a_{w12} and a_{w23} is obtained, but the operating pressure angles of the external α_{w12} and the internal α_{w23} meshing are different:

$$a_{w12} = m\frac{z_1 + z_2}{2} \cdot \frac{\cos\alpha}{\cos\alpha_{w12}} = a_{w23} = m\frac{z_3 - z_2}{2} \cdot \frac{\cos\alpha}{\cos\alpha_{w23}}. \tag{4.9}$$

And most often the new center distance a_w is larger than the reference one, i.e.,

$$a_w = a_{w12} = a_{w23} > a = a_{12} = a_{23}. \tag{4.10}$$

Ultimately, the coaxiality condition can be expressed as follows:

$$\boxed{\frac{z_1 + z_2}{\cos\alpha_{w12}} = \frac{z_3 + z_2}{\cos\alpha_{w23}},} \tag{4.11}$$

where the operating pressure angles α_{w12} and α_{w23} are determined by the following dependencies:

$$\boxed{\cos\alpha_{w12} = \frac{a_{12}}{a_w}\cos\alpha} \text{ and } \boxed{\cos\alpha_{w23} = \frac{a_{23}}{a_w}\cos\alpha.} \tag{4.12}$$

The reduction in the number of teeth z_2 of planets 2 is possible because it does not affect the basic speed ratio $i_0 = z_3/z_1$ and hence the speed ratios i of the PGT in its various modes of operation. This reduction Δz_2 is possible, however, only in narrow limits because of geometrical reasons—an unacceptable reduction in contact ratio $\varepsilon_{\alpha 12}$ of external meshing between the sun gear 1 and the planets 2.

4.3 Adjacent Condition

This condition is related to the premise that planets 2 lie in one plane (the most common case) and, of course, they should not interfere. As can be seen from Figure 4.2, this condition is expressed as follows:

$$\boxed{d_{a2} < 2a_w \cdot \sin\frac{\pi}{k},} \tag{4.13}$$

where d_{a2} is the tip diameter of planets 2, and a_w is the center distance.

Obviously, this condition limits the maximum possible basic speed ratio i_0 for PGT with three or more planets.

Figure 4.3 shows the maximum basic speed ratio $|i_0|_{max}$ for different number of planets k, i.e., dependence $|i_0|_{max} = f(k)$. The opposite task is also possible—looking for the maximum number of planets k_{max} that can be used for a particular basic speed ratio, i.e., $k_{max} = f(|i_0|)$.

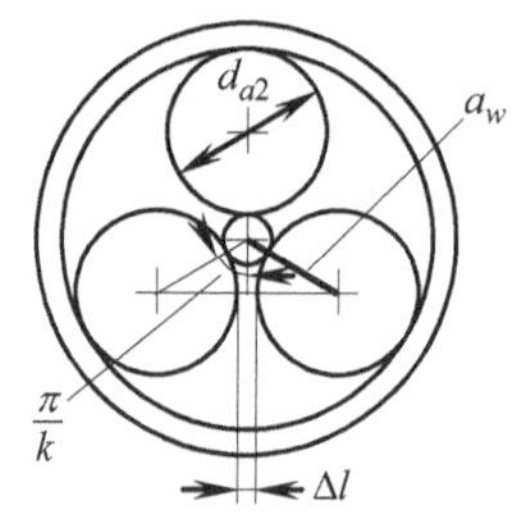

FIGURE 4.2
Planets spacing without interference (adjacent condition).

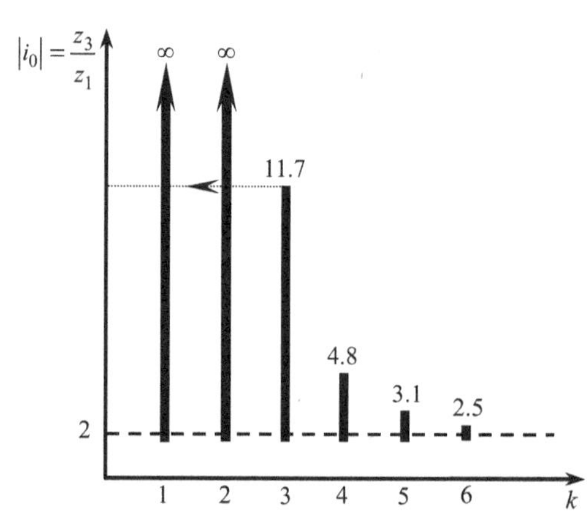

FIGURE 4.3
Maximum basic speed ratio i_0 depending on the number of planets k in case of one-plane planet spacing.

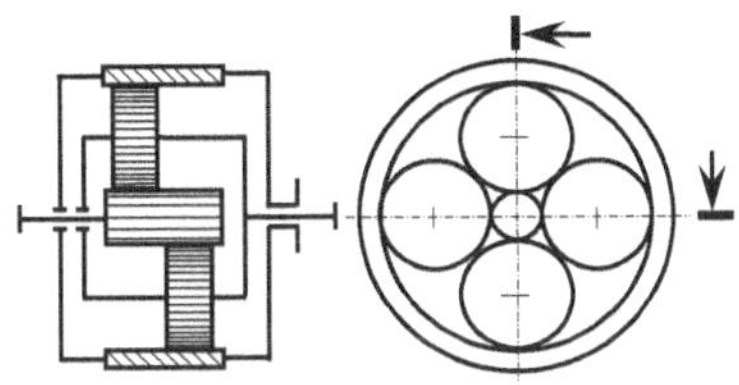

FIGURE 4.4
Displaced planets spacing in two planes to avoid their interference.

If it is possible to space planets in two planes (displaced spacing) (Figure 4.4), this third condition may also be disregarded. Of course, this can happen with an even number of planets $k = 4, 6, 8$, etc.

From the above, it can be concluded that the three specific conditions of PGTs should not be conceived as a dogma, but can be circumvented in many cases, which reveals good design possibilities.

5

Meshing Geometry of $\overline{AI}$-Planetary Gear Train and Its Peculiarities

5.1 General

In an $\overline{AI}$-planetary gear train ($\overline{AI}$-PGT), planets 2 are involved in two meshings—an external one with sun gear 1 and an internal one with ring gear 3. In this case, the gear parameters—module m, number of teeth z_1, z_2, and z_3, and profile shift coefficients x_1, x_2, and x_3, *cannot* be selected independently of each other.

In general, external meshing is the weak point of the PGT, because in case of nonhardened gears, its load capacity is lower than that of the internal gear (as a rule, this is the contact load capacity but in some cases the bending one, too). However, things are changing drastically when using a hardened sun gear 1 and planets 2.

The tooth geometry of the external and internal meshing has been examined most extensively in the textbooks on machine elements [58, 175], in monographs [46, 120, 158, 192, 208] and in standards [64, 65]. Profile shift coefficients x of the external and internal toothing can be selected within certain limits. Diagrams in Figure 5.1 show these limits [64, 65].

It is noteworthy that there is significantly more interference in internal meshing than in external one. The literature [46, 120, 171] presents the very useful, so-called *blocking contours*, which give the allowable limits of variation of profile shift coefficients x_1 and x_2 of

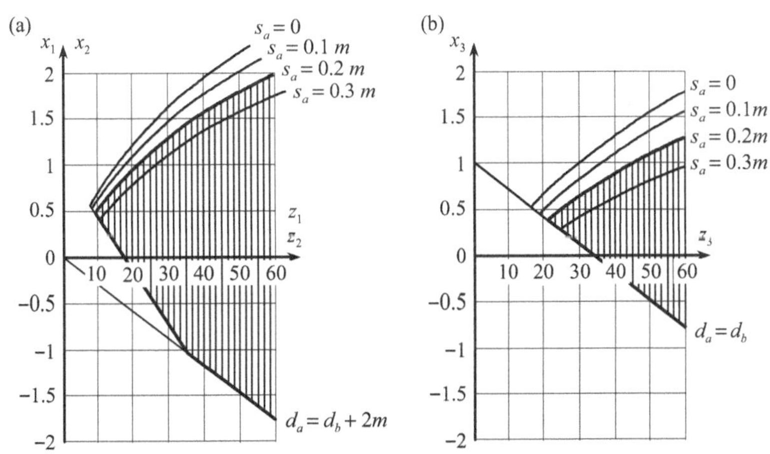

FIGURE 5.1
Certain limits of profile shift coefficients x_1, x_2, and x_3 [64, 65]: (a) External toothing (sun gear 1 and planets 2); (b) internal toothing (ring gear 3).

external, respectively, x_2 and x_3 of internal meshing, about the concrete combination of the number of teeth z_1 and z_2 (z_2 and z_3, respectively), and the number of teeth z_0 of used shaper cutter. Nevertheless, it is advisable not to go beyond the following moderate limits for coefficients x_1, x_2, and x_3 [65, 175] for reasons explained further in the text (see Section 5.5).

$$0 < x_1 < +0.65$$

$$-0.65 < x_2 < +0.65$$

$$0 < x_3 < +0.65.$$

In this book tooth geometry of spur gears ($\beta = 0$) is considered only. It is determined by the following parameters:

Module m
Number of teeth z
Profile shift coefficient x

as well as by the parameters of the basic rack:

Profile angle α
Addendum coefficient (factor) h_a^*
Clearance coefficient (factor) c^*
Coefficient of root fillet radius ρ_f^*

For strength calculations, the tooth face width b should also be taken into account.

As for the types of *meshing*, there are three external and three internal ones [120, 175]:

- Null meshing (non-shifted)
- Equal-shifted
- Shifted

which are characterized in the following ways:

1. By profile shift coefficients of the gears x_1, x_2, and x_3:
   Null meshing:
   $x_1 = x_2 = 0$, respectively, $x_1 + x_2 = 0$—external meshing;
   $x_3 = x_2 = 0$, respectively, $x_3 - x_2 = 0$—internal meshing;
   Equal-shifted meshing:
   $x_1 = -x_2$, respectively, $x_1 + x_2 = 0$—external meshing;
   $x_3 = x_2 = 0$, respectively, $x_3 - x_2 = 0$—internal meshing;
   Shifted meshing:
   $x_1 \neq -x_2$, respectively, $x_1 + x_2 \neq 0$—external meshing;
   $x_3 \neq x_2$, respectively, $x_3 - x_2 \neq 0$—internal meshing;

2. By operating center distances a_{w12} and a_{w23}:
   $a_{w12} = a_{12}$ and $a_{w23} = a_{23}$—null and equal-shifted meshing
   $a_{w12} > a_{12}$ and $a_{w23} < a_{23}$, accordingly $a_{w12} \equiv a_{12} = a_w$—only for shifted meshing
3. By operating pressure angles α_{w12} and α_{w23}:
   $\alpha_w = \alpha = 20°$—null and equal-shifted meshing
   $\alpha_w \neq= 20°$—only for shifted meshing
4. By operating circle diameters d_{w1}, $d_{w2(1)}$, $d_{w2(3)}$, and d_{w3}:
   $d_{w1} \equiv d_1;\ d_{w2(1)} = d_{w2(3)} \equiv d_2;\ d_{w3} \equiv d_3$—null and equal-shifted meshing
   $d_{w1} > d_1;\ d_{w2(1)} > d_2;\ d_{w2(3)} < d_2;\ d_{w3} < d_3$—only for shifted meshing
   It is necessary to explicitly point out that the planets 2 have two operating circles with
   $d_{w2(1)}$—in mesh with sun gear 1
   $d_{w2(3)}$—in mesh with ring gear 3

In the case of profile shifts, there are the so-called forward and inverse tasks.

5.2 External Meshing—The Forward and the Inverse Task

In case of *external meshing*, the data of the ***forward task*** are

Given
- Module m
- Number of teeth z_1 and z_2
- Reference center distance a_{12}
- Profile shift coefficients x_1 and x_2, respectively, $x_{\Sigma 12} = x_1 + x_2$
- Reference pressure angle (rack profile angle) $\alpha = 20°$
- Addendum coefficient (factor) $h_a^* = 1.0$
- Clearance coefficient (factor) $c^* = 0.25$

Sought (results)
- Operating pressure angle α_{w12}
- Operating center distance a_{w12}
- Gears shift coefficient y_{12}
- Return shift coefficient (tooth tip shortening coefficient) Δy_{12}
- Addendums h_{a1} and h_{a2}, and dedendums h_{f1} and h_{f2} of gears, respectively, the heights of their teeth h_1 and h_2
- Reference circle diameters d_1 and d_2
- Tip circle diameters d_{a1} and d_{a2}, which eventually need tooth tip shortening $\Delta y \cdot m$ (Figure 5.2) [175, 198]
- Root circle diameters d_{f1} and d_{f2}
- Operating (pitch) circle diameters d_{w1} and $d_{w2(1)}$

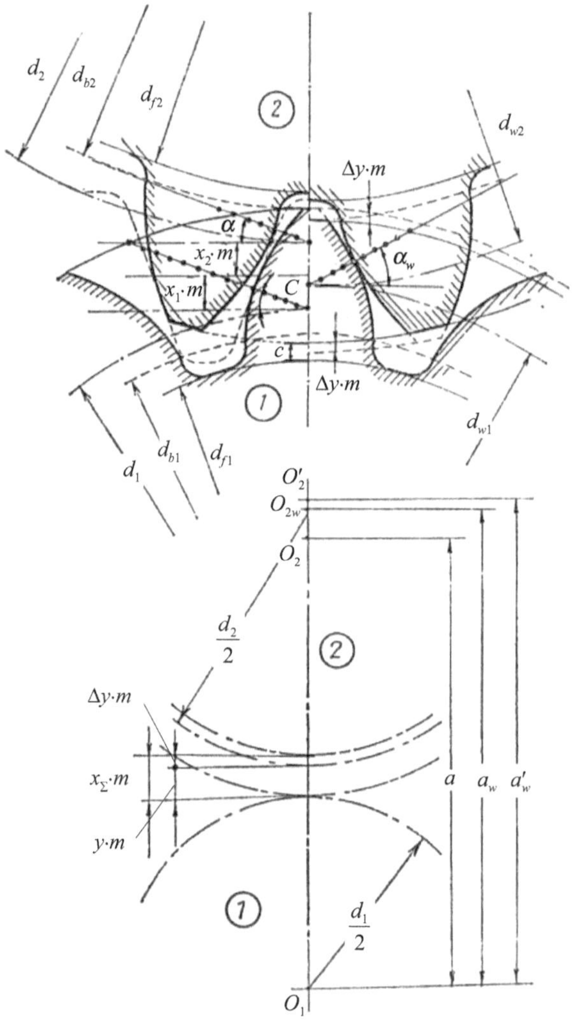

FIGURE 5.2
Necessary return shift $\Delta y_{12}{\cdot}m$ at external meshing.

Base circle diameters d_{b1} and d_{b2}

Tip circle profile angles α_{a1} and α_{a2}

The forward task solution is as follows:

Operating pressure angle

$$\boxed{\operatorname{inv}\alpha_{w12} = 2\frac{x_1 + x_2}{z_1 + z_2}\tan\alpha + \operatorname{inv}\alpha} \tag{5.1}$$

with recommended values within the limits $\alpha_{w12} = 24° \div 27°$ [147–149].

Operating center distance after shifting

$$a_{w12} = a_{12}\frac{\cos\alpha}{\cos\alpha_{w12}} = m\frac{z_1 + z_2}{2}\cdot\frac{\cos\alpha}{\cos\alpha_{w12}}. \tag{5.2}$$

Gears shift coefficient

$$y_{12} = \frac{a_{w12} - a_{12}}{m} = \frac{a_{12}}{m}\left(\frac{\cos\alpha}{\cos\alpha_{w12}} - 1\right) = \frac{z_1 + z_2}{2}\left(\frac{\cos\alpha}{\cos\alpha_{w12}} - \right). \tag{5.3}$$

Return shift coefficient

$$\Delta y_{12} = x_{\Sigma 12} - y_{12} = (x_1 + x_2) - \frac{z_1 + z_2}{2}\left(\frac{\cos\alpha}{\cos\alpha_{w12}} - 1\right). \tag{5.4}$$

Apropos, this is the tip shortening coefficient of the teeth of mated gears.

Another formula suitable for this purpose is

$$\Delta y_{12} = \frac{1}{m}\left[a_{12} + m(x_1 + x_2) - a_{w12}\right]. \tag{5.5}$$

Addendums and dedendums of both gears, and the heights of their teeth, respectively,

$$\left.\begin{aligned} h_{a1} &= m\left(h_a^* + x_1 - \Delta y_{12}\right) \\ h_{f1} &= m\left(h_a^* + x_1 + c^*\right) \end{aligned}\right\} h_1 = h_{a1} + h_{f1} = m\left(2h_a^* + c^* - \Delta y_{12}\right); \tag{5.6}$$

$$\left.\begin{aligned} h_{a2} &= m\left(h_a^* + x_2 - \Delta y_{12}\right) \\ h_{f2} &= m\left(h_a^* - x_2 + c^*\right) \end{aligned}\right\} h_2 = h_{a2} + h_{f2} = m\left(2h_a^* + c^* - \Delta y_{12}\right). \tag{5.7}$$

Reference circle diameters

$$\begin{aligned} d_1 &= m\cdot z_1, \\ d_2 &= m\cdot z_2 \end{aligned} \tag{5.8}$$

Tip circle diameters

$$\begin{aligned} d_{a1} &= d_1 + 2h_{a1} = m\left(z_1 + 2h_a^* + 2x_1 - 2\Delta y_{12}\right), \\ d_{a2} &= d_2 + 2h_{a2} = m\left(z_2 + 2h_a^* + 2x_2 - 2\Delta y_{12}\right). \end{aligned} \tag{5.9}$$

Root circle diameters

$$\begin{aligned} d_{f1} &= d_1 - 2h_{f1} = m\left[z_1 - 2\left(h_a^* + c^*\right) + 2x_1\right], \\ d_{f2} &= d_2 - 2h_{f2} = m\left[z_2 - 2\left(h_a^* + c^*\right) + 2x_2\right]. \end{aligned} \tag{5.10}$$

Operating circle diameters of the sun gear 1 and the planets 2 in their meshing with the sun gear (external mesh)

$$d_{w1} = d_1 \frac{\cos\alpha}{\cos\alpha_{w12}} = m \cdot z_1 \frac{\cos\alpha}{\cos\alpha_{w12}}, \tag{5.11}$$

$$d_{w2(1)} = d_2 \frac{\cos\alpha}{\cos\alpha_{w12}} = m \cdot z_2 \frac{\cos\alpha}{\cos\alpha_{w12}}. \tag{5.12}$$

The following formulae are also applicable:

$$d_{w1} = 2a_w \frac{z_1}{z_1 + z_2} = 2a_w \frac{1}{u_{12} + 1}, \tag{5.13}$$

$$d_{w2(1)} = 2a_w \frac{z_2}{z_1 + z_2} = 2a_w \frac{u_{12}}{u_{12} + 1} = u_{12} \cdot d_{w1}, \tag{5.14}$$

where $u_{12} = z_2/z_1$ is the teeth ratio of mated gears 1 and 2.

Base circle diameters

$$\begin{aligned} d_{b1} &= d_1 \cdot \cos\alpha, \\ d_{b2} &= d_2 \cdot \cos\alpha \end{aligned} \tag{5.15}$$

Tip circle profile angles

$$\begin{aligned} \alpha_{a1} &= \arccos\frac{d_{b1}}{d_{a1}}, \\ \alpha_{a2} &= \arccos\frac{d_{b2}}{d_{a2}} \end{aligned} \tag{5.16}$$

The data of the ***inverse task*** are

Given

m, z_1, z_2, respectively, a_{12}, a_{w12}, $\alpha = 20°$, $h_a^* = 1.0$, and $c^* = 0.25$.

Sought (results)

α_{w12}, $x_{\Sigma 12} = x_1 + x_2$ and their distribution, y_{12}, Δy_{12}, h_{a1}, h_{a2}, h_{f1}, h_{f2}, h_1, h_2, d_1, d_2, d_{f1}, d_{f2}, d_{w1}, $d_{w2(1)}$, d_{b1}, and d_{b2}.

The corresponding formulae are

Operating pressure angle

$$\boxed{\cos\alpha_{w12} = \frac{a_{12}}{a_{w12}} \cos\alpha = \frac{m\dfrac{z_1 + z_2}{2}}{a_{w12}} \cos\alpha.} \tag{5.17}$$

Sum of profile shift coefficients

$$\boxed{x_{\Sigma 12} = x_1 + x_2 = \frac{z_1 + z_2}{2} \cdot \frac{\operatorname{inv}\alpha_{w12} - \operatorname{inv}\alpha}{\tan\alpha}.} \tag{5.18}$$

Formulae for the other dimensions y_{12}, Δy_{12}, h_{a1}, h_{a2}, h_{f1}, h_{f2}, h_1, h_2, d_{a1}, d_{a2}, d_{f1}, d_{f2}, d_{w1}, $d_{w2(1)}$, d_{b1}, and d_{b2} are the same as above. The choice of the profile shift coefficients x_1 and x_2 from their sum $x_{\Sigma 12}$ as determined by formula (5.15) also depends on the parameters of the internal meshing, respectively, on the choice of x_3, and hence on x_2. Usually $x_1 > x_2$ is chosen because the number of teeth most often is $z_1 < z_2$. Thus, the sun gear 1 has the largest number of load cycles N_1, and from this point of view, it is the most loaded gear.

It should be borne in mind that profile shifts are effective mostly with a small number of teeth, e.g., $z<20$ (25). With a greater number of teeth, the profile shift does not provide a significant benefit for the increase of tooth strength.

The diagram in Figure 5.3 [162] presents recommended semi-sum values of the profile shift coefficients depending on the semi-sum of the number of teeth of mated gears, i.e.,

$$\frac{x_1 + x_2}{2} = f\left(\frac{z_1 + z_2}{2}\right).$$

When the point defined by both semi-sums lies in the gray area, favorable, balanced results for the gears are obtained in respect of their geometry, kinematics, and strength.

5.3 Internal Meshing—The Forward and the Inverse Task

The situation of *internal meshing* is similar, of course, with the respective differences.[1] The ***forward task*** data are

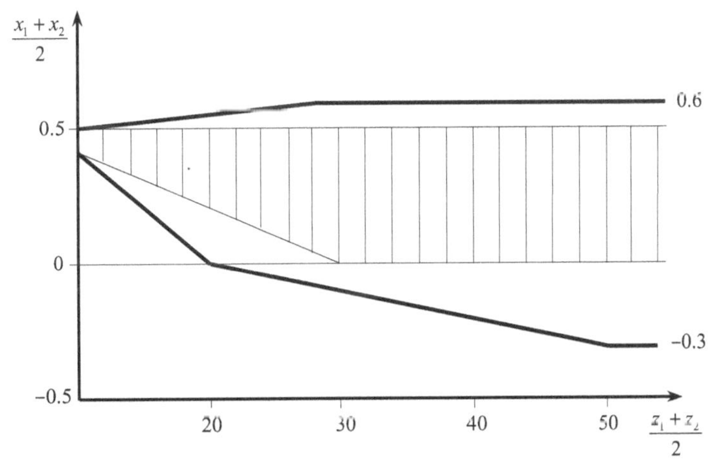

FIGURE 5.3
Recommended values for $(x_1 + x_2)/2$ depending on $(z_1 + z_2)/2$ at $\alpha = 20°$. (Redrawn from MAAG-Taschenbuch. 2. Auflage. Zürich: MAAG-Zahnräder AG, 1985.)

[1] In order to use the same formulae for both external and internal meshing (very appropriate for software), the teeth number of internal gear z_3 is conventionally assumed as negative; i.e., the respective data are negative, too: d_3, d_{a3}, d_{b3}, d_{f3}, d_{w3}, a_{23}, a_{w23}, u_{23} [65, 175]. The algebraic sign of the profile shift coefficient x_3 also changes. However, in order to keep the reader from confusion, in this book, this conditionality *is not accepted*.

Given

m, z_2, z_3, a_{23}, x_2, and x_3, respectively, $x_{d23} = x_3 - x_2, \alpha = 20°, h_a^* = 1.0$, and $c^* = 0.25$ or $c^* = 0.30$ [57].

Sought (results)

α_{w23}, a_{w23}, internal tooth tip shortening coefficient Δh_3^*, h_{a3}, h_{f3}, h_3, d_{a3}, d_{f3}, d_{w3}, $d_{w2(3)}$, d_{b3}, and α_{a3}.

The corresponding formulae are

Operating pressure angle

$$\boxed{\operatorname{inv}\alpha_{w23} = 2\frac{x_3 - x_2}{z_3 - z_2}\tan\alpha + \operatorname{inv}\alpha} \quad (5.19)$$

with recommended values within the limits $\alpha_{w23} = 17° \div 19°$ [147–149].

Operating center distance after shifting

$$\boxed{a_{w23} = a_{23}\frac{\cos\alpha}{\cos\alpha_{w23}} = m\frac{z_3 - z_2}{2}\cdot\frac{\cos\alpha}{\cos\alpha_{w23}}.} \quad (5.20)$$

Internal tooth tip shortening

$$\Delta h_3 = \Delta h_3^* \cdot m = 7.6\frac{(1 - x_3)}{z_3}m\cdot \quad (5.21)$$

The formula (5.18) applies to null (zero) and equally shifted meshing [147–149], and GOST 19274 even recommends $\Delta h_3^* = 0.2 = \text{const}$ in all cases. The large contact ratio $\varepsilon_{\alpha 23}$ of the internal meshing allows for shortening Δh_3. This shortening of the addendum of the internal tooth (Figure 5.4) avoids the interference between the tip of ring gear 3 and the flank of planets 2 (Figure 5.7). It has nothing in common with the coefficient Δy_{12} of the externally mated gears 1 and 2.

Addendums and dedendums of ring gear 3, and the heights of its tooth, respectively,

$$\left.\begin{aligned} h_{a3} &= m\left(h_a^* - x_3 - \Delta h_3^*\right) \\ h_{f3} &= m\left(h_a^* + x_3 + c^*\right) \end{aligned}\right\} h_3 = h_{a3} + h_{f3} = m\left(2h_a^* + c^* - \Delta h_3^*\right). \quad (5.22)$$

Reference circle diameters

$$d_3 = m \cdot z_3. \quad (5.23)$$

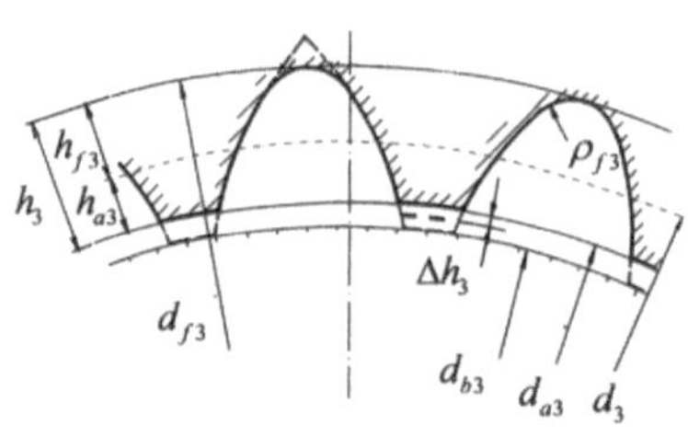

FIGURE 5.4
Internal tooting of ring gear 3.

Tip circle diameter

$$d_{a3} = d_3 - 2h_{a3} = m\left(z_3 - 2h_a^* + 2x_3 + 2\Delta h_3^*\right). \tag{5.24}$$

Root circle diameter

$$d_{f3} = d_3 + 2h_{f3} \approx m\left(z_3 + 2h_a^* + 2c^* + 2x_3\right). \tag{5.25}$$

The dependence (5.21) is approximate because the diameter d_{f3} depends not only on the ring gear parameters (z_3 and x_3) but also on the current parameters of the cutter (z_0 and x_0). The profile shift coefficient varies after each resharpening, and hence, the root diameter d_{f3} varies, too. This diameter may actually be of interest only in a case of thin-rim ring gear, made for greater compliance. In general, the exact determination of d_{f3} is not of interest. If this is still necessary, it is done as follows:

First, the pressure angle of gear tooling mesh α_{w03} is defined by the following formula:

$$\boxed{\operatorname{inv}\alpha_{w03} = 2\frac{x_3 - x_0}{z_3 - z_0}\tan\alpha + \operatorname{inv}\alpha,} \tag{5.26}$$

and through it, the center distance a_{w03} of the same mesh is determined:

$$\boxed{a_{w03} = a_{03}\frac{\cos\alpha}{\cos\alpha_{w03}} = m\frac{z_3 - z_0}{2}\cdot\frac{\cos\alpha}{\cos\alpha_{w03}}.} \tag{5.27}$$

For determination of the root diameter d_{f3} of ring gear 3, it is still necessary to know (or measure) the current tip diameter d_{a0} of the tool after its corresponding resharpening. From Figure 5.5, the radius r_{f3} and the sought root diameter d_{f3}, respectively, are obtained by the following formula

$$d_{f3} = 2a_{w03} + d_{a0}. \tag{5.28}$$

It should not be forgotten that d_{f3} is not invariable, i.e., $d_{f3} \neq \text{const}$, but depends on the current tip diameter d_{a0} and on the profile shift coefficient x_0 of the cutter, respectively, after each resharpening, i.e., $d_{f3} = f(x_0)$ in invariable m, z_3, x_3, and z_0.

Operating diameters of ring gear 3 and of planets 2 in their meshing with ring gear 3 (internal meshing) are

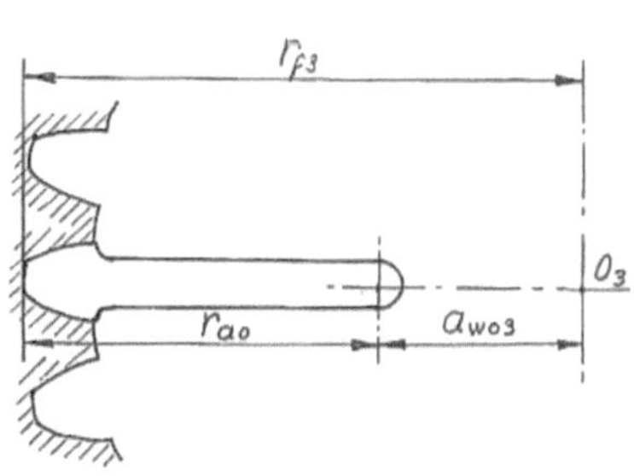

FIGURE 5.5
Determining the root diameter d_{f3} of ring gear 3.

$$d_{w3} = d_3 \frac{\cos\alpha}{\cos\alpha_{w23}} = m \cdot z_3 \frac{\cos\alpha}{\cos\alpha_{w23}}, \tag{5.29}$$

$$d_{w2(3)} = d_2 \frac{\cos\alpha}{\cos\alpha_{w23}} = m \cdot z_2 \frac{\cos\alpha}{\cos\alpha_{w23}}. \tag{5.30}$$

Obviously, in general, when the operating pressure angles of external and internal meshing are different, i.e., $\alpha_{w12} \neq \alpha_{w23}$, the planets have two operating circles with diameters $d_{w2(1)}$ and $d_{w2(3)}$.

For internal meshing, the operating diameters determination can be done by analogous formulae, as in the external meshing:

$$d_{w2(3)} = 2a_w \frac{z_2}{z_3 - z_2} = 2a_w \frac{1}{u_{23} - 1}, \tag{5.31}$$

$$d_{w3} = 2a_w \frac{z_3}{z_3 - z_2} = 2a_w \frac{u_{23}}{u_{23} - 1} = u_{23} \cdot d_{w2(3)}, \tag{5.32}$$

where $u_{23} = z_3/z_2 > 1$ is the teeth ratio of mated gears 2 and 3.

Base circle diameters

$$d_{b3} = d_3 \cdot \cos\alpha. \tag{5.33}$$

The data of the ***inverse task*** are

Given

m, z_2, z_3, respectively, a_{23}, a_{w23}, $\alpha = 20°$, $h_a^* = 1.0$, and $c^* = 0.25$ or $c^* = 0.30$.

Sought (results)

α_{w23}, $x_{d23} = x_3 - x_2$ and their distribution Δh_3, h_{a3}, h_{f3}, d_{a3}, d_{f3}, $d_{w2(3)}$, and d_{w3}.

The corresponding formulae are

Operating pressure angle

$$\boxed{\cos\alpha_{w23} = \frac{a_{23}}{a_{w23}} \cos\alpha = \frac{m \dfrac{z_3 - z_2}{2}}{a_{w23}} \cos\alpha.} \tag{5.34}$$

Difference of profile shift coefficients

$$\boxed{x_{d23} = x_3 - x_2 = \frac{z_3 - z_2}{2} \cdot \frac{\operatorname{inv}\alpha_{w23} - \operatorname{inv}\alpha}{\tan\alpha}.} \tag{5.35}$$

The choice of profile shift coefficients x_2 and x_3 from their difference $x_{d23} = x_3 - x_2$ is important too.

Tip circle profile angle

$$\alpha_{a3} = \arccos\frac{d_{b3}}{d_{a3}}, \tag{5.36}$$

Formulae for the other dimensions Δh_3, h_{a3}, h_{f3}, h_3, d_3, d_{a3}, d_{f3}, d_{w3}, and $d_{w2(3)}$ are the same as above.

5.4 Cases of Profile Shift Corrections (Modifications)

By using profile shift corrections (modifications), different meshing cases are possible (Figure 5.6). Here, only some relatively *simple cases of profile shifting* are given. In these cases, it is assumed that the number of teeth of planets 2 is reduced, i.e., $z_2 < \frac{z_3 - z_1}{2}$, respectively, $z_2 = \frac{z_3 - z_1}{2} - \Delta z_2$, where Δz_2 may be equal to 0.5, 1, or 1.5 (see Section 4.2 and formula 4.8).

1. Case

Shifted external meshing $(x_1 + x_2 > 0)$

Null internal meshing: $x_3 - x_2 = 0$

This case is possible and necessary when the number of teeth of sun gear 1 is small, within the limits of $z_1 = 10 \div 20$, so that it is necessary or appropriate to apply a profile shift $(x_1 > 0)$.

1. Case
Shifted external meshing $(x_1 + x_2 > 0)$;
Null internal meshing: $x_3 - x_2 = 0$.
$x_1 > 0$; $x_2 = 0$; $x_3 = 0$
$a_w = a_{23} > a_{12}$
$\alpha_{w12} > \alpha = 20°$; $\alpha_{23} = \alpha = 20°$
$d_{w1} > d_1$; $d_{w2(1)} > d_2$; $d_{w2(3)} \equiv d_2$; $d_{w3} \equiv d_3$

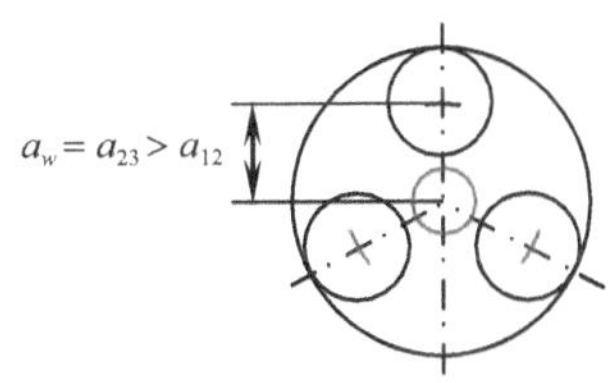

2. Case
Shifted external meshing $(x_1 + x_2 > 0)$;
Equal-shifted internal meshing $(x_3 - x_2 = 0)$.
$x_1 > 0$; $x_2 > 0$; $x_3 = x_2 > 0$
$a_w = a_{23} > a_{12}$
$\alpha_{w12} > \alpha = 20°$; $\alpha_{23} = \alpha = 20°$
$d_{w1} > d_1$; $d_{w2(1)} > d_2$; $d_{w2(3)} \equiv d_2$; $d_{w3} \equiv d_3$

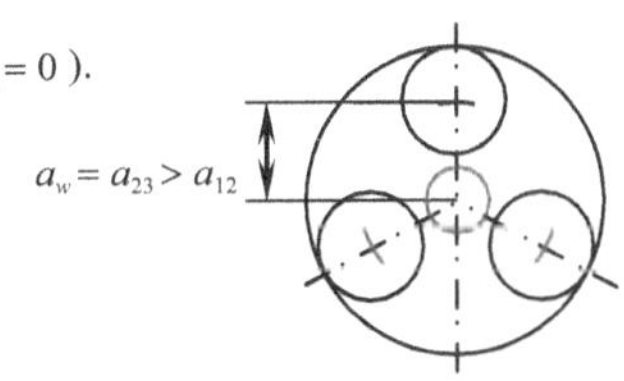

3. Case
Shifted external meshing $(x_1 + x_2 > 0)$;
Shifted internal meshing $(x_3 - x_2 < 0)$.
$x_1 > 0$; $x_2 > 0$; $x_3 \geq 0$; $x_3 < x_2$
$a_{12} < a_w < a_{23}$
$\alpha_{w12} > \alpha = 20°$; $\alpha_{w23} < \alpha = 20°$
$d_{w1} > d_1$; $d_{w2(1)} > d_2$; $d_{w2(3)} < d_2$; $d_{w3} < d_3$

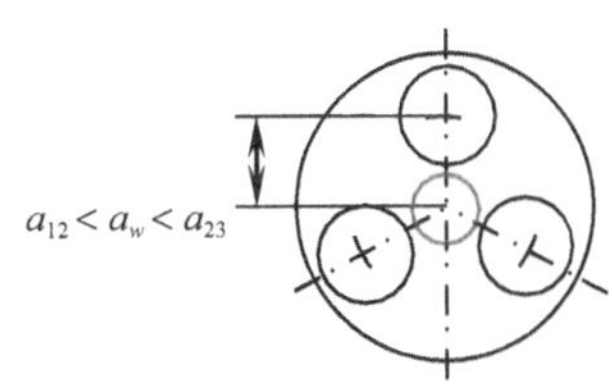

4. Case
Null external meshing $(x_1 = x_2 = 0)$;
Null internal meshing $(x_2 = x_3 = 0)$.
$x_1 = x_2 = x_3 = 0$
$a_{12} = a_{23}$
$\alpha_{12} = \alpha_{23} = \alpha = 20°$
$d_{w1} \equiv d_1$; $d_{w2(1)} \equiv d_{w2(3)} \equiv d_2$; $d_{w3} \equiv d_3$

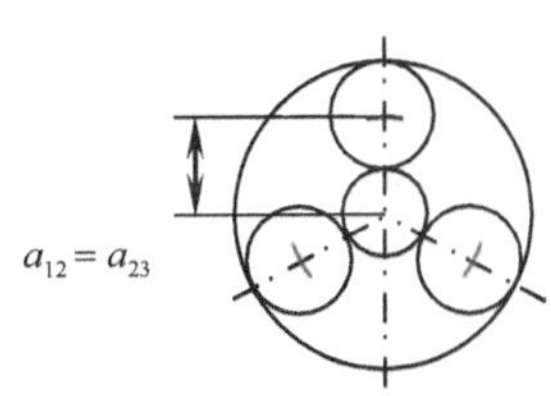

FIGURE 5.6
Possible cases of meshing depend on profile shift corrections (modifications).

Whereas the planets 2 and the ring gear 3 have large number of teeth $z_2 > 40$ and $z_3 > 100$, and there is no need of correction in respect of the strength. Operating pressure angle of external mesh α_{w12} depends on planets teeth number reduction—$\Delta z_2 = 0.5$, 1, or 1.5. Both the strength of the teeth and the size of the contact ratio $\varepsilon_{\alpha 12}$ depend on this angle. Operating pressure angle is important for the noise level of the gear train, and its appropriate values are [147–149] $\alpha_{w12} = 24° \div 27°$.

2. Case

$$\text{Shifted external meshing}\,(x_1 + x_2 > 0)$$
$$\text{Equally shifted internal meshing}\,(x_3 - x_2 = 0)$$

This case can also be used for a small number of teeth, both of the sun gear and planets, $z_2 = 15 \div 20$, e.g., so that their profile shift is appropriate. This case is relatively simple and is therefore often used with a larger teeth number of planets. In this case, the recommended values of operating pressure angle are the same as in the previous one.

3. Case

$$\text{Shifted external meshing}\,(x_1 + x_2 > 0)$$
$$\text{Shifted internal meshing}\,(x_3 - x_2 < 0)$$

This is the most general and the most complex case. It has the greatest design freedom, but geometric computations are more complex. The following operating pressure angles are recommended [147–149]:

- For external meshing: $\alpha_{w12} = 24° \div 27°$
- For internal meshing: $\alpha_{w23} = 17° \div 19°$

4. Case

$$\text{Null external meshing}\,(x_1 = x_2 = 0)$$
$$\text{Null internal meshing}\,(x_2 = x_3 = 0)$$

This case without shifting is appropriate when the number of teeth of all gears z_1, z_2, and z_3, in particular z_1, is very large ($z_1 > 25 \div 30$ and $z_2 > 25 \div 30$), so it is pointless, from the point of view of strength, to shift the profile because the effect is negligible.

What has been said so far about the gear teeth number and profile shift should not be conceived as a dogma, but as one orientation only. In PGTs design, besides strength, there are other considerations that a designer must keep in mind—kinematic, constructive, production, and operational.

The cases mentioned here are only a few of the possibilities. However, they are simple and convenient to work, and are therefore recommended. The simplest one to use is the second case.

When choosing the profile shift coefficients, in particular those of the externally meshed gears, in addition to adhering to the permissible limits of x_1 and x_2 (Section 5.1), care must be taken to obtain a sufficient contact ratio $\varepsilon_{\alpha 12}$, because the quiet operation of the gear train depends on it. This means that one should not overdo the corrections, i.e., large values of x_1 and x_2 (Section 5.1), hence the pressure angle α_{w12}. In order to avoid heating due to the compactness and hence the small cooling surface of the PGTs, the aim is to obtain the highest possible efficiency η by a profile shift (see Chapters 11 and 12). This is always important, but especially in the powerful, high-speed PGTs. In low-speed, heavily loaded trains in which tooth wear comes to the fore, the aim is to obtain a leveling of specific teeth sliding, especially of the external mesh, i.e., of sun gear 1 and of planets 2 [23, 33, 34].

5.5 Internal Teeth Cutting and Its Peculiarities

Unlike external mesh, there are more interference hazards in internal one [57, 65, 120, 158, 175, 197]. Particularly at $\overline{AI}$-PGT, the number of possible interferences is lower than that of some other types of PGTs. The following two interferences are possible (Figure 5.7):

- Interference of internal tooth (ring gear 3) tip with external tooth (planet 2) fillet (Figure 5.7a)
- Interference of planet 2 tooth tip with internal tooth (ring gear 3) fillet (Figure 5.7b)

In the figure, the boundary points G_2 and G_3 between the involute flank and the fillet curve are given.

The first interference can be avoided by reducing the height of the internal tooth with $\Delta h_3 = \Delta h_3^* \cdot m$. That is why this reduction is made.

The second interference can be avoided (often but not always!) by using a cutter with the number of teeth z_0 larger than that of the planets 2, i.e., $z_0 > z_2$ [175]. However, this is not always possible in the case of a large teeth number of planets ($z_2 > z_0$), as well as in a very large module m for which there is no suitable cutter and a gear-cutting machine, too. In such case, for low-accuracy gears mainly, the following options may be used:

- Vertical slotting by shaped tool
- Form milling by shaped wheel-type mill gear cutter
- Form milling by module end mill gear cutter

Form milling may be accomplished in standard milling machines with the corresponding attachments. Profile tools, of course, must help to avoid interference, that is, to respond to the specific case. Apropos, planets addendum shortening with $\Delta y_{12} \cdot m$ also helps to avoid the second interference (Figure 5.7b).

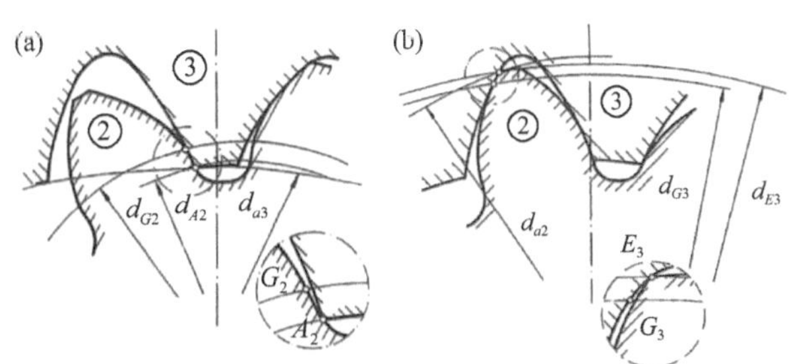

FIGURE 5.7
Cases of interference with the fillet curve of planets 2 (a) and of ring gear 3 (b).

5.6 Interference with the Filet Curve of Ring Gear

Figure 5.8 shows interference with the fillet curve of ring gear 3. The presence of such interference is determined in the following way. Boundary point G_3 of involute profile of ring gear 3 is determined by the following formula [46]:

$$\boxed{\tan\alpha_{G3} = \tan\alpha_{w03} + \frac{z_0}{z_3}(\tan\alpha_{a0} - \tan\alpha_{w03}) = \frac{z_3 - z_0}{z_3}\tan\alpha_{w03} + \frac{z_0}{z_3}\tan\alpha_{a0},} \tag{5.37}$$

where α_{w03} is the pressure angle of the machining meshing, determined by formula (5.26), and α_{a0} is the profile angle of current tip circle with diameter d_{a0} of the cutter after its corresponding resharpening which is determined as follows:

$$\cos\alpha_{a0} = \frac{d_{b0}}{d_{a0}} = \frac{m \cdot z_0 \cdot \cos\alpha}{d_{a0}}. \tag{5.38}$$

Boundary point E_3 of the meshing between planets 2 and ring gear 3 is obtained by [46]

$$\boxed{\tan\alpha_{E3} = \tan\alpha_{w23} + \frac{z_2}{z_3}(\tan\alpha_{a2} - \tan\alpha_{w23}) = \frac{z_3 - z_2}{z_3}\tan\alpha_{w23} + \frac{z_2}{z_3}\tan\alpha_{a2},} \tag{5.39}$$

where α_{w23} is the pressure angle of the internal meshing, determined by formula (5.19), and α_{a2} is the profile angle of tip circle with diameter d_{a2} of the planets 2 which is determined by

$$\cos\alpha_{a2} = \frac{d_{b2}}{d_{a2}} = \frac{m \cdot z_2 \cdot \cos\alpha}{d_{a2}}, \tag{5.40}$$

where d_{b2} is the diameter of the base circle of planets 2.

Interference *will not occur* if the condition is kept (Figure 5.8a):

$\tan\alpha_{E3} < \tan\alpha_{G3}$, respectively, $d_{E3} = \dfrac{d_{b3}}{\cos\alpha_{E3}} < d_{G3} = \dfrac{d_{b3}}{\cos\alpha_{G3}}$

And vice versa—interference *will occur* if

$\tan\alpha_{E3} > \tan\alpha_{G3}$, respectively, $d_{E3} = \dfrac{d_{b3}}{\cos\alpha_{E3}} > d_{G3} = \dfrac{d_{b3}}{\cos\alpha_{G3}}$.

When using these formulae, it is necessary not to work with the nominal values of the quantities but to take into account the tolerances of gears and the clearances of planet bearings, i.e., to work with a safety factor.

5.7 Obtaining Filet Curve Radius of Ring Gear Equal to Zero

Apart from interference, in the internal teeth, there is another danger as a result of the machining, which unfortunately is often forgotten in practice. This danger concerns the fillet curve of the internal tooth, which is an *extended (very rarely—shortened) hypocycloid.*

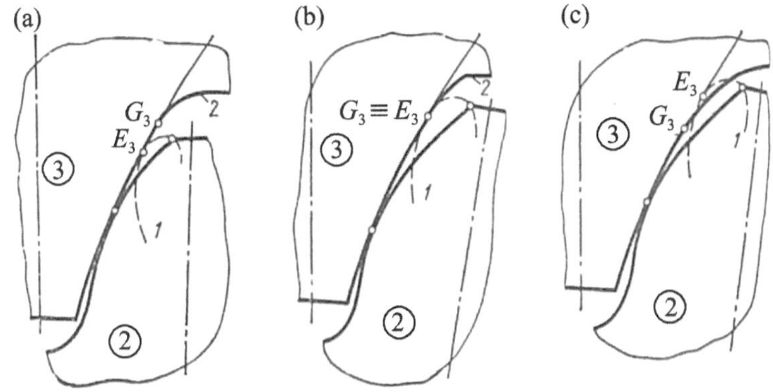

FIGURE 5.8
Interference of planets tooth tip with the filet curve of ring gear (1—trajectory of the sharp of planets tooth; 2—filet curve of ring gear tooth): (a) No interference; (b) boundary case; (c) interference.

At each of its points, the radius of curvature ρ_f is different. For bending strength, the radius of the curvature in the critical section (Figure 5.9a) is important, and its decreasing or degeneration into a point ($\rho_{f3} = 0$, Figure 5.9b) can lead to unacceptable concentration of the bending stress σ_F. This occurs during the machining operating (pitch) circle (i.e., machining centroid) of the cutter when the diameter d_{w0} coincides with its current tip circle with diameter d_{a0}, i.e., when becomes

$$\boxed{d_{w0} = d_{a0}.}$$

This can be obtained by unfavorably combining the profile shift coefficients x_3 of the ring gear 3 and x_0 of the cutter. More in particular, this occurs when the difference of the profile shift coefficients $x_{d03} = x_3 - x_0$ becomes very large, and hence, the machining operating (pitch) circle diameter (i.e., the machining centroid) d_{w0} of the gear increases to such an extent that it equals the diameter of the current tip circle d_{a0}. Since the cutters are most often without a curvature of their addendum edge (sharp-cornered cutter), i.e., $\rho_{a0} = 0$ (as opposed to hobs and racks), their sharp corner directly reproduces onto the ring gear 3.

When there is a coincidence of the machining operating circle (i.e., the machining centroid) with the current tip circle of the cutter after its resharpening, its tooth tip point

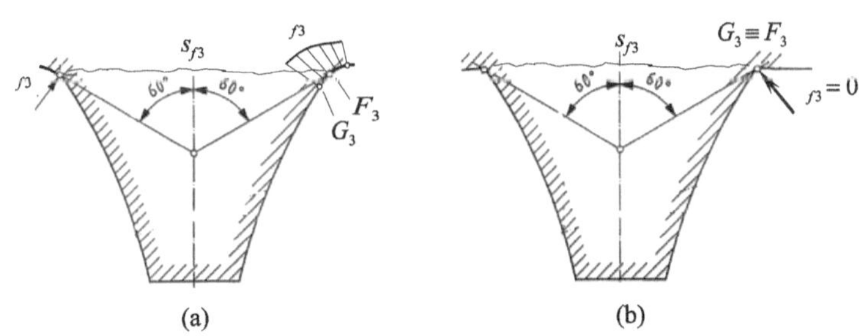

FIGURE 5.9
Curvature radius of the filet curve ρ_{f3} at the critical section of an internal tooth (a) and its degeneration into a point (b).

describes a non-elongated (or truncated) hypocycloid but a common hypocycloid, whose inflex point corresponds with the fillet curve degenerated in the point.

On the basis of its features, the cutter for spur teeth can be conventionally understood in the following ways (Figure 5.10):

1. As a cylindrical spur gear with linearly adjustable profile shift $x_0 \cdot m$ along the tooth face width. Therefore, at each resharpening, its coefficient x_0 decreases, and from being positive in the beginning ($x_0 > 0$) after the final resharpening, it becomes even negative ($x_0 < 0$).
2. Its addendum is increased with $c = 0.25m$ or $c = 0.3m$ to create a radial clearance of the mated gears.
3. There is no curvature of its addendum edge (sharp-cornered cutter), i.e., $\rho_{a0} = 0$, as opposed to hobs and racks.

Check for $\rho_{f3} = 0$ is as follows:

Current profile shift coefficient x_0 of the cutter after corresponding resharpening can be determined, based on its current tip d_{a0} diameter:

$$d_{a0} = m\left(z_0 + 2h_a^* + 2c^* + 2x_0\right). \tag{5.41}$$

The coefficient x_0 is

$$x_0 = \frac{d_{a0} - m\left(z_0 + 2h_a^* + 2c^*\right)}{2m} = \frac{1}{2}\left(\frac{d_{a0}}{m} - z_0 - 2h_a^* - 2c^*\right). \tag{5.42}$$

Especially in the case of prefinishing cutters, x_0 determination is more accurate if it results from the current span measurement W_0 of the tool

$$x_0 = \frac{\dfrac{W_0}{m \cdot \cos\alpha} - \pi(z_W - 0.5) - z_0 \cdot \operatorname{inv}\alpha}{2\tan\alpha}, \tag{5.43}$$

where z_W is the number of the spanned teeth (see Section 15.4).

The profile angle α_{a0} at the current tip circle with diameter d_{a0} of the cutter can be determined by formula (5.38):

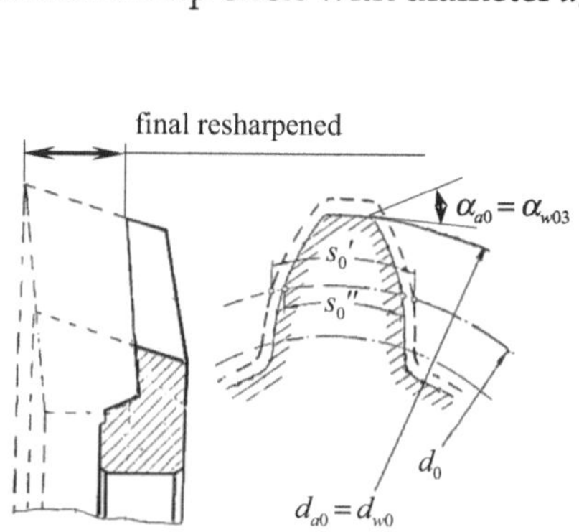

FIGURE 5.10
Maximally resharpened shaper cutter.

$$\cos\alpha_{a0} = \frac{d_{b0}}{d_{a0}} = \frac{m \cdot z_0 \cdot \cos\alpha}{d_{a0}}.$$

The operating pressure angle of tooling mesh α_{w03} is determined by formula (5.26)

$$\operatorname{inv}\alpha_{w03} = 2\frac{x_3 - x_0}{z_3 - z_0}\tan\alpha + \operatorname{inv}\alpha.$$

If these two angles coincide or are very close, i.e.,

$$\boxed{\alpha_{w03} \approx \alpha_{\alpha 0},}$$

it follows that $d_{w0} \approx d_{a0}$, and the fillet curve of the internal tooth of ring gear 3 is degenerate at a point (radius of curvature $\rho_{f3} \approx 0$).

There is also another way to check whether $\rho_{f3} = 0$ is obtained, which provides the designer with a tool to detect the results, which is not unimportant.

From the intended angle of the tooling mesh α_{w03}, its center distance is determined by formula (5.27)

$$a_{w03} = a_{03}\frac{\cos\alpha}{\cos\alpha_{w03}} = m\frac{z_3 - z_0}{2} \cdot \frac{\cos\alpha}{\cos\alpha_{w03}}.$$

The diameter of the machining operating (pitch) circle (i.e., the machining centroid) d_{w0} of the tool can be determined in two ways, which also gives the designer a good opportunity:

$$d_{w0} = m \cdot z_0 \frac{\cos\alpha}{\cos\alpha_{w03}} \tag{5.44}$$

or

$$d_{w0} = 2a_{w03}\frac{z_0}{z_3 - z_0} = 2a_{w03}\frac{1}{u_{03} - 1}, \tag{5.45}$$

where $u_{03} = z_3/z_0 > 1$ is the teeth ratio of tooling mesh.

Knowing this diameter d_{w0}, as well as the current tip diameter d_{a0}, it is possible to determine the current machining operating tooth addendum h_{aw0} of the toothed gear:

$$\boxed{h_{aw0} = \frac{d_{a0} - d_{w0}}{2}.} \tag{5.46}$$

This value is important for obtaining the smallest radius of the fillet curvature at point F_3 of the critical section of the tooth (Figure 5.9a). If $h_{aw0} \approx 0$, then $\rho_{f3} \approx 0$. Therefore, it is advisable to be within the limits:

$$h_{aw0} = (0.6 \div 1.0)m.$$

The increase of the tooling mesh pressure angle α_{w03} occurs with the increase of x_3 and with increasing (by absolute value) the negative profile shift coefficient x_0 of the cutter

($x_0 < 0$). Then, the difference $x_{d03} = x_3 - x_0$ becomes greatest, the tooling mesh center distance a_{w03} increases, and the diameter d_{w0} of the machining operating circle (machining centroid) of the cutter approaches its tip circle diameter d_{a0}.

From the above, it is not difficult to estimate that the danger of $\rho_{f3} \approx 0$ increases when working with a maximally resharpened cutter (Figure 5.10) and with a high value for profile shift coefficient x_3 of the ring gear 3. This explains why it is advisable (Section 5.1) to choose x_3 within moderate limits [65, 175]

$$x_3 = 0 \div +0.65.$$

From the above, it can also be conceived that in the case of fixed z_3 and x_3, if with a particular gear, $\rho_{f3} \approx 0$ is obtained, a different cutter with other geometric parameters must be used.

All this calls for special attention to the fact that it is best for the designer to work closely with the tooting technologist.

5.8 Peculiarities of Internal Tooting and Internal Meshing

Finally, it is appropriate to briefly summarize and clearly formulate the peculiarities of internal tooting and internal meshing (Figure 5.4).

1. The tooth profile of both the external and the internal teeth consists of four sections: a tip circle, an involute flank, a fillet curve, and a root circle.
2. The addendum of internal teeth h_{a3} of ring gear 3 is reduced with $\Delta h_3 = \Delta h_3^* \cdot m$ in order to avoid interference with fillet curve of planet 2.
3. The dedendum h_{f3} of internal teeth is not constant, but depends not only on the geometrical parameters (z_3, x_3) of ring gear 3 but also on the current status of the geometrical parameters of the cutter (z_0, x_0) after each resharpening, i.e.,

$$h_{f3} = f(m, z_3, x_3, z_0, x_0) \neq \text{const};$$

4. It follows from the above that the root diameter is also not constant, i.e.,

$$d_{f3} = f(m, z_3, x_3, z_0, x_0) \neq \text{const},$$

   and hence the height of the internal teeth

$$h_3 = h_{a3} + h_{f3} \neq \text{const};$$

5. The fillet curve of the internal teeth is an extended (very rarely—shortened) hypocycloid whose radius of curvature ρ_{f3} is different at each point.
6. The beginning of the fillet curve, i.e., the boundary point G_3 (Figures 5.8 and 5.9), also depends on the abovementioned parameters—m, z_3, x_3, z_0, x_0, and changes with each resharpening of the cutter, as the point G_3 is shifted to a larger diameter d_{G3}, where the involute flank extends and the fillet curve shortens.

7. The same applies to the curvature radius of the fillet curve ρ_{f3}, which is not a constant but depends on the parameters mentioned above:

$$\rho_{f3} = f\left(m, z_3, x_3, z_0, x_0\right) \neq \text{const}$$

and with their unfavorable combination, $\rho_{f3} = 0$ may be obtained.

6

Forces and Torques in $\overline{AI}$-Planetary Gear Train

6.1 Preconditions for Forces and Torques Determination

The two tangential forces F_{t12} and F_{t32} that load planets 2 in both meshings—the external one between sun gear 1 and planets 2, and the internal one between planets 2 and ring gear 3 (Figure 6.1), can be determined under the following idealized conditions:

1. A stationary load is assumed; i.e., all gears (1, 2, and 3) and carrier H rotate evenly, so there are no additional inertial forces, i.e.,
   $\omega_1 = \text{const}$, $\omega_2 = \text{const}$, $\omega_3 = \text{const}$ (most often $\omega_3 = 0$), and $\omega_H = \text{const}$.
2. It is assumed that all planets are loaded evenly, i.e., the mesh load factor K_γ (Chapter 8), which represents the unevenness of their load, is $K_\gamma = 1$.
3. It is assumed that planetary gear train (PGT) operates without losses; i.e., its basic efficiency η_0 (Chapter 11) with a fixed carrier H and power transmitting in both directions—from sun gear 1 to ring gear 3 and vice versa, is

$$\eta_0 = \eta_{13(H)} = \eta_{31(H)} = 1.$$

In these assumptions, the tangential forces F_{t12} and F_{t32} acting on planets 2 are calculated as follows, starting from the torque T_1 acting on sun gear 1:

$$F_{t12} = F_{t32} = F_t = 2000\frac{T_1}{k \cdot d_1} = \text{const}, \tag{6.1}$$

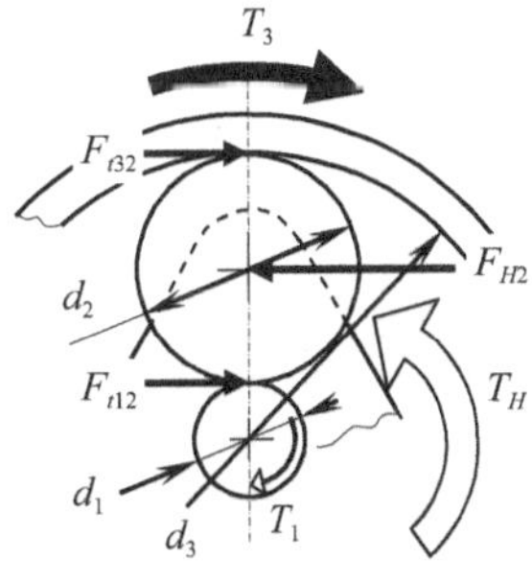

FIGURE 6.1
A planet loaded by tangential forces F_{t12} and F_{t32}.

where T_1 is the torque of sun gear 1, Nm, k is the number of planets, and d_1 is the reference diameter of sun gear 1, mm.

These two forces can be designated as *nominal.* They do not take into account the dynamic effects as well as the losses in the gear train. They can also be called *ideal,* unlike the real forces that consider losses. Of course, in an operating gear train, as it will be seen later (Chapters 8 and 11), neither $\eta_0 = 1$ nor $K_\gamma = 1$, and in the nonstationary mode, the angular velocities ω are constant.

Apart from the two tangential forces F_{t12} and F_{t32}, the planets 2 are loaded with their axes and with the carrier H, respectively, with force F_{H2}. This force is opposite to the tangential forces F_{t12} and F_{t32}, and in absolute value (though approximately in shifted mesh but sufficiently precise), it is equal to their sum, i.e.,

$$F_{H2} = F_{t12} + F_{t32} = 2F_t. \tag{6.2}$$

6.2 Ideal External Torques

From the equilibrium of the two tangential forces $F_{t12} = F_{t32} = F_t$ acting on different radii r_1 and r_3 of the reference circles of sun gear 1 and ring gear 3, the torque of ring gear T_3 is determined at a certain torque of sun gear T_1. It can, of course, be determined by basic speed ratio i_0:

$$T_3 = +\frac{r_3}{r_1}T_1 = +\frac{z_3}{z_1}T_1 = -i_0 \cdot T_1. \tag{6.3}$$

As can be seen from Figure 6.1 (as in Figures 2.5 and 3.1), these two torques are unidirectional (because $i_0 < 0$!), and the third torque T_H of carrier H is opposite, and in absolute value, it is equal to their sum, i.e.,

$$T_H = -(T_1 + T_3). \tag{6.4}$$

Here follows the condition (2.2):

$$\boxed{\sum T_i = T_1 + T_3 + T_H = 0,} \tag{6.5}$$

given in Section 2.4, which is available regardless of the operating mode of the gear train. In size the torques are arranged as follows (2.4):

$$T_1 < T_3 < |T_H|. \tag{6.6}$$

It is noteworthy that under the precondition ($\eta_0 = 1$), the *ideal external torques* are meant here, not real ones (see Section 10.2).

6.3 Torque Ratio

For convenience in further work, including the facilitation of kinematic analysis, here a new ratio is defined, called *torque ratio* t [13], which is the ratio of the unidirectional ideal external torques, namely, the larger one to the smaller one, so that it is always $t>+1$:

$$\boxed{t = \frac{T_3}{T_1} = +\frac{z_3}{z_1} = -i_0 > +1.} \tag{6.7}$$

As can be seen, the torque ratio t of $\overline{AI}$-PGT in absolute value is equal to the basic speed ratio i_0. This new ratio, which is a power characteristic of the train, complements the other PGT characteristics [120, 175] very well:

- *Speed ratio* $i = \omega_A/\omega_B$, where ω_A and ω_B are the input and output angular velocities. This ratio is a *kinematic characteristic* of each train (not only of the gear or PGT!).
- *Teeth ratio*—in case of $\overline{AI}$-PGT, they are two—of the external meshing $u_{12} = z_2/z_1 > 1$ (less often $u_{21} = z_1/z_2 > 1$, when $z_2 < z_1$) and of the internal meshing $u_{23} = z_3/z_2 > 1$, and they are *geometric characteristics* of mating gear pairs *only* (!) [120, 175].

In this way, a set of three gear train characteristics is obtained:

- Kinematic i
- Geometric u_{12} and u_{23}
- Power t

which characterize it to the fullest extent.

Through the torque ratio t, three ideal external torques T_1, T_3, and T_H can be expressed in the following way [see also Section 7.4.1 and formula (7.26)]:

$$\boxed{T_3 = +t \cdot T_1} \text{ and } \boxed{T_H = -(T_1 + T_3) = -(1+t)T_1.} \tag{6.8}$$

They are at definite invariable ratio. This question is explained in the next Chapter 7, where the kinematics of the PGT is examined through the torque method—a very convenient method not only for the kinematic analysis but also for efficiency determination (see Chapter 12).

7

Kinematics of $\overline{AI}$-Planetary Gear Train

Various methods for the kinematic analysis of planetary gear trains (PGTs) exist [147, 151, 167, 194, 237, 248]. Only the most commonly used will be displayed here. It should be borne in mind that the number of planets k does not affect the planetary gear kinematics, i.e., its gear ratio $i \neq f(k)$.

7.1 Analytical Method of Willis

7.1.1 Essence of the Method

In this method [248], the relative movement of the gears—sun 1 and ring 3—with their relative angular velocities $\omega_{1\,rel}$ and $\omega_{3\,rel}$ with respect (relative) to the carrier H is considered. The carrier itself undergoing inversion (setting the reverse angular velocity ω_H to all central elements) is considered stationary. In this way, PGT converts to non-planetary (fixed-axis gear train), and its analysis is simplified. Thus, the basic ratio i_0 is determined by angular velocities ω (or rotation speeds n) according to the following formula:

$$\boxed{i_{13(H)} = \frac{\omega_{1\,rel}}{\omega_{3\,rel}} = \frac{\omega_1 - \omega_H}{\omega_3 - \omega_H} = i_0 = -\frac{z_3}{z_1} < 0.} \tag{7.1}$$

This formula is fundamental for the method of Willis and can be transformed to

$$\boxed{\omega_1 - i_0 \cdot \omega_3 - (1 - i_0)\omega_H = 0.} \tag{7.2}$$

Special attention should be paid here to a vicious practice which unfortunately is spread widely in Bulgaria (and not only). This is the place of the index of fixed (stationary) gear train element in the symbol of speed ratio i. The vicious practice is to place the index where in mathematics it is explicitly defined as an exponent. The record i_{1H}^3 means nothing more than the speed ratio is raised to the third power, instead of meaning that the ring gear 3 is fixed. Therefore, in this book, the fixed gear train element is correctly marked in the brackets—where it belongs, i.e.,

$i_{1H(3)}$, not i_{1H}^3.

In this book, the following indices are used:

A—input shaft (element)

B—output shaft (element)

C—fixed (stationary) shaft (element)

Below are given possible modes of work of PGT, with $F = 1$ and $F = 2$ degrees of freedom.

7.1.2 Work of PGT with $F = 1$ Degree of Freedom

At $F = 1$ degree of freedom, three cases of fixed element (sun gear 1, ring gear 3, or carrier H) are possible. Each of them has two modes of work—as a reducer or as a multiplier (Figure 3.2).

At $\omega_3 = 0$—the most often case

- Work as a reducer

$$i_{1H(3)} = \frac{\omega_1}{\omega_H} = \frac{\omega_A}{\omega_B} = 1 - i_0 = 1 + \frac{z_3}{z_1} > +1; \tag{7.3}$$

- Work as a multiplier

$$i_{H1(3)} = \frac{\omega_H}{\omega_1} = \frac{\omega_A}{\omega_B} = \frac{1}{1 - i_0} = \frac{1}{1 + \frac{z_3}{z_1}} < +1. \tag{7.4}$$

At $\omega_1 = 0$

- Work as a reducer

$$i_{3H(1)} = \frac{\omega_3}{\omega_H} = \frac{\omega_A}{\omega_B} = 1 - \frac{1}{i_0} = \frac{i_0 - 1}{i_0} = 1 + \frac{z_1}{z_3} > +1; \tag{7.5}$$

- Work as a multiplier

$$i_{H3(1)} = \frac{\omega_H}{\omega_3} = \frac{\omega_A}{\omega_B} = \frac{1}{1 - \frac{1}{i_0}} = \frac{i_0}{i_0 - 1} = \frac{1}{1 + \frac{z_1}{z_3} < +1.} \tag{7.6}$$

At $\omega_H = 0$

- Work as a reducer

$$i_{13(H)} = \frac{\omega_1}{\omega_3} = \frac{\omega_A}{\omega_B} = i_0 = -\frac{z_3}{z_1} < -1, \text{resp.} \left|i_{13(H)}\right| > 1, \tag{7.7}$$

- Work as a multiplier

$$i_{31(H)} = \frac{\omega_3}{\omega_1} = \frac{\omega_A}{\omega_B} = \frac{1}{i_0} = -\frac{z_1}{z_3} > -1, \text{resp.} \left|i_{31(H)}\right| < 1. \tag{7.8}$$

Obviously, in this third case, the PGT does not work as a planetary one and the term *pseudo-PGT* is appropriate.

7.1.3 Work of PGT with $F = 2$ Degrees of Freedom

Willis's formula, of course, can also be used when the PGT works with $F = 2$ degrees of freedom (as a differential)—as a summation or as a division gear train. Different combinations of the two drives or driven elements (shafts) are possible. However, the most common cases are known from practice, so they will mostly be considered here.

When a PGT operates with $F = 2$ degrees of freedom, one cannot talk about speed ratio i. In this case, it is of interest to determine the angular velocities ω (absolute and relative) of the gears (7.11) and carrier. This issue at $F = 1$ degree of freedom is not a difficulty, and therefore, it is not discussed above.

Work as a summation PGT

The most often case is when both sun gear 1 and ring gear 3 are drives; i.e., their shafts are input ones. The shaft of the carrier H is an output one. Its angular velocity ω_H at given angular velocities ω_1 and ω_3 is defined by the following formula, starting from the formula (7.2)

$$\boxed{\omega_H = \frac{\omega_1 - i_0 \cdot \omega_3}{1 - i_0}.} \tag{7.9}$$

The same result, moreover, occurs if the *superposition method* is used—formulae (7.20) and (7.21) (see Section 7.3.3)

$$\omega_H = i_{H1(3)} \cdot \omega_1 + i_{H3(1)} \cdot \omega_3 = \frac{\omega_1}{i_{1H(3)}} + \frac{\omega_3}{i_{3H(1)}}.$$

When the two drive elements are other, e.g., 1 and H or 3 and H, the corresponding formulae are

$$\begin{aligned} \omega_3 &= \frac{\omega_1 - (1 - i_0)\omega_H}{i_0} = i_{31(H)} \cdot \omega_1 + i_{3H(1)} \cdot \omega_H = \frac{\omega_1}{i_{13(H)}} + \frac{\omega_3}{i_{H3(1)}}, \\ \omega_1 &= i_0 \cdot \omega_3 + (1 - i_0)\omega_H = i_{13(H)} \cdot \omega_3 + i_{1H(3)} \cdot \omega_H = \frac{\omega_3}{i_{13(H)}} + \frac{\omega_3}{i_{H3(1)}}. \end{aligned} \tag{7.10}$$

Knowing the angular velocity of the carrier ω_H, the *relative angular velocities* of the gears and bearings (of interest for their calculation) can be determined

$$\begin{aligned} \omega_{1rel} &= \omega_{1(H)} = \omega_1 - \omega_H, \\ \omega_{3rel} &= \omega_{3(H)} = \omega_3 - \omega_H, \\ \omega_{2rel} &= \omega_{2(H)} = -\omega_{1rel}\frac{z_1}{z_2} = +\omega_{3rel}\frac{z_3}{z_2}. \end{aligned} \tag{7.11}$$

Work as a division PGT

Here, things are more special. As shown in Figure 7.1, PGT can be used, e.g., for the opposite rotation of the propellers of twin-propeller aircraft engine, wherein one of them is connected with the ring gear 3, and the other one—with the carrier H. Since the torques of ring gear 3 and carrier H are different in size ($T_3 < |T_H|$), and the two propellers are with the same shape, and here, they can be assumed to have the same mechanical characteristic $T_B = f(\omega)$ (regardless of the disturbed air flow), the angular velocities ω_3 and ω_H will necessarily be different, as shown in the figure $\omega_3 < \omega_H$. This case is discussed in Section 7.4.3 using the torque method, which is more suitable for this purpose.

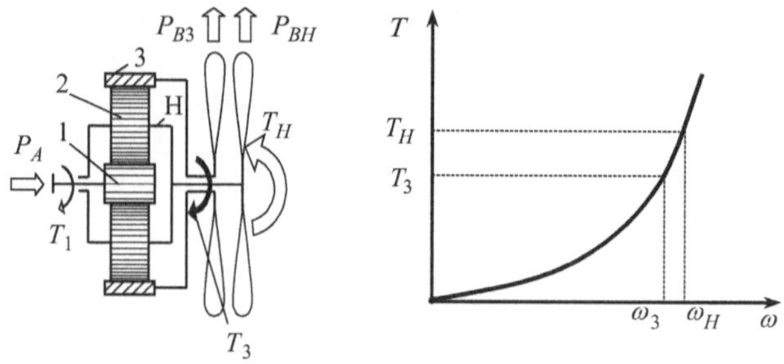

FIGURE 7.1
Work of PGT with $F = 2$ degrees of freedom as a division differential.

7.1.4 Method Estimation

Willis's method is undoubtedly accurate and versatile, and that is his great advantage. It has proven its qualities for nearly two centuries since its creation [248]. However, it is deprived of any kind of clarity. Besides, its application to complicated compound multi-carrier PGTs becomes difficult requires significant computational work with the risk of error, and there is no simple way to verify results. The most meticulous repetition of the calculations is necessary.

7.2 Graphical Method of Kutzbach

7.2.1 Essence of the Method

The graphical method is newer than that of Willis [151]. Unlike Willis's method, which uses angular velocities ω (or rotational speeds n), the graphical method works with a diagram of the peripheral velocities of the gears.

7.2.2 Work of PGT with $F = 1$ Degree of Freedom

Figure 7.2 shows diagrams of velocities and corresponding speed ratios for three possible modes of work of the train with $F = 1$ degree of freedom. The speed ratios are expressed by the ratio of the segments that are plotted on the *measuring line* (Figure 7.2) and are proportional to the angular velocity of the respective gear ($v = r \cdot \omega \approx r \cdot \tan\alpha$). In Figures 7.2 and 7.3 are given both the absolute and the relative angular velocities $\omega_{1(H)} \equiv \omega_{1rel}$, $\omega_{2(H)} \equiv \omega_{2rel}$ and $\omega_{3(H)} \equiv \omega_{3rel}$ of the three gears 1, 2, and 3 with respect to the carrier H.

7.2.3 Work of PGT with $F = 2$ Degrees of Freedom

Figure 7.3 shows a velocity diagram of PGT with $F = 2$ degrees of freedom when two angular velocities are known—ω_1 and ω_3.

7.2.4 Method Estimation

Unlike the analytical method, the graphical method is visual, and this is its great advantage. However, working with segments (Figures 7.2 and 7.3) cannot provide accurate

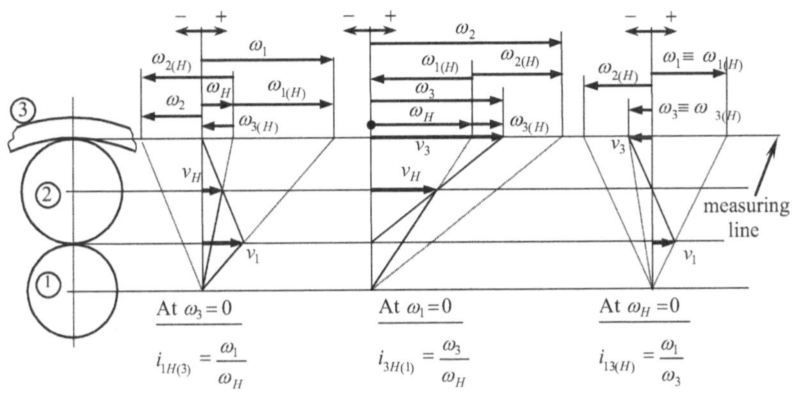

FIGURE 7.2
Velocity diagrams of a PGT with $F = 1$ degree of freedom.

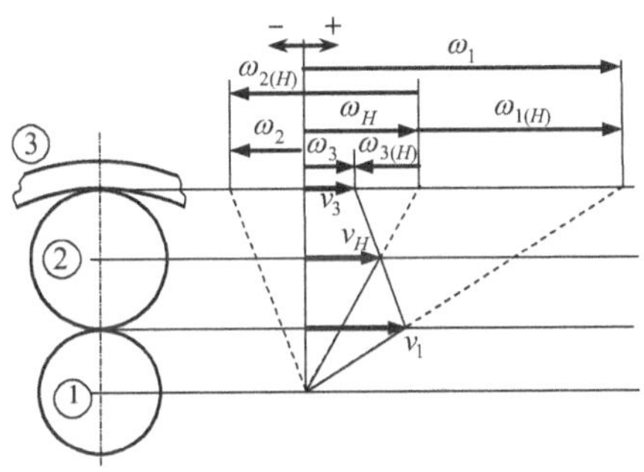

FIGURE 7.3
Velocity diagram of a PGT with $F = 2$ degrees of freedom.

results. Therefore, the graphical method is often converted into grapho-analytical, using the radius of the gears, which in turn is proportional to the corresponding number of teeth.

Nowadays, the graphical and grapho-analytical methods have a very limited application, more often for illustration, thanks to their clarity. At complex multi-carrier PGTs, such a tangle of lines is produced that working with this method is not only difficult, but it is simply impossible.

7.3 Superposition Method, Table Method of Swamp

7.3.1 Essence of the Method

In this method [236], the motion of each of the PGT elements is represented by the sum of two movements (rotations, see Section 2.1):

- *Coupling movement* of PGT which rotates as a whole (as a rigid coupling) without relative rotation of gears to the carrier
- *Relative movement* of gears and their shafts with respect to the carrier

In the tables, the different possible modes of work of PGT with both $F=1$ and $F=2$ degrees of freedom are investigated.

Initially, the entire gear train is rotated in a single turn as a whole (the first movement). Then, depending on which element is fixed (most often $\omega_3=0$, less often $\omega_1=0$), this element rotates one turn at a fixed carrier H (already rotated to a single turn), whereby the third element rotates further. This is the second relative movement mentioned above. In this way, the fixed element returns to its starting position, as though from the very beginning it was invariably immobile. This is illustrated in the tables, bearing in mind that the angular velocities ω_1, ω_3, and ω_H (also the rotational speeds n_1, n_3, and n_H) are proportional to the angles of rotation φ_1, φ_3, and φ_H of the elements 1, 3, and H of the gear train at the same time.

7.3.2 Work of PGT with $F=1$ Degree of Freedom

At $\omega_3=0$—the most often case

Element	φ_1	φ_3	φ_H
Coupling movement	+1	+1	+1
Relative movement	$+\frac{z_3}{z_1}$	−1	0
Result	$1+\frac{z_3}{z_1}$	0	+1

- Work as a reducer

$$i_{1H(3)}=\frac{\omega_1}{\omega_H}=\frac{\omega_A}{\omega_B}=\frac{\varphi_1}{\varphi_H}=\frac{1+\frac{z_3}{z_1}}{1}=1+\frac{z_3}{z_1}=1-i_0>+1; \tag{7.12}$$

- Work as a multiplier

$$i_{H1(3)}=\frac{\omega_H}{\omega_1}=\frac{\omega_A}{\omega_B}=\frac{\varphi_H}{\varphi_1}=\frac{1}{i_{1H(3)}}\frac{1}{1+\frac{z_3}{z_1}}=\frac{1}{1-i_0}<+1. \tag{7.13}$$

At $\omega_1=0$

Element	φ_1	φ_3	φ_H
Coupling movement	+1	+1	+1
Relative movement	−1	$+\frac{z_1}{z_3}$	0
Result	0	$1+\frac{z_1}{z_3}$	+1

- Work as a reducer

$$i_{3H(1)}=\frac{\omega_3}{\omega_H}=\frac{\omega_A}{\omega_B}=\frac{\varphi_3}{\varphi_H}=\frac{1+\frac{z_1}{z_3}}{1}=1+\frac{z_1}{z_3}=1-\frac{1}{i_0}=\frac{i_0-1}{i_0}>+1; \tag{7.14}$$

- Work as a multiplier

$$i_{H3(1)} = \frac{\omega_H}{\omega_3} = \frac{\omega_A}{\omega_B} = \frac{\varphi_H}{\varphi_3} = \frac{1}{i_{3H(1)}} = \frac{1}{1+\frac{z_1}{z_3}} = \frac{1}{1-\frac{1}{i_0}} = \frac{i_0}{i_0 - 1} < +1. \tag{7.15}$$

At $\omega_H = 0$

In this case, it was initially known that

- Work as a reducer

$$i_{13(H)} = \frac{\omega_1}{\omega_3} = \frac{\omega_A}{\omega_B} = \frac{\varphi_1}{\varphi_3} = i_0 = -\frac{z_3}{z_1} < -1, \text{resp.} \left|i_{13(H)}\right| > 1; \tag{7.16}$$

- Work as a multiplier

$$i_{31(H)} = \frac{\omega_3}{\omega_1} = \frac{\omega_A}{\omega_B} = \frac{\varphi_3}{\varphi_1} = \frac{1}{i_0} = -\frac{z_1}{z_3} > -1, \text{resp.} \left|i_{31(H)}\right| < 1, \tag{7.17}$$

so here superposition is not necessary.

7.3.3 Work of PGT with $F = 2$ Degrees of Freedom

If sun gear 1 and ring gear 3 are assumed to be driving, $\omega_1 \neq 0$ and $\omega_3 \neq 0$ are given. The angular velocity of the carrier ω_H is as follows:

It is initially assumed, e.g., that $\omega_3 = 0$ and sun gear 1 is driving ($\omega_1 \neq 0$), whereby it is obtained:

$$\omega_H = i_{H1(3)} \cdot \omega_1 = \frac{\omega_1}{i_{1H(3)}} = \frac{1}{1-i_0} \cdot \omega_1 = \frac{1}{1+\frac{z_3}{z_1}} \cdot \omega_1. \tag{7.18}$$

Similarly, it is assumed $\omega_1 = 0$ and obtained:

$$\omega_H = i_{H3(1)} \cdot \omega_3 = \frac{\omega_3}{i_{3H(1)}} = \frac{i_0}{i_0 - 1} \cdot \omega_3 = \frac{1}{1+\frac{z_1}{z_3}} . \omega_3, \tag{7.19}$$

or ultimately as a sum is obtained:

$$\boxed{\omega_H = i_{H1(3)} \cdot \omega_1 + i_{H3(1)} \cdot \omega_3 = \frac{\omega_1}{i_{1H(3)}} + \frac{\omega_3}{i_{3H(1)}}.} \tag{7.20}$$

Taking into account from above that $i_{H1(3)} = \frac{1}{1-i_0}$ and $i_{H3(1)} = \frac{i_0}{i_0 - 1}$ finally, the following expression is obtained:

$$\boxed{\omega_H = \frac{1}{1-i_0} \cdot \omega_1 + \frac{i_0}{i_0 - 1} \cdot \omega_3 = \frac{\omega_1 - i_0 \cdot \omega_3}{1 - i_0},} \tag{7.21}$$

which is the same as that determined by the method of Willis (7.9).

7.3.4 Method Estimation

The Swamp's method has the advantage that it clearly separates the movement of the PGT elements into two constituent movements: coupling and relative. This is also very useful for some of the further considerations here (Section 10.1). Generally, however, this method is rarely used.

7.4 Torque Method

7.4.1 Essence of the Method

In contrast to the methods of Willis and Kutzbach which use the angular velocities ω (or the rotation speed n) and peripheral velocities v, for the analysis of the PGTs, with the method revealed later, the torques of the three central PGT elements are used. These are the ideal external torques (without consideration of the gear train internal losses) T_1, T_3, and T_H and the real external torques (with consideration of the losses) T_1', T_3', and T_H' of the sun gear 1, the ring gear 3, and the carrier H. As it will be seen later, the ideal external torques are used for the determination of the speed ratio i, and the real one—for the determination of the gear train efficiency η.

Nature of the torque method developed and presented originally in [13], and also further developed and presented in consequence in series of publications of the authors [18, 22–26, 30, 31, 130, 137] is the following:

1. The well-known *symbol of Wolf* [254] is used (Figure 7.4), at which the simple PGT due to its circular form is displayed by a circle and the three shafts coming out of it—with lines. With the aim of achieving maximal visualization, here the symbol of Wolf is used, but not in its initial appearance, but *modified*, as the three PGT shafts are marked where appropriate with different lines, as follows:
   - Both shafts with unidirectional ideal external torques T_1 and T_3 (of the sun gear 1 and the ring gear 3) are marked with single lines, but with different thickness, according to the size of the corresponding torque. As $T_1 < T_3$, the shaft of the sun gear 1 is marked with a thin single line and the shaft of the ring gear 3—with a thick single line.
   - The shaft of the carrier H is marked with a double line, because its torque T_H is the biggest one [254]. It is in opposite direction of the other two and equal in an absolute value to the sum (see Chapter 6).

$$T_H = -(T_1 + T_3). \tag{7.22}$$

   In this case, the three *ideal external torques* are arranged in size in the following way (2.3), (6.6):

$$T_1 < T_3 < |T_H|. \tag{7.23}$$

2. As defined in Chapter 6 (6.7), *torque ratio* t of the unidirectional torques is used:

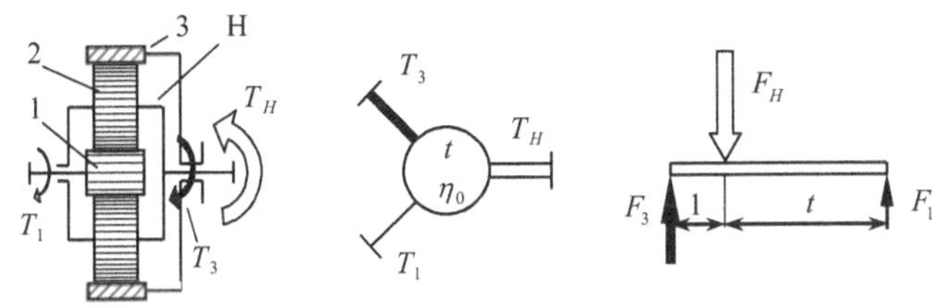

Precondition: $\eta_0 = \eta_{13(H)} \approx \eta_{31(H)} = 1$

$$t = \frac{T_3}{T_1} = \frac{F_3}{F_1} > +1$$

Torques:	Forces:
$T_1 < T_3 < \lvert T_H \rvert$	$F_1 < F_3 < \lvert F_H \rvert$
$T_1 : T_3 : T_H = +1 : +t : -(1+t)$	$F_1 : F_3 : F_H = +1 : +t : -(1+t)$
$\sum T_i = T_1 + T_3 + T_H = 0$	$\sum F_i = F_1 + F_3 + F_H = 0$

FIGURE 7.4
$\overline{AI}$-PGT—kinematic scheme, ideal external torques, modified symbol of Wolf, and lever analogy.

$$t = \frac{T_3}{T_1} = +\frac{z_3}{z_1} = -i_0 > +1. \tag{7.24}$$

It is necessary specifically to underline that the relation $t = -i_0$ is not valid with all types of PGTs and this is not a general relation. It is valid for the negative-ratio PGTs but not for the positive-ratio ones (see Sections 2.3–2.5 and Figure 2.3)!

The so-defined *torque ratio t* (see Section 6.1) appears to be very convenient to work with (not only at $\overline{AI}$-PGT). The ideal external torques can be expressed by it as follows (6.8):

$$\boxed{T_3 = +t \cdot T_1} \text{ and } \boxed{T_H = -(T_1 + T_3) = -(1+t)T_1}. \tag{7.25}$$

If we assume that $T_1 = +1$, the so-called *specific torques* are defined [193] as follows:

$$T_3 = +t \text{ and } T_H = -(1+t), \tag{7.26}$$

and it is very convenient to work with them.

Using these relations, it is very easy to establish the equilibrium of the sum of the three ideal torques and particularly:

$$\boxed{\sum T_i = T_1 + T_3 + T_H = 1 + t - (1+t) = 0}. \tag{7.27}$$

This relationship provides a good opportunity to check the accuracy of the calculations, which possibility is missing in the methods of Willis and Kutzbach.

3. It is important also to keep in mind that the three ideal external torques T_1, T_3, and T_H are always in a *permanent ratio*:

$$T_1 : T_3 : T_H = T_1 : +t \cdot T_1 : -(1+t)T_1 = +1 : +t : -(1+t), \tag{7.28}$$

irrespective of

- With how many degrees of freedom F is running the gear train, with $F = 1$ or $F = 2$
- Which element of the PGT is fixed at $F = 1$ degree of freedom
- What is the direction of power transmission, i.e., whether the PGT works as a speed reducer or a multiplier when $F = 1$, or as a summation or a division PGT at $F = 2$, i.e., as a differential
- Whether the PGT works independently, or as a part of a compound multi-carrier PGT

4. From those relationships, for the three ideal external torques of the PGT shafts, it is not difficult to establish that there is an analogy of the gear train with a lever, loaded with three forces, i.e., there is a *lever analogy*. In Figure 7.4, there is a comparison of the PGT with a straight lever, in which the complete analogy can be seen. This lever analogy with its visibility is very useful for ease of understanding and insight into the work of the PGTs.
5. *The ideal external torques*—input T_A, output T_B, and reactive T_C (of the fixed element)—are used to determine the speed ratio i in the work of the PGT with $F = 1$ degree of freedom (not with $F = 2$!). Here, the law of conservation of energy is considered with the acceptance that the gear train works without losses; i.e., the basic coefficient of efficiency is $\eta_0 = 1$. Denoting the power input with P_A, and the output one with P_B, the condition of conservation of energy is expressed as follows:

$$\boxed{\sum P_i = P_A + P_B = T_A \cdot \omega_A + T_B \cdot \omega_B = 0}. \tag{7.29}$$

Here, for the *speed ratio* is obtained:

$$\boxed{i = \frac{\omega_A}{\omega_B} = -\frac{T_B}{T_A}}. \tag{7.30}$$

6. *The real external torques* T_1', T_3', and T_H', respectively, T_A' and T_B', with the consideration of the losses in the gear train, are determined depending on the direction of the relative power P_{rel} (see Chapter 10):

- Whether P_{rel} is transmitted from the sun gear 1 through the planets 2, to the ring gear 3, or vice versa
- From the ring gear 3 to the sun gear 1

These real external torques enable the determination of the efficiency of the gear train [161, 237]. This is performed by first determining the so-called *torque transmit ratio* i_T (*torque transformation*) of the real external torques of the input A and the output B of the gear train:

$$\boxed{i_T = \frac{T_B'}{T_A'}}. \tag{7.31}$$

This ratio is used for the determination of the *coefficient of efficiency* η (see Section 12.2, formula 12.6):

$$\boxed{\eta = \frac{-i_T}{i}}. \tag{7.32}$$

Unlike the torque transmit ratio i_T, the speed ratio i is also called the *kinematic transmit ratio* and, in some cases, for reasons of clarity, is denoted by the index k, i.e., i_k.

Formula (7.32) is derived from formula (12.5):

$$\sum P_i = \eta \cdot P_A + P_B = \eta \cdot T'_A \cdot \omega_A + T'_B \cdot \omega_B = 0, \tag{7.33}$$

from where formula (7.32) is obtained

$$\boxed{\eta = -\frac{P_B}{P_A} = -\frac{T'_B \cdot \omega_B}{T'_A \cdot \omega_A} = -\frac{T'_B / T'_A}{\omega_A / \omega_B} = -\frac{i_T}{i_k}}. \tag{7.34}$$

7.4.2 Work of PGT with $F = 1$ Degree of Freedom

In Figure 7.5, the formulae are given for determining the speed ratios of the six possible cases of work of the gear train with $F = 1$ degree of freedom. For better visibility, the lever analogy is displayed with the corresponding direction of power transmission.

Figure 7.6 shows the change of the speed ratio i of the six cases and the limits, which may vary depending on the torque ratio $t = z_3/z_1$ and on the basic speed ratio i_0, respectively.

Figures 7.5 and 7.6 show that for the motion transmission between the shafts with unidirectional torques, the speed ratio is negative, and for transmitting the movement between the shafts with diverging torques—positive. This is seen very clearly at the modified symbol of Wolf (Figure 7.7).

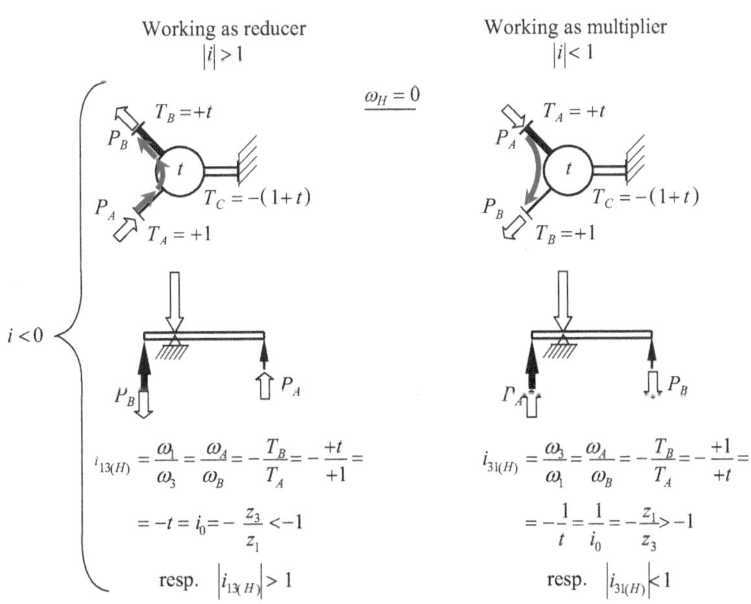

FIGURE 7.5
Speed ratios of $\overline{AI}$-PGT working with $F = 1$ degree of freedom.

(Continued)

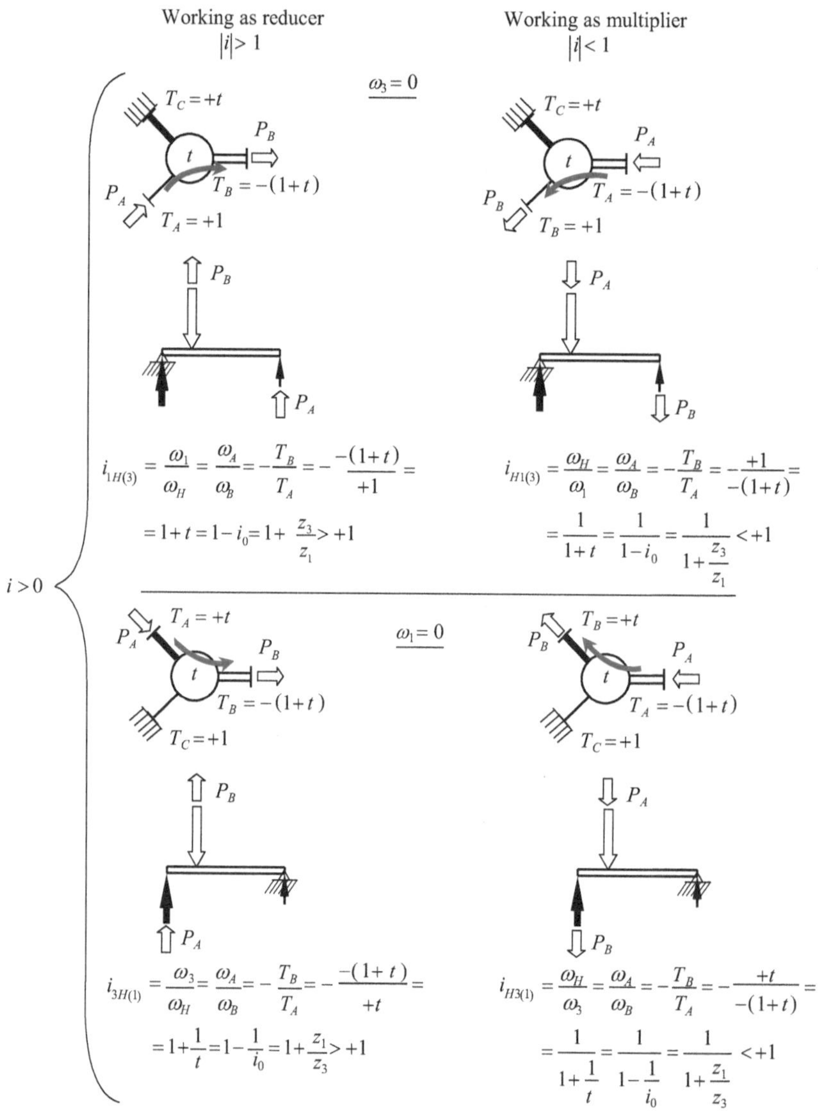

FIGURE 7.5 (CONTINUED)
Speed ratios of $\overline{\text{AI}}$-PGT working with $F = 1$ degree of freedom.

7.4.3 Work of PGT with F = 2 Degrees of Freedom

Work of PGT as a summation one

At given angular velocities ω_1 and ω_3 of the sun gear 1 and the ring gear 3, the angular velocity ω_H of the carrier is defined by the condition of the sum of the input P_{A1}, P_{A3}, and the output P_B powers, which is given by

$$\sum P_i = P_{A1} + P_{A3} + P_B = T_1 \cdot \omega_1 + T_3 \cdot \omega_3 + T_H \cdot \omega_H = 0. \tag{7.35}$$

From here, the following can be defined:

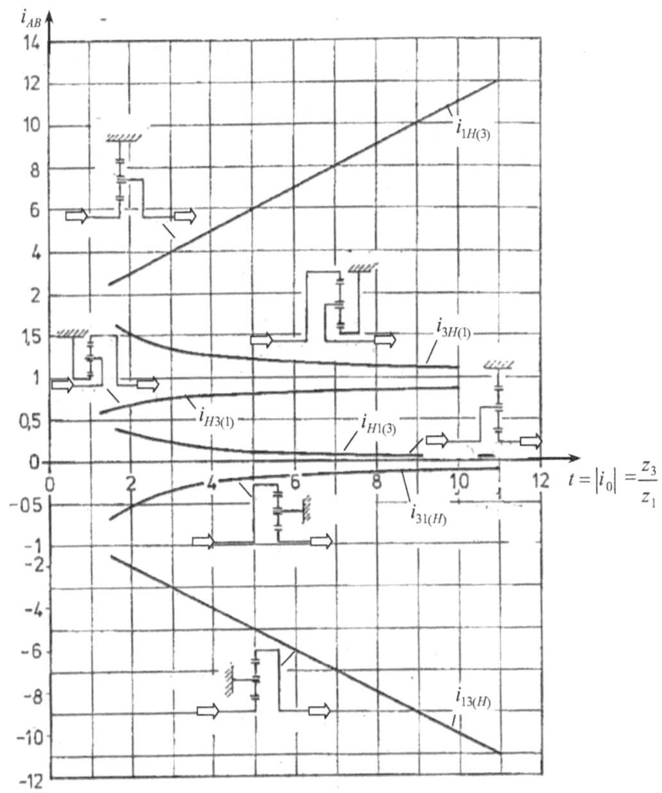

FIGURE 7.6
Speed ratio i_{AB} of $\overline{AI}$-PGT working with $F = 1$ degree of freedom depending on the torque ratio t, on the number of teeth z_1 and z_3, and the basic speed ratio i_0 respectively.

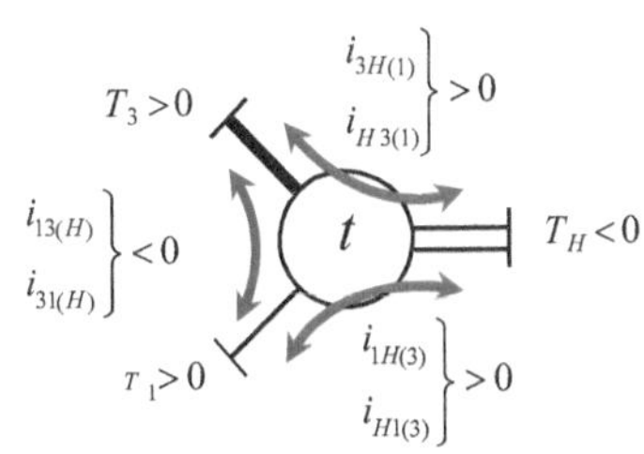

FIGURE 7.7
A practical way to determine the algebraic sign of the speed ratio i_{AB}.

$$\boxed{\omega_H = \omega_B = -\frac{T_1 \cdot \omega_1 + T_3 \cdot \omega_3}{T_H} = i_{H1(3)} \cdot \omega_1 + i_{H3(1)} \cdot \omega_3}, \tag{7.36}$$

which is the same formula as formula (7.20) determined by the superposition method (Swamp). The result is obviously the same as formula (7.9), determined by the method of Willis. Thus, the designer has three ways to establish the correctness of the results.

A formula for the output angular velocity ω_B can be derived in a similar way at a different combination of the given angular velocities.

As it is known, the angular velocity ω_H of the carrier, determining the relative angular velocities of the gear wheels $\omega_{1rel} = \omega_{1(H)}$, $\omega_{3rel} = \omega_{3(H)}$, and $\omega_{2rel} = \omega_{2(H)}$, is not a problem (see formula 7.11 of Willis's method).

In the case of twin-motor driving of sun gear 1 and ring gear 3, considered in Section 7.3.3, four speed ratios can be realized depending on whether each of the motors is switched on or off. Given that the two motors have the same angular velocities $\omega_1 = \omega_3 = \omega_A$, four speed ratios are obtained by formula (7.36) as follows:

First case: Both motors work in the same direction, i.e., $\omega_1 = \omega_3 = \omega_A$:

$$\omega_H = \omega_B = \frac{\omega_1}{i_{1H(3)}} + \frac{\omega_3}{i_{3H(1)}} = \left(\frac{1}{1+\frac{z_3}{z_1}} + \frac{1}{1+\frac{z_1}{z_3}} \right) \omega_A = f\left(|i_0| = \frac{z_3}{z_1} \right).$$

Obviously, $\omega_B = \omega_A$ is obtained and

$$i = \frac{\omega_B}{\omega_A} = +1.$$

Second case: Only the motor driving ring gear 3 works at fixed sun gear 1, i.e., $\omega_3 = \omega_A$:

$$\omega_H = \omega_B = \frac{\omega_3}{i_{3H(1)}} = \frac{\omega_A}{1+\frac{z_1}{z_3}} = f\left(|i_0| = \frac{z_3}{z_1} \right).$$

In this case, $\omega_H < \omega_A$ is obtained; i.e., the carrier rotates more slowly than the motor and

$$i = \frac{\omega_A}{\omega_B} = 1 + \frac{z_1}{z_3} > +1.$$

Third case: Only the motor driving sun gear 1 works at fixed ring gear 3, i.e., $\omega_1 = \omega_A$:

$$\omega_H = \omega_B = \frac{\omega_1}{i_{1H(3)}} = \frac{\omega_A}{1+\frac{z_3}{z_1}} = f\left(|i_0| = \frac{z_3}{z_1} \right).$$

In this case, the lower carrier velocity ω_H is obtained from the previous case and an even higher speed ratio:

$$i = \frac{\omega_A}{\omega_B} = 1 + \frac{z_3}{z_1} > +1.$$

Fourth case: Both motors work in the opposite directions, i.e., $\omega_1 = \omega_3$.

$$\omega_H = \omega_B = \frac{\omega_1}{i_{1H(3)}} + \frac{\omega_3}{i_{3H(1)}} = \left(-\frac{1}{1+\frac{z_3}{z_1}} + \frac{1}{1+\frac{z_1}{z_3}} \right) \omega_A = f\left(|i_0| = \frac{z_3}{z_1} \right).$$

In this case, the lowest carrier velocity ω_H and the highest speed ratio are obtained: $i = \omega_A/\omega_B > +1$.

By appropriately choosing the number of teeth z_1 and z_3, respectively, and the basic speed ratio i_0, desired speed ratios i, respectively, output velocities ω_H can be obtained (Figure 7.8). If the angular velocities of the motors are changed (e.g., by changing the number of poles), the number of possible output velocities increases.

Work of PGT as a division one

Figure 7.9a is an example of the PGT operation with $F = 2$ degrees of freedom as a differential division gear train for a stirrer (mixer) with two propellers, driven by an electric motor. In this case, the task is to identify the three angular velocities ω_1, ω_3, and ω_H of the three central PGT elements—sun gear 1, ring gear 3, and carrier H.

The approach to this task is the following. The mechanical characteristics of the engine are considered which drives the sun gear 1:

$$T_A \equiv T_1 = f(\omega_A),$$

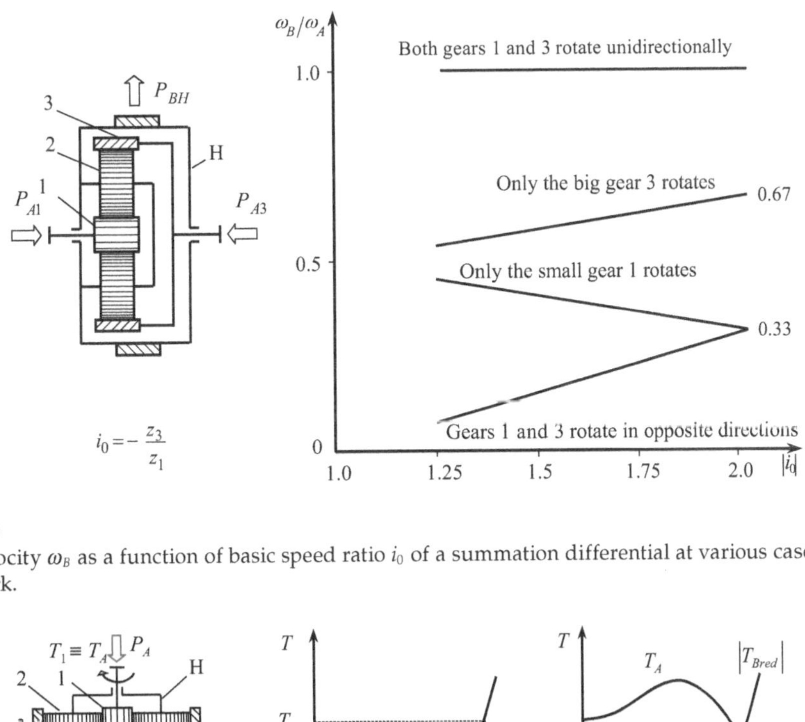

FIGURE 7.8
Output velocity ω_B as a function of basic speed ratio i_0 of a summation differential at various cases of driving motors work.

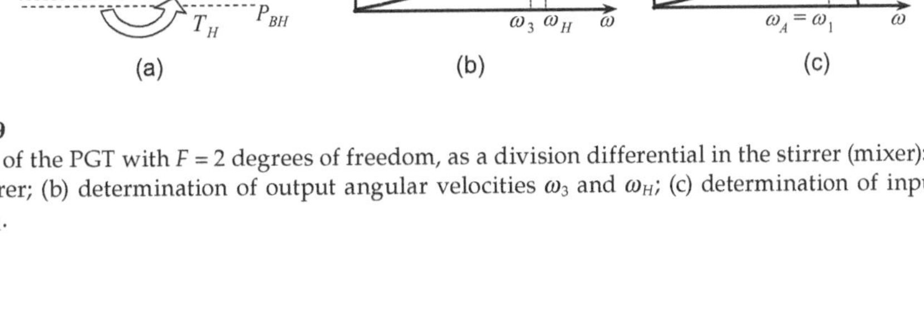

FIGURE 7.9
Operation of the PGT with $F = 2$ degrees of freedom, as a division differential in the stirrer (mixer): (a) Sketch of the stirrer; (b) determination of output angular velocities ω_3 and ω_H; (c) determination of input angular velocity ω_1.

and the mechanical characteristics of the two working bodies (propellers) associated with the ring gear 3 and the carrier H:

$$T_3 = f(\omega_3) \text{and} T_H = f(\omega_H)$$

(Figure 7.9b). These three mechanical properties can be set graphically or analytically. It is necessary that the mechanical characteristics T_3 and T_H should be adduced to the shaft of the sun gear 1, and to the engine, respectively, as a common mechanical (working) characteristic depending on the angular velocity $\omega_A = \omega_1$ of the engine $T_{B\,red} = f(\omega_A)$ (see Figure 7.1). This mechanical characteristic is the total working load of both working bodies, adduced to the input gear train shaft. Both mechanical characteristics $T_A = f(\omega_A)$ of the engine and $T_{B\,red} = f(\omega_A)$ of the working bodies are applied in a common chart, and their point of intersection defines first the angular velocity ω_1 of the sun gear 1 (Figure 7.9c) and then that of the ring gear 3 of the carrier H.

When solving the given task, we have to work in the following sequence:

We have to accept a specific numerical value for the torque T_3 of the ring gear 3 and to determine its angular velocity (it is possible to begin from carrier H) from the diagram:

$$\omega_3 = f(T_3).$$

Both torques T_3 and T_H are in a definite relation through the torque ratio t (7.26), (7.28):

$$T_H = -T_3\frac{1+t}{t}.$$

The angular velocity ω_H of the carrier H is determined by the torque T_H from the mechanical characteristic of the coupled propeller:

$$\omega_H = f(T_H).$$

The overall working load $T_{B\,red} = f(\omega_A)$ of the two working bodies reduced to the shaft of the sun gear 1 is defined by the well-known condition for equilibrium of the gear train torques—formulae (2.2), (6.5), and (7.27):

$$T_{B\,red} \equiv T_1 = T_H - T_3 \text{ or directly from } T_{B\,red} \equiv T_1 = \frac{T_3}{t}.$$

Angular velocities ω_3 and ω_H, which are different, where $\omega_H > \omega_3$, define the angular velocity ω_1 of the sun gear 1 (i.e., the motor) determined by the formula (7.10):

$$\omega_A = \omega_1 = i_{13(H)} \cdot \omega_3 + i_{1H(3)} \cdot \omega_H.$$

This will yield a point from the overall mechanical working characteristic $T_{B\,red} = f(\omega_A)$. Repeating the above calculations several times, point by point, the common mechanical characteristic (total loading) is defined:

$$T_{B\,red} = f(\omega_A).$$

The intercept point of both mechanical characteristics $T_A = f(\omega_A)$ and $|T_{B\,red}| = f(\omega_A)$ allows the determination of the torque T_1 of the sun gear 1 and its angular velocity ω_1. The torques T_3 and T_H are determined from T_1 of the working bodies using the known formulae:

$$T_3 = t \cdot T_1 \text{ and } T_H = -(1+t)T_1.$$

Knowing these torques and using the corresponding diagrams, the angular velocities ω_3 and ω_H are determined:

$$\omega_3 = f(T_3) \text{ and } \omega_H = f(T_H).$$

Knowing the torques T_1, T_3, and T_H, and the angular velocities ω_1, ω_3, and ω_H allows for the determination of the powers P_1, P_3, and P_H of the three central elements. Also, the same applies for the relative angular velocities ω_{1rel}, ω_{2rel}, and ω_{3rel}. Thus, the given task is fully solved.

Besides graphically, the given task can be solved analytically, when the mechanical characteristics of the engine and the working body are set analytically.

7.4.4 Method Estimation

From the foregoing, it can be concluded that the torque method is a *simple and practical one*, as it combines the *accuracy* of the method of Willis with the *clarity* of the method of Kutzbach, which can be found separately in each of them. This is its greatest advantage. Furthermore, except the speed ratio, the method allows for the definition of efficiency (see Chapter 12), so that it possesses more possibilities than the methods of Willis and Kutzbach. The method also has other advantages that occur and are particularly useful in the study of complex compound multi-carrier PGTs, where other methods are either difficult to apply or totally impractical. And last but not least, unlike other methods, the torque method allows for easy and quick check of the final result by the sum of the ideal external torques (7.27)—an advantage that it is not negligible. We have a similar case with the real external torques in determining the efficiency (see Chapter 12).

8

Load Distribution between the Planets in $\overline{AI}$-Planetary Gear Train, Its Unevenness, and Equalization

8.1 Causes for Uneven Load Distribution

A specific feature of all split torque gear trains [these are the planetary gear trains (PGTs) with over $k = 1$ planets] is the more or less pronounced unevenness of loading of the parallel power paths, i.e., of the individual planets in the PGT. Consequently, the problem of uneven load distribution[1] (load sharing imbalance[2]) in PGTs is their cardinal problem, and its solution is not so easy.

The reasons for the uneven load distribution are the following:

1. Inevitable manufacturing anomalies within the tolerances of gear train elements,
2. Its static indeterminacy in case of more planets, i.e., with $k > 1$.

8.2 Uneven Load Distribution Changing and Its Quantitative Expressions

Figure 8.1 shows an exemplary variation of the load F_{tj} (tangential forces) of the individual planets (j = I, II, III, … k) for a kinematic cycle of loading of the gear train, after which the load variation is repeated. Kinematic gear cycle is determined by the number of teeth z_1, z_2, and z_3, and finding their least common multiple. After each kinematic cycle, the reciprocal position of the gears is repeated.

From current loads—tangential forces F_{tj} on individual planets (j = I, II, III, …k) in every moment, and attitude (rotation angle) φ, respectively, the maximum of all the j values is taken, and thus, the second diagram for the *mesh load factor* K_γ (so-called coefficient of unequal distribution of the load between planets) is obtained. This coefficient can be expressed as follows:

$$K_\gamma = \frac{F_{tj}(\varphi)}{F_{tm}} = K_{\gamma 0} + K'_\gamma + K''_\gamma = f(\varphi) > 1. \tag{8.1}$$

[1] ISO term

[2] AGMA term

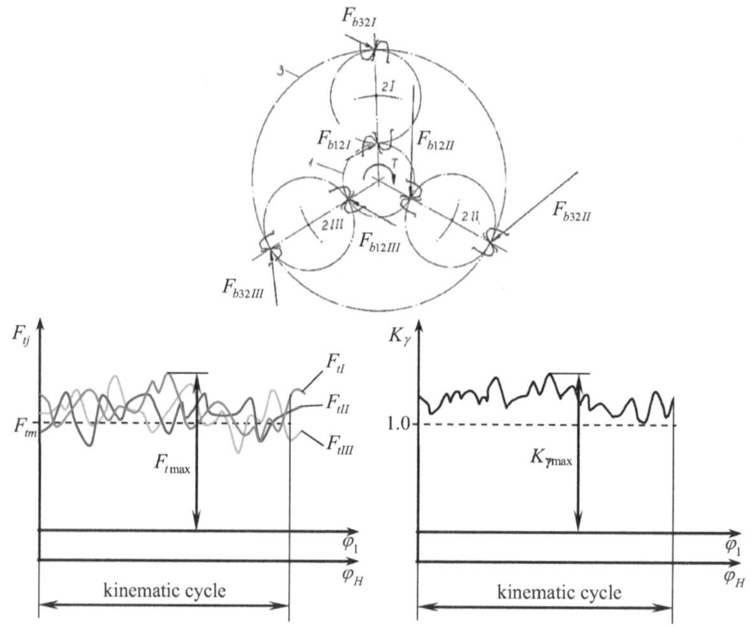

FIGURE 8.1
Variation of tangential force F_{tj} of each planet and mesh load factor K_γ.

This dependence consists of a constant value $K_{\gamma 0} = \text{const}$, a low frequency value K'_γ, and a high frequency value K''_γ. The F_{tm} is the average (nominal) value of the load (when it is evenly distributed between the planets). From this dependence, the maximal value is taken:

$$K_{\gamma max} = \frac{F_{tj\,max}}{F_{tm}} > 1, \tag{8.2}$$

that appears at all, and it is used in the calculations, although it would be right to work with the equivalent load on the most loaded tooth of any planet. Since in the design of a PGT, the exact unevenness of the load on the planets is not known, and even less their load spectrum, which is different for each gear, and even for each tooth, and hence their equivalent load, in the calculations, it is accepted to work with the maximum load coefficient in question. Obviously, this is a conscious mistake, but it is made for the sake of safety.

For this factor $K_{\gamma\,max}$, either numerical values from specific experimental results [9, 11, 14, 15, 37, 38, 47, 48, 90, 122] or recommended regulatory values [5, 237] are used.

8.3 Influencing Factors

The uneven load distribution between planets (load imbalance) is determined by various factors:

- Negative influencing
- Positive influencing

- Noninfluencing (neutral)
- Other factors

It is still here to pay special attention to the fact that the problem of uneven load distribution, the influence of various factors, and the equalization of unevenness are extremely extensive, complex, and convoluted intricate. Here, things will be dealt in a somewhat simplified and shorter manner, but still quite sufficient for practical use (for more details, see the specialized literature given at the end of Section 8.6.6).

8.3.1 Negative Influencing Factors

For individual elements of the gear train, these factors are given as follows (Figure 8.2):

Gear wheels: run-out F_r (eccentricity f_e) (see Figure 15.5) of *all* gears; difference Δs_2 of the planets tooth thickness s_2.

Carrier H: tangential misalignment f_{tHj} of the planet axes (planet centerlines) positioning, and a different central angle $\psi \neq 2\pi/k$ between them, respectively; misalignment f_{1H} between the bore of sun gear bearing and carrier bearings sites.

Planet bearings: non-equal radial clearances $j_{L2I} \neq j_{L2II} \neq j_{L2III}$—it has the same effect as the tangential misalignment f_{tHj} of the position of planet axes on the carrier H.

It is necessary to pay particular attention to the fact that the tangential misalignment f_{tHj} of the planet axes has a very negative impact, which necessitates the very precise fabrication of the carrier.

8.3.2 Positive Influencing Factors

Compliance of all the elements of the gear train: gear meshings, bearings, planets pins, shaft of the sun gear 1—bending, ring gear 3—radial compliance (provided that the deformations do not result in an uneven load distribution across the entire width of the tooth face).

Bearing clearance of the central elements—j_{L1} of sun gear 1, j_{LH} of carrier H, and j_{L3} in case of rotating ring gear 3.

8.3.3 Noninfluencing (Neutral) Factors

These are the radial misalignment f_{rHj} of the planet centerlines (I, II, III … k) positioning, and the misalignments Δa_j of center distances $a_{w12} = a_{w23} = a_w$ of meshings—a_{w12} of

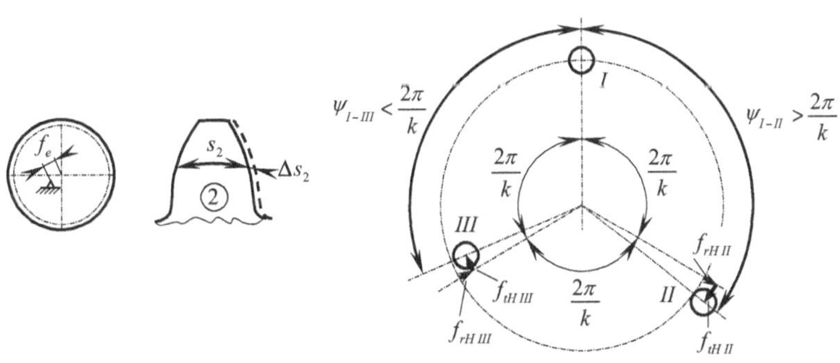

FIGURE 8.2
Negative influencing factors of the gear wheels and the carrier.

sun gear 1 with planets 2 and a_{w23} of planets 2 with ring gear 3, respectively. This is true, however, in identical pressure angles α_{w12} and α_{w23} of both meshings—external and internal. This is also true, provided that the thickness of the teeth s_1, s_2, and s_3 is such that at the available center distance a_w, there is a sufficient backlash j_{12} and j_{23} in both meshings (external and internal) in order to prevent the nondriving side of the teeth to make contact and rattle (or self-lock).

8.3.4 Other Influencing Factors

These are factors that generally affect the load distribution but cannot be influenced by gear design and fabrication.

Load—tangential force F_t on gear meshing, and input torque T_A or output torque T_B of the gear train, respectively.

Angular velocity—most often it is given ω_1 of the sun gear, rarely ω_H of the carrier.

Attitude (rotation angle)—angle φ, of the input (φ_A) or output (φ_B) of the gear train.

Friction forces—they are important for kinematic equalizing (distribution [88]; allaying [29]) devices because they reduce their sensitivity ξ (see Section 8.6.5).

In PGTs without a kinematic equalizing (distribution) device, the principal change of the mesh load factor $K_{\gamma\,max}$ for some of the influence factors is shown in Figure 8.3 [42]. It provides useful information to a designer.

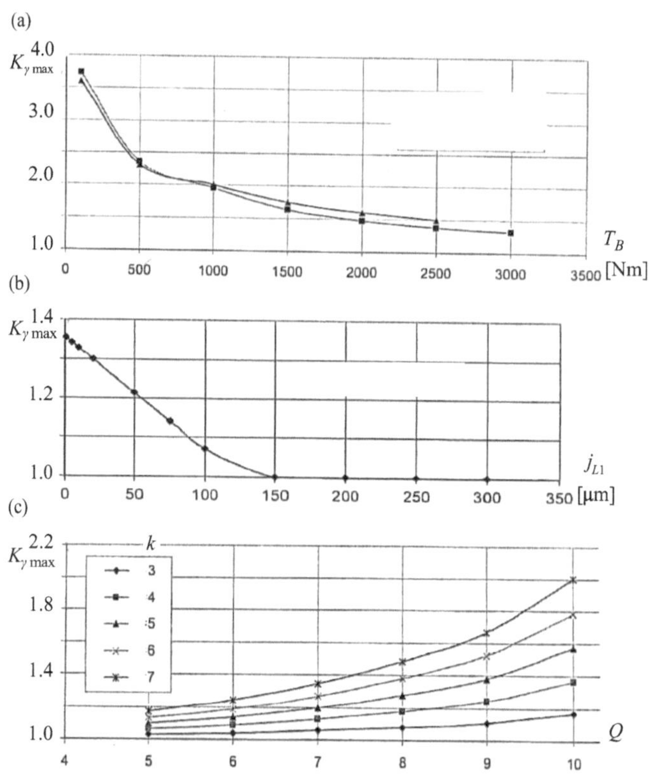

FIGURE 8.3
Influence of the main factors on mesh load factor $K_{\gamma\,max}$: (a) Output torque T_B; (b) clearance j_{L1} of sun gear 1 bearings; (c) accuracy grade Q and number of planets k. (Courtesy of Baumann, F. *Theoretische Untersuchungen zur Lastaufteilung und Lastverteilung in Planetengetrieben*. Dissertation. Dresden: Technische Universität, 2012.)

For equalizing (distribution) devices acting on a kinematic principle, the combination of the following factors is important [6, 7]:

Mass—force—path (displacement).

It concerns the mass floating under a certain active force, whereby the mass has to make a certain displacement determined by the tolerances (allowances) of the dimensions of the gear train elements in order to achieve load equalization.

For kinematic equalizing (distribution) devices, their sensitivity ξ of the equalization process is also significant (see Sections 8.6.5 and 8.17).

8.4 Possibilities for Theoretical Determination of the Unevenness of Load Distribution

In designing PGTs, of course, it is highly desirable that the designer can determine the mesh load factor $K_{\gamma\, max}$ in this case in a short, simple, and reliable way. With the large number of factors listed above, and their previously unknown size in the said case, and their unknown combination, this can be done neither in short nor in simple, but often not in a credible way as well. Therefore, in this short book, this option is not considered. More details one can find in specialized literature [42–44, 118, 147, 165, 181, 196, 211, 233]. Because of abovementioned reasons, the experimental determination of the unevenness of load sharing and of the load mesh factor $K_{\gamma\, max}$ remains the most reliable. Therefore, together with the use of the accumulated experience, it is recommended in the design.

8.5 Ways for Experimental Determination of the Unevenness of Load Distribution

Experimental studies are mainly done for two reasons:

- Need for fundamental research to give general recommendations [9, 45, 105, 122, 159, 199].
- Ascertaining the actual load unevenness and mesh load factor $K_{\gamma\, max}$ of a concrete PGT for the purpose of verifying the reliability of the calculations and the expected durability of the gears and gear train (Section 8.7 and Figure 8.20). Such tests are done with particularly responsible PGTs, or when the customer explicitly asks for testing and such is enclosed in the gear train manufacturing agreement.

While in fundamental research, this is done for the various influencing factors: load, angular velocity, choice of the floating element, different quality (different geometric misalignments) of the gear train elements, sensitivity of the equalizing (distribution) device, compliances, etc. in the study of a concrete already manufactured PGT for the practice, one can only investigate the impact of two factors: the gear train load T and the angular velocity ω.

8.5.1 Test Rigs (Benches)

Different test rigs exist. The simplest one consists of an electric motor, a tested gear train, and a friction brake (Figure 8.4a). This rig is suitable for testing low-power gear trains because all drive power is lost.

Another possibility is to combine two gear trains with the same speed ratio, where the tested PGT works as a reducer and the other (which may be non-planetary) as a multiplier (Figure 8.4b). In this case, the output angular velocity of the multiplier is equal to input velocity of the tested PGT. Multiplier's output shaft is connected with an electric generator or with a hydraulic brake.

A third possibility is a rig with closed power loop (Figure 8.4c). Two identical PGTs are used. The loading can be done either by using a torsion shaft between the two sun gears or by mounting the second gear train to work as a multiplier and creating a load in the connecting shafts between the sun gears and carriers. This method is very economical because the motor's power covers only the losses in the power loop.

8.5.2 Ways for Uneven Loading Registration

As shown in Figure 8.5, there are different ways to detect the unevenness of load distribution. In general, the stress in the various elements of the PGT is used and only rarely their deformation (deflection):

- Of the ring gear outside
- Of the ring gear teeth
- Of the planet pins

In the first method, the stress state on the outer surface of the ring gear is recorded (Figure 8.6), where strain gauges are attached (Figure 8.5a). In this way, only individual momentary values of the load (when the individual planets pass through the respective location) are recorded, without them being recorded as a continuous function of the time t or the attitude (position of rotation) φ. Moreover, the measurement results affect the load

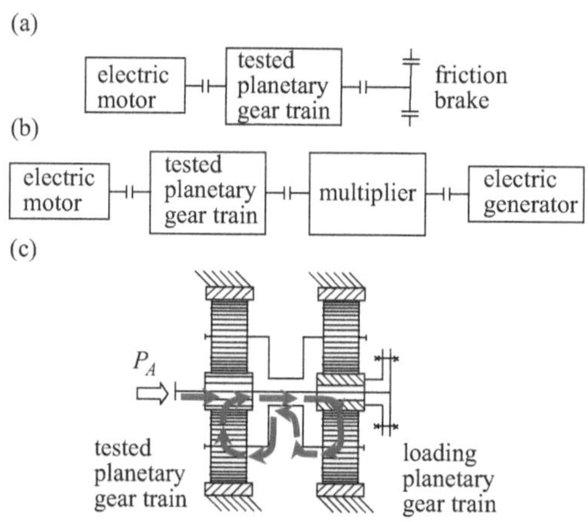

FIGURE 8.4
Test rigs for PGTs with different ways of loading: (a) With friction brake (Prony brake); (b) with electric generator; (c) with closed power loop.

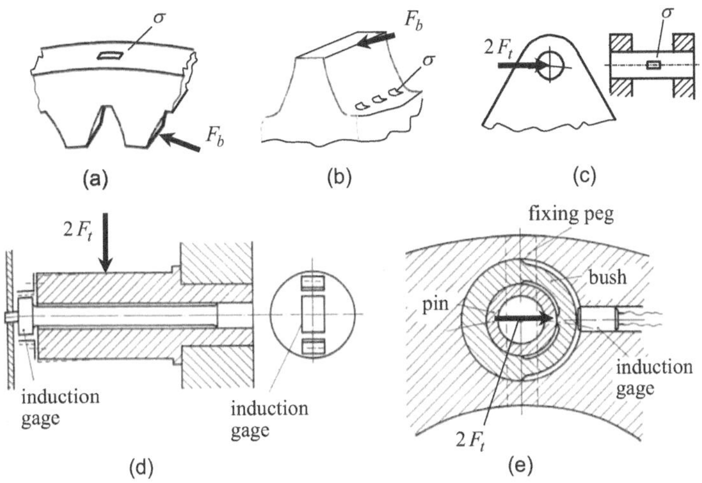

FIGURE 8.5
Experimental methods for determining the load distribution: (a) Stress in ring gear outside [122]; (b) stress in ring gear teeth [191]; (c) stress in planet pins [9, 45]; (d) deformation (strain) of planet pins; (e) deformation of an additional deflective bush of planet pins.

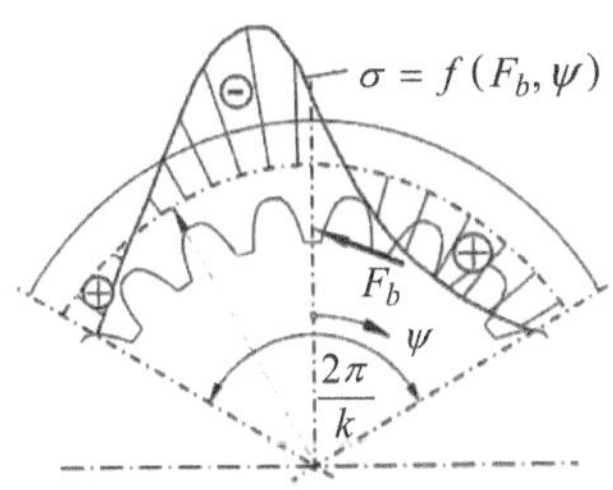

FIGURE 8.6
Stress in the ring gear.

on the ring gear from other planets. The method is applicable only to thin-rim (flexible) fixed ring gears, and only when there is an access to their outer surface. Obviously, the method is not universal.

The same can be said for the measurement of the stress in the fillet of the ring gear teeth (Figure 8.5b). In this way, the load does not occur as a continuous function of the time t or the attitude φ, too. Only momentary values are recorded. Another issue is the difficult strapping of the strain gauges in the confined space. This method is not universal, too.

The third method uses the stress state (bending) of the planet pins (Figure 8.4c). Bending stress, or load of the particular planet, respectively, is obtained as a continuous function of the time t, respectively, of the attitude φ, which is a very substantial advantage. This also applies to the last two methods in which the deformation of the planet pins by an induction gauge (Figure 8.5g) or the deformation of deflective bushes placed at the points of attachment of the planet pins to the carrier (Figure 8.5d) is measured. The methods in which the load on the planet pins is measured can be considered universal, convenient for practical application, and hence, only they will be taken into account. Indeed, in these methods, the influence of the internal dynamic forces both in the two meshings (external and internal) of an individual planet is registered. In principle, these are taken into account in the load capacity calculations by dynamic load factor K_v (see Chapter 17 and Section 17.2). The measurements register the actual load on the planets (which leads to failures!). In other words, the joint influence of the two factors used in the load capacity calculations, namely the product of these factors $K_{\gamma\,max} \cdot K_v$ (see Section 17.2), is practically determined.

8.5.3 Planet Pins Calibration

This is a very important and delicate problem in practice. The difficulty lies in the fact that they do not have to or cannot be calibrated on their own, outside the gear, and then mounted, because the calibration results will be distorted and not reliable. It is best to calibrate them after assembling the gear train.

One original and already tested solution of this problem [12] is shown in Figure 8.7. A so-called *calibration sun gear* with the same tooth geometry as the original sun gear is applied. The specific thing here is that it has partial toothing $z_t \le z_1/k$. Thus, the planets come consequentially one by one in mesh with the calibration sun gear and always only one planet can be loaded (Figure 8.8). Thus, a calibration line for every planet is obtained (Figure 8.9). The loading is repeated several times to achieve the repeatability of the results and the dispersion of the corresponding calibration line, respectively.

8.5.4 Experiments Conducting

As noted above, in the test of a particular PGT, only two dimensions can be changed: the input T_A (or output T_B) load and the input ω_A (or output ω_B) angular velocity. In order to achieve a possible nonlinear dependence of the mesh load factor $K_{\gamma\,max}$ on these parameters,

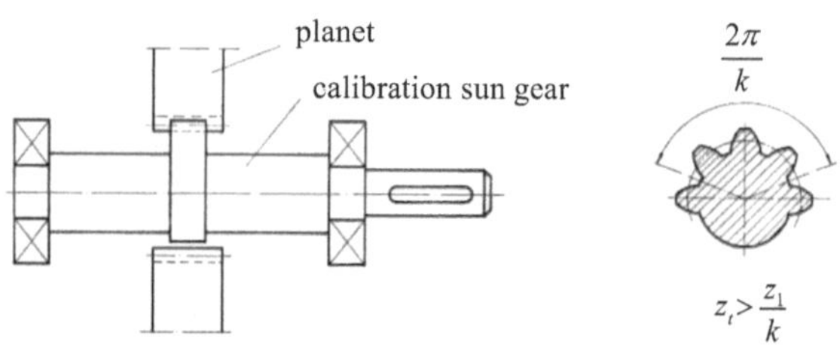

FIGURE 8.7
Calibration sun gear with partial toothing [12].

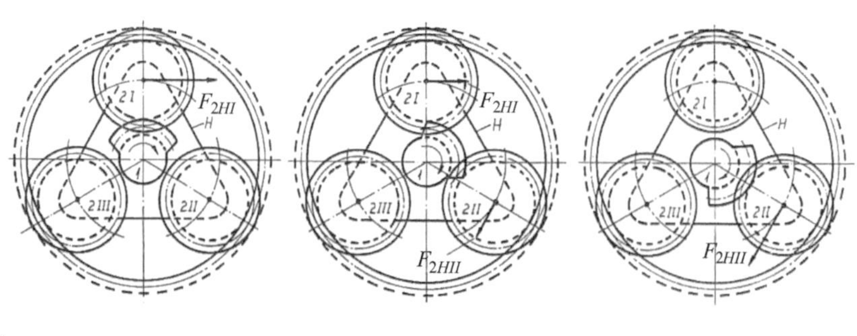

FIGURE 8.8
Various positions of the calibrating sun gear and loading of only one planet.

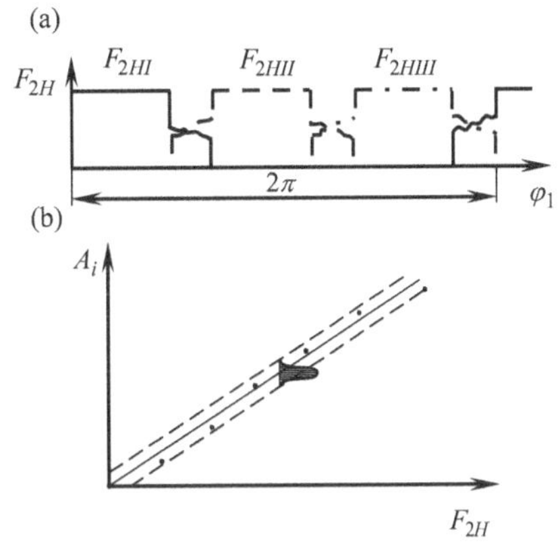

FIGURE 8.9
Sequential loading of individual planets (a) and obtaining a calibration line (b).

it is advisable to make the measurements at least for three values for each of them, e.g., 1/3, 2/3, and 1 of the nominal, and better, e.g., five. On the other hand, in order to achieve the repeatability of the results, and their dispersion, respectively, the measurements must also be repeated at least three times or more.

The minimum required number of measurements can be determined in the planning of the experiment.

Figure 8.10 depicts the cross section of an original research PGT,[3] allowing for fundamental studies of the influence of the various factors [9–11]:

- Change the floating element: sun gear 1, planets 2, ring gear 3 or carrier H.
- Change the ring gear compliance.
- Gear train load T_B.
- Rotation speed, and angular velocity ω_A, respectively.
- Change the mass of some of the floating elements—sun gear and ring gear.
- Change the sensitivity ξ of the equalizing (distribution) devices of some floating elements (planets and ring gear) through changing the friction in devices by hydraulic mode.
- Measuring the equalizing displacement of the sun gear, planets, and ring gear by specially developed induction gauges.

The original in these experimental studies is that the ring gear with its fixed geometrical deviations within the tolerances, which is initially made pliable (Figure 8.11), is made nonpliable by the strain of a maximally lightweight stiffening washer. Another interesting and original thing at the ring gear is that its connecting cylindrical part has axial slots which allow its local deformation at the point where it meshes with a planet at a given moment.

Figure 8.12 shows a part of test results [9, 199] under following influence factors (except for the input speed n_A):

[3] Developed by the author K. Arnaudov as a PhD student [9].

- Different floating[4] elements: sun gear, planets, ring gear, and carrier
- Different compliances α of pliable[5] element or its mounting (ring gear and carrier)
- Different load T_B of the gear train

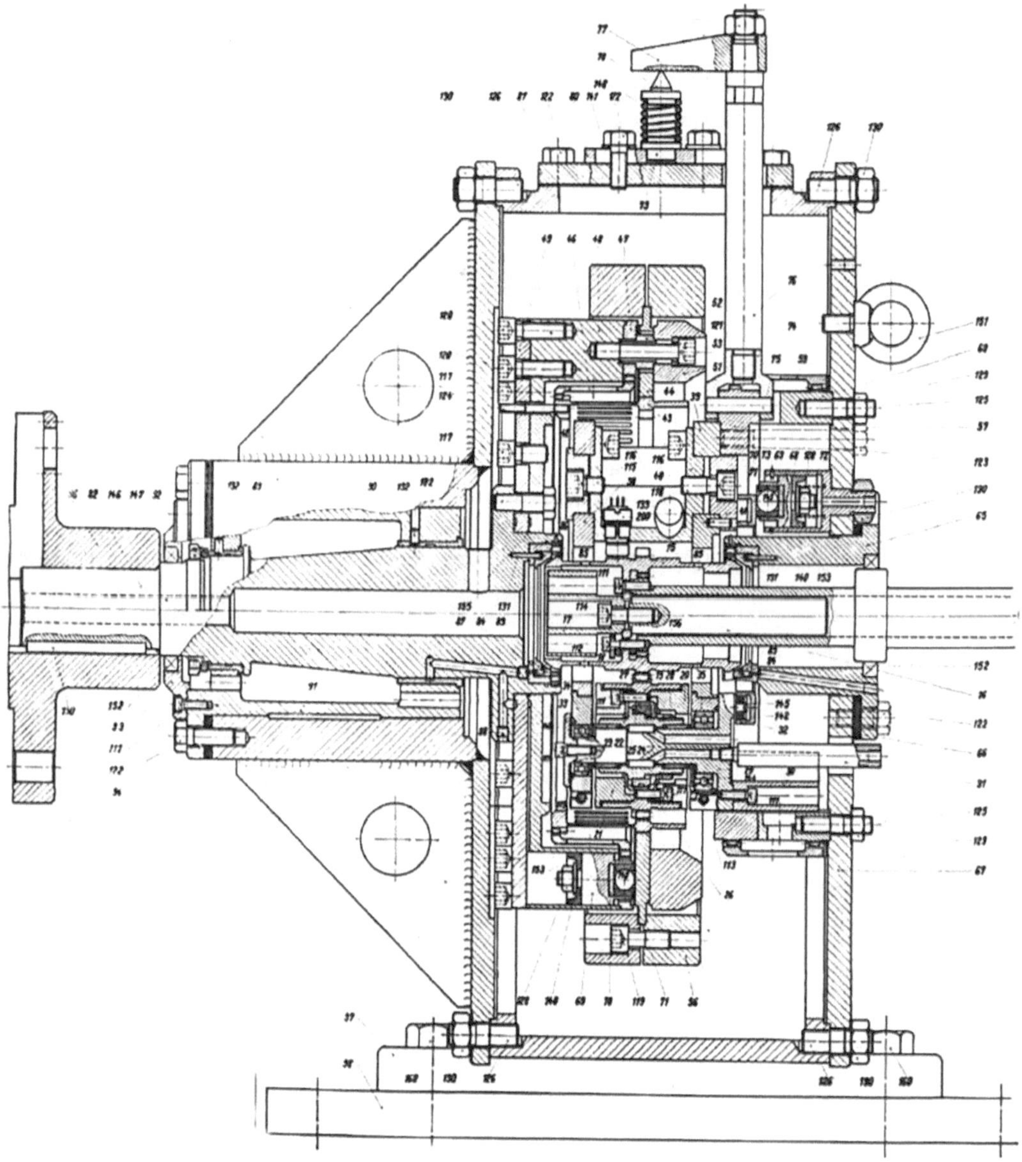

FIGURE 8.10
Cross section of an original research PGT [9], with which the experimental studies described herein have been performed.

---

[4] Displacement due to kinematic flexibility (because of clearances in mounting)
[5] Displacement due to deformations

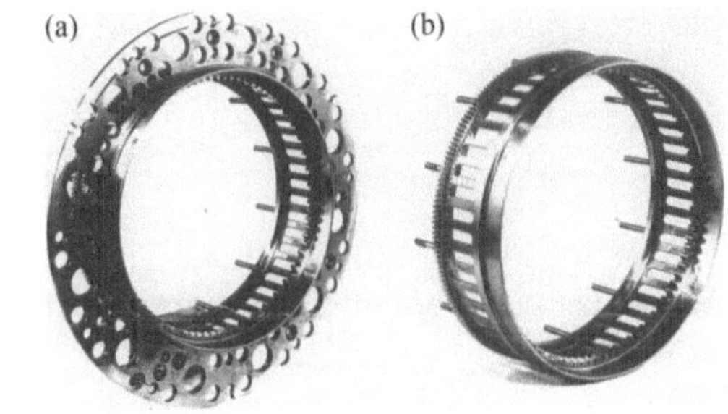

FIGURE 8.11
Pliable ring gear of the research PGT from Figure 8.10 with (a) and without (b) a stiffening washer.

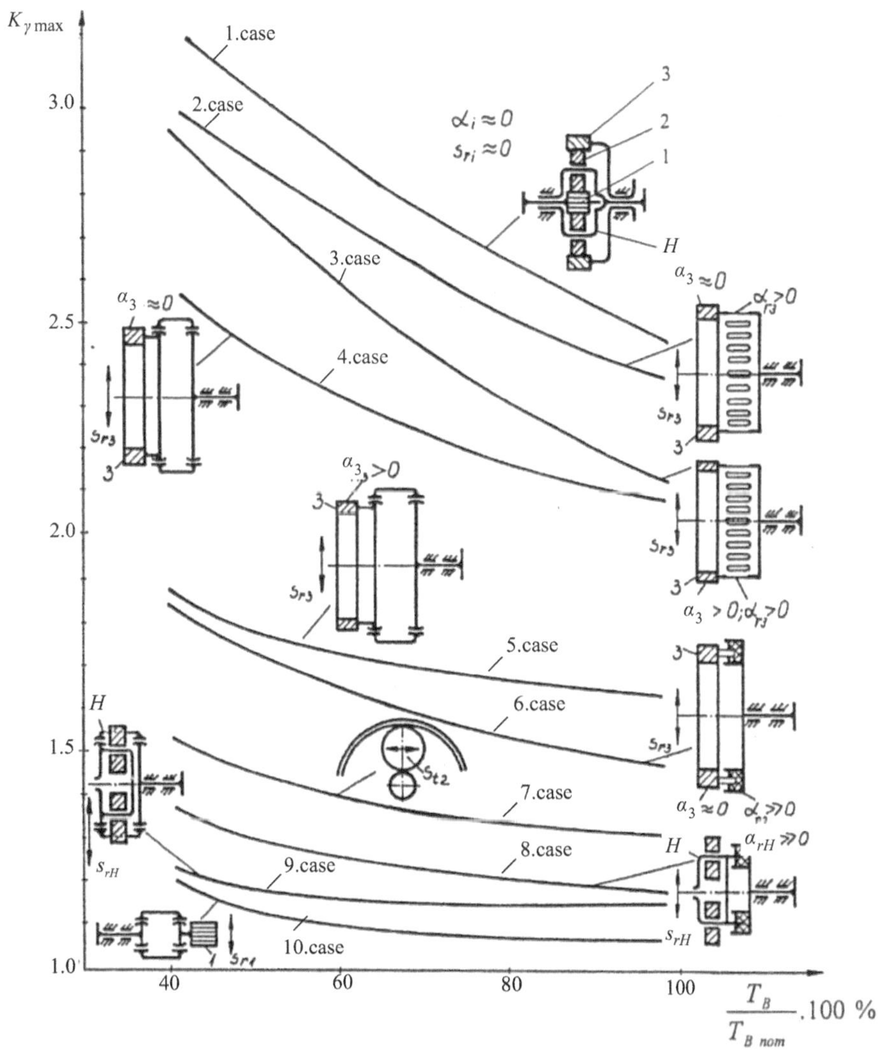

FIGURE 8.12
Values of mesh load factor $K_{\gamma\,max}$ depending on: the pliable element and the kind of the kinematic equalizing (distribution) device; compliance α_i of gear train elements; gear train load T_B, respectively, F_t.

Figure 8.13 shows two typical test records—with good and bad equalizing of load distribution between the planets.

The results of experimental studies, depicted in Figure 8.12, concern the following ten cases [9–11]:

1. *Case*: There are not any actions to equalize the load. The compliance of gear train elements is $\alpha_i \approx 0$. There are no kinematic equalizing (distribution) devices and radial floating, i.e., $s_{ri} \approx 0$.
2. *Case*: Ring gear 3 is stiff, non-pliable (with mounted stiffening washer), i.e., $\alpha_3 \approx 0$, but it has any radial compliance $\alpha_{r3} > 0$ thanks to the thin connecting cylindrical part with axial slots.
3. *Case*: Like the previous case, but with pliable ring gear, i.e., $\alpha_3 > 0$, and the axial slots of the thin connecting part allow for local deformation in the points of meshing.
4. *Case*: Ring gear 3 is stiff, non-pliable, i.e., $\alpha_3 \approx 0$, but with radial floating, i.e., $s_{r3} > 0$, thanks to a double-articulated gear coupling. Its sensitivity is not very high because of the small ratio L/d (Figure 8.17).
5. *Case*: Like the previous case, but with a pliable ring gear, i.e., $\alpha_3 > 0$.

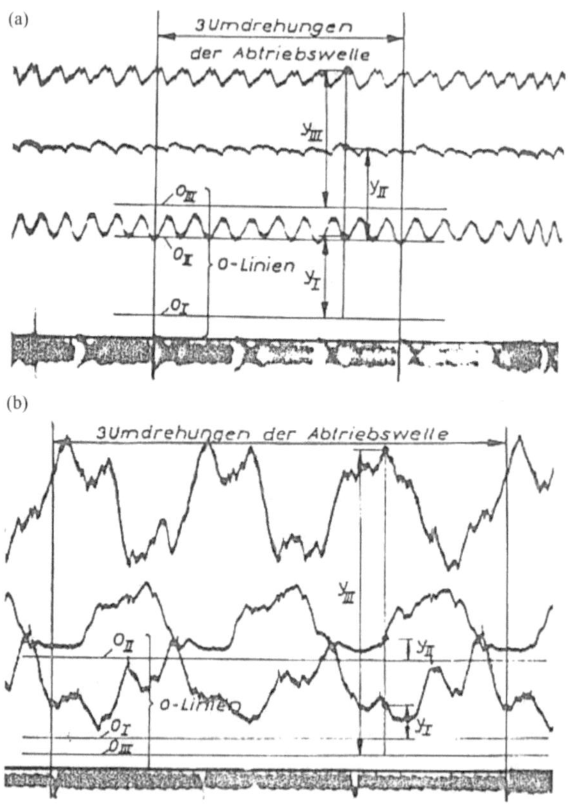

FIGURE 8.13
Typical test records [9–11]—with good (a) and bad (b) equalizing of load distribution (sharing) between the planets.

6. *Case*: Ring gear 3 is stiff, non-pliable $\alpha_3 \approx 0$, but it is mounted by elastic rubber pads with big radial compliance, i.e., $\alpha_{r3} \gg 0$. This allows for radial floating, i.e., $s_{r3} > 0$.
7. *Case*: Planets 2 float tangentially, i.e., $s_{t2} > 0$, thanks to a lever equalizing (distribution) device (not shown in Figure 8.12).
8. *Case*: Carrier H with planets 2 is mounted by elastic rubber pads with big radial compliance, i.e., $\alpha_{rH} \gg 0$. This allows for radial floating, i.e., $s_{rH} > 0$.
9. *Case*: Carrier H with planets 2 is mounted by a double-articulated gear coupling. This allows for radial floating, i.e., $s_{rH} > 0$.
10. *Case*: Sun gear 1 is mounted by a double-articulated gear coupling. This allows for radial floating, i.e., $s_{r1} > 0$.

To these results can be made the following:

Short comments:

Compliance, as in Cases 3, 6, and 8, favorably affects the equalization of load distribution (sharing), which is not unexpected.

As can be seen from Cases 4 and 5, when the non-pliable ring gear is mounted by an equalizing (distribution) device with a double-articulated gear coupling, this contributes to reducing the load imbalance and $K_{\gamma\, max}$, respectively.

Surprisingly, the ninth Case with floating carrier (distribution device with double-articulated gear coupling) produces a favorable result (reducing load imbalance and $K_{\gamma\, max}$, respectively), although the carrier with the planets has the largest mass of all the central elements of the gear train.

It is also surprising that the variation of the input speed in the low revolution range $n_A = 800 \div 1500\ \text{min}^{-1}$ does *not affect* the load imbalance and does *not determine* a relation like

$$K_{\gamma\, max} = f(n_A), \text{ respectively.}$$

In fact, other own rig tests, allowing a higher input speed of up to $n_A = 3{,}000\ \text{min}^{-1}$, as well as studies by other authors, establish the same. Such dependence is only observed at higher peripheral velocities—$v > 10 \div 12\,\text{m/s}$

The influence of the sensitivity ξ of equalizing (distribution) devices is very pronounced in Cases 4, 5, 9, and 10. When ratio L/d of gear coupling is big, as in Case 10 with floating sun gear, the sensitivity ξ of equalizing (distribution) device is high, load distribution (sharing) equalizing is good, and mesh load factor $K_{\gamma\, max}$ has a low value. And vice versa, in case of small ratio L/d, as in Cases 4, 5, and 9, the sensitivity ξ of equalizing (distribution) devices is low, the load imbalance is heavily expressed, and the mesh load factor $K_{\gamma\, max}$ has high values.

For completeness, it should be noted that with two floating elements of the gear train—the sun gear and ring gear, very favorable results are obtained, as in Case 10.

In all cases, the most characteristic and general is that the load imbalance is most dependent on the load of the gear train, i.e.,

$$K_{\gamma\, max} = \frac{\text{const}}{T_B/T_{B\,nom}} = \frac{C(f_i, \alpha_i, j_{Li}, \xi)}{T_B/T_{B\,nom}}. \tag{8.3}$$

In this formula, the constant C is specific for each particular PGT with its

- Geometric misalignments f_i within the tolerances by size and combination
- Compliance α_i of particular elements of the train
- Bearing clearance j_{Li}, especially of the central elements
- Sensitivity ξ of equalizing (distribution) device, if any

Obviously, all of these influential factors are in a different combination in size in each PGT. This explains the great complexity and ambiguity of the problem. Therefore, experimental studies are authoritative and dominant. They can provide results usable in practice.

It is worth noting that the experimental results presented here cast wide-ranging light on the different aspects of the problem, and because of their practical orientation, they have found appropriate acceptance and application [11, 155].

Other interesting and important later studies are described in [139]. There is a partial coincidence between their research program and that of [9], which provides a convenient comparison option. The coincidence of the two different studies concerns the following:

- Sun gear can be borne or not, i.e., floating by a double-articulated gear coupling.
- Ring gear can be non-pliable or pliable.
- Floating elements are sun gear and ring gear—either individually or together.

Unlike the [9], in [139], the load distribution on the tooth face width (factor K_β; see Section 17.1) is investigated by selecting different planet bearings—self-aligning or non-self-aligning bearings.

Unlike the [9] in [139], however, it has not been investigated:

- Floating of the PGT elements—sun gear, planets, ring gear, and carrier
- Different loading T_B of the PGT
- Different rotation speed n_A, which in [139] is constant and very low ($n_A = 40\ \text{min}^{-1}$), so researches are actually static

The conclusions that are made in the [139] are important and in line with the results of the [9] research:

- The best equalization of the load distribution, expressed by factors $K_{\gamma\, max}$ and K_β, is obtained by floating sun gear and self-aligning planet bearings.
- Floating ring gear is less effective.

Fundamentally in PGTs, there is not enough space and the ring gear hanging is with a little ratio L/d of equalizing device (see Section 8.6.5 and Figure 8.17), resulting in its low sensitivity ξ and high value of mesh load factor $K_{\gamma\, max}$(the second conclusion). On the contrary, the sun gear has plenty of space to achieve a high ratio L/d and better equalization of the load distribution, as noted in the first conclusion.

It is also worth noting that in the [139], research planets loading is not a continuous function of the position (attitude of the gears), as in [9]. The load of several individual teeth is measured, which determines the complex processing of the measurement records.

Article [45] describes a contemporary rig for measuring planet load sharing by means of strain gauges mounted on the planet pins and by measuring the displacement of the sun gear.

8.6 Ways of Problem-Solving

There are different ways of solving (or not solving) the problem of reducing the unevenness of load distribution (sharing), i.e., load imbalance, and its consequences in the process of a PGT design:

- No purposive actions
- Purposive accuracy
- Purposive compliance
- Kinematic flexibility (radial float)
- Complex way

For the designer, it is important to know what values of load mesh factor $K_{\gamma\, max}$ to use in its calculations in the different ways discussed below for the most common case with $k = 3$ planets.

8.6.1 No Purposive Actions

In PGTs, where no special measures are aimed at reducing the load imbalance and the mesh load factor $K_{\gamma\, max}$, respectively, the latter varies widely depending on the accuracy, bearing clearances and compliance of the gear train elements, and can reach very high values. For this reason, very roughly, it may be considered:

$$K_{\gamma\, max} \approx 1.5 \div 2.0(2.5).$$

This approach can lead to big misfortunes and should be avoided.

8.6.2 Purposive Accuracy

This is, in principle, the most correct way because it affects the causes, i.e., the misalignment within the tolerances of the gear train elements and not the consequences of them—the load imbalance. This approach, however, is expensive, cannot always be applied, and cannot ensure full uniformity of the load sharing $\left(K_{\gamma\, max} = 1\right)$. Therefore, it can be accepted approximately:

$$K_{\gamma\, max} \approx 1.3 \div 1.5.$$

It is particularly important to point out that the impact of the inaccurate planet pins location, especially the tangential misalignment f_{tHj} (Figure 8.2), is very strong and unfavorable. Therefore, the holes for the planet pins in the carrier have to be made very precisely by a jig boring machine.

8.6.3 Purposive Compliance

Compliance also positively affects the load sharing. Various elements of the gear train can be made pliable (Figures 8.14 and 15.1):

(a)

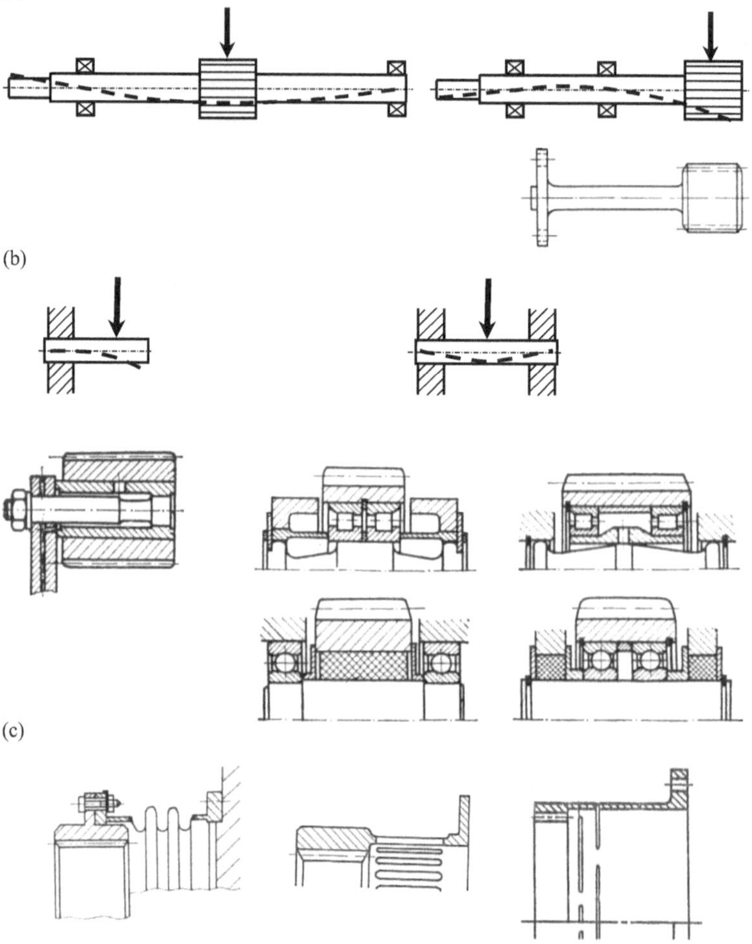

(b)

(c)

FIGURE 8.14
Increased compliance of the PGT elements, and the elastic mounting, respectively: (a) Shaft of the sun gear; (b) pins of the planets; (c) ring gear and its mounting.

- Shaft of the sun gear 1
- Pins of the planets 2
- Ring gear 3 and its mounting [42, 116, 211]

Additional elastic elements such as rubber pads and springs can also be used. However, it should not be forgotten that the change in compliance also affects the amplitude–frequency response of the gear train [213, 214], which is important in high-speed PGTs.

Principally in this approach, a complete equalization of load sharing is not possible, but only partial. Of course, the variations in the dimensions of the gear train elements (resp. their size) which are compensated by the deformations (of one or more gear train elements) are of importance, too.

Approximately, it could be accepted:

$$K_{\gamma\, max} \approx 1.2 \div 1.4.$$

Using this method, however, care must be taken to ensure that the gears do not tilt in the deformations, which will lead to an uneven load on their face width, i.e., to a strong increase in the face load factor $K_\beta > 1$ (load distribution factor [5]), taking into account this unevenness in the load capacity calculations (Chapter 17). It is appropriate not only in these cases but in general, the use of only one planet bearing (of course, with sufficient load capacity) allowing for some self-aligning, e.g., a single-row ball bearing, a double-row self-aligning (spherical) ball, or a roller bearing. In case of small diameter of planets, because of the lack of space, needle bearings are used.

8.6.4 Kinematic Flexibility

Kinematic flexibility (radial float) of the individual elements of a PGT is created using a corresponding equalizing (distribution) device. A great number of patented equalizing (distribution) devices exist, and many of them are not used in practice. The most common are the single or double-articulated gear couplings to provide the desired displacement (Figure 8.15). With a properly selected and designed equalizing device (i.e., under certain conditions), this method results in *the best equalization of the load sharing.*

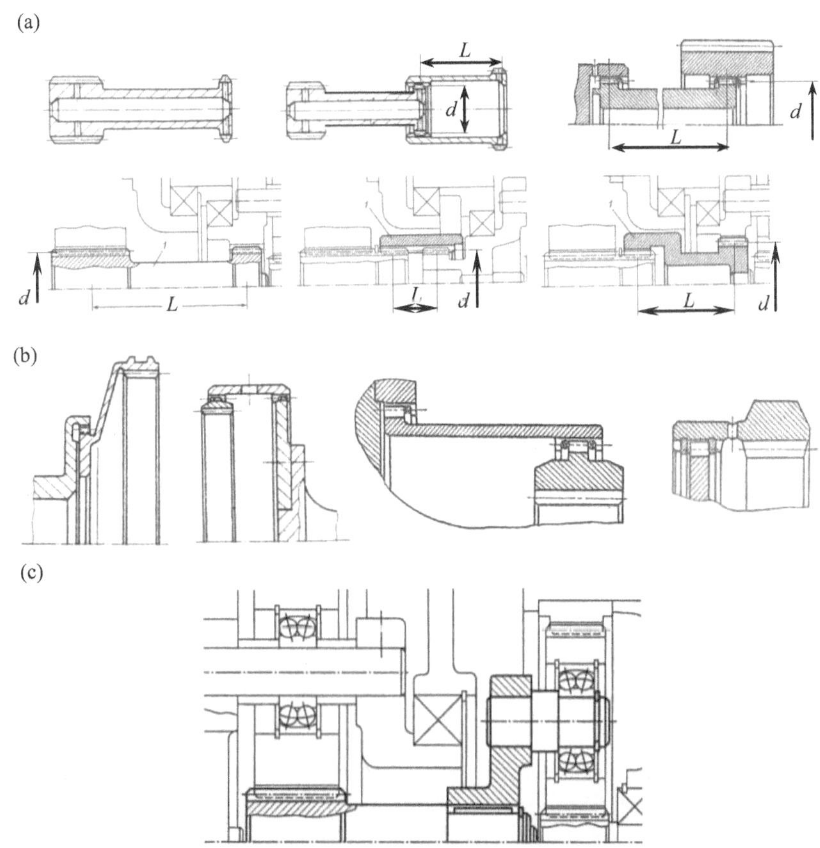

FIGURE 8.15
The most commonly used kinematic equalizing (distribution) devices on the: (a) Sun gear; (b) ring gear; (c) carrier.

The best equalization is achieved when a PGT is with $k = 3$ planets (since three points determine the position of a circle), which is, moreover, the most common case, making one of the central elements—the sun gear 1 (most often), the ring gear 3, or the carrier (most rarely), to float. The best result is achieved with a floating sun gear, which is most often with the smallest mass m_1 and therefore with the lowest inertia. It is best to use a double-articulated gear coupling rather than a single (Figure 8.15), because the single one leads to a greater tilting of the sun gear and an uneven load on the tooth face width (factor K_β!). This tilt in a short coupling (single or double articulated) can be so great, and the factor K_β to increase so much, that the favorable effect of the equalization (factor $K_{\gamma\,max}$) is canceled. Due to the relationship between the factors K_β and $K_{\gamma\,max}$, their product has to be low. It is also very important to use the longest possible sleeve between the two coupling tooth pairs (large ratio L/d—Figure 8.15) to ensure maximum sensitivity of the equalizing device. Moreover, it is best if the coupling teeth are crowned. Otherwise, the width of the teeth face should be as small as possible.

When the number of planets is $k = 3$, it could be accepted:

$$K_{\gamma\,max} \approx 1.1 \div 1.2.$$

In cases of a floating sun gear, the following values of mesh load factor may be taken according to the number of planets $K_{\gamma\,max} = f(k)$, based on [5]

k	3	4	5	6	7	8
$K_{\gamma\,max}$	1.1 ÷ 1.2	1.15 ÷ 1.3	1.2 ÷ 1.4	1.25 ÷ 1.5	1.3 ÷ 1.6	1.35 ÷ 1.7

The use of a floating sun gear has two additional advantages. Its bearings are eliminated (simplified and cheaper product), which are the most high-speed and cause the greatest losses, especially when they are plain bearings.

8.6.5 Equalizing Devices Characteristics

When it comes to equalizing (distribution) devices, it is very important to pay attention to the properties they possess [6, 7, 259]:

1. Inertia
2. Dynamism (fast action)
3. Sensitivity
4. Compensating ability

1. *Inertia* directly depends on the mass m_i of the element that performs the equalizing movements. As can be seen from Figure 8.16, they are quite different and depend on the basic speed ratio $i_0 = z_3/z_1$, and on the torque ratio t, respectively. Naturally, inertial forces counteract the equalizing process.
2. In relation to the inertia determined by the mass m_i of the moving element, the *dynamics (fast action)* is determined by the active force F_{ac} that causes the equalization process. For sun gear 1 and ring gear 3, this active force is proportional to the normal forces F_b acting in the gear teeth, because it is caused by their difference at a given moment, i.e.,

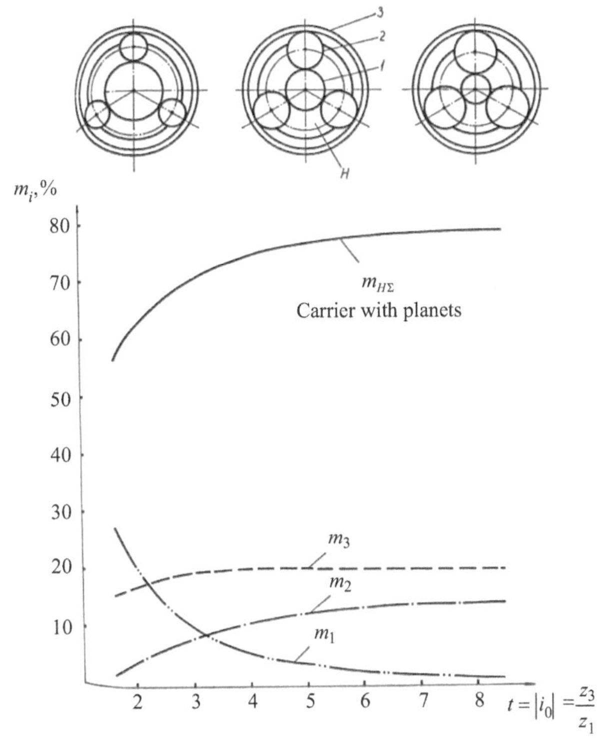

FIGURE 8.16
Relative mass m_i of the main elements of a PGT depending on the basic speed ratio $|i_0| = z_3/z_1$.

$$F_{ac} = \Delta F_b. \tag{8.4}$$

For the carrier, the active force Γ_{ac} is approximately proportional to the double normal force, i.e.,

$$F_{ac} = \Delta(2F_b). \tag{8.5}$$

This follows from the loading of planets pins.

3. A very important characteristic of equalizing devices is their *sensitivity* [15]. It is in close connection with the friction forces F_R that, like the inertial forces, counteract the equalization process. Especially for the most commonly used gear coupling, the sensitivity ξ depends on the ratio L/d (Figure 8.17):

$$\xi = f\left(\frac{L}{d}\right). \tag{8.6}$$

The sensitivity of equalizing devices is, moreover, defined as follows, taking into account the friction force in them:

$$\xi = 1 - \frac{F_R}{k \cdot F_{bm}} = 1 - \frac{F_R}{\sum_{j=1}^{k} F_{bj}}, \tag{8.7}$$

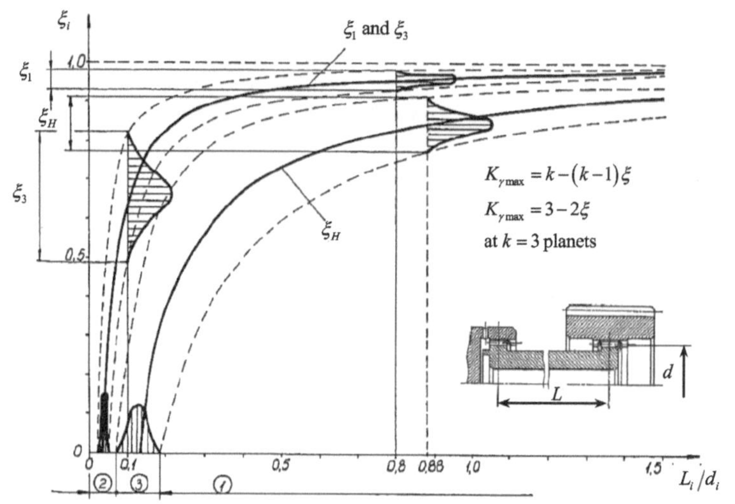

FIGURE 8.17
Sensitivity ξ_i of an equalizing (distribution) device with a double-articulated gear coupling on a central element of PGT depending on the ratio L_i/d_i.

where F_R is the friction force in equalizing device, F_{bm} is the average normal force in meshing, and F_{bj} is the nominal force in individual (the j) meshing.

It is easy to ascertain that this formula meets the boundary conditions:

- At $F_R = 0$, sensitivity is $\xi = 1$, and load sharing (in static conditions) is completely uniform, i.e., $K_\gamma = 1$.
- At $F_R = k \cdot F_{bm}$, sensitivity is $\xi = 0$, and there is no load sharing, i.e., $K_\gamma = k$ (only one planet takes all load).

It is clear that for high sensitivity ξ, a large ratio L/d has to be chosen. Conversely, at low value of the ratio L/d, high sensitivity ξ and good equalization of load sharing cannot be expected, because the equalization process will be severely hampered. In the case of low sensitivity, it is necessary to create a big difference ΔF_b between the normal forces in the meshing in order to overcome the friction forces F_R in the equalizing device and only then it starts to act. This means a great load imbalance and a high value of the mesh load factor $K_{\gamma\, max}$. In Figure 8.17, the three zones of operation of the equalizing device of the carrier H are clearly visible. They are the result of the dispersion of the friction coefficient μ ($\mu = 0.05 \pm 50\% = 0.025 \div 0.075$) and hence of the dispersion of the friction forces F_R and the sensitivity:

Zone 1—equalizing (distribution) device is certainly functioning.

Zone 2—equalizing (distribution) device is certainly not functioning.

Zone 3—the functioning of the equalizing device is in question (transition zone).

In Figure 8.18 are given two computational examples [16] of the sensitivity ξ of the three central elements of PGT—sun gear 1, ring gear 3, and carrier H for two different basic speed ratios $i_0 = -2$ and $i_0 = -10$. The results are very indicative.

For the sake of completeness, it should be noted that in the underlying low-speed PGTs (e.g., the last stage of a multistage PGT), the sensitivity ξ of the equalizing device (Figure 8.17) plays a dominant role in the equalizing process. Conversely, in high-speed PGTs (e.g., turbine gear trains), the dynamic effects and inertia are

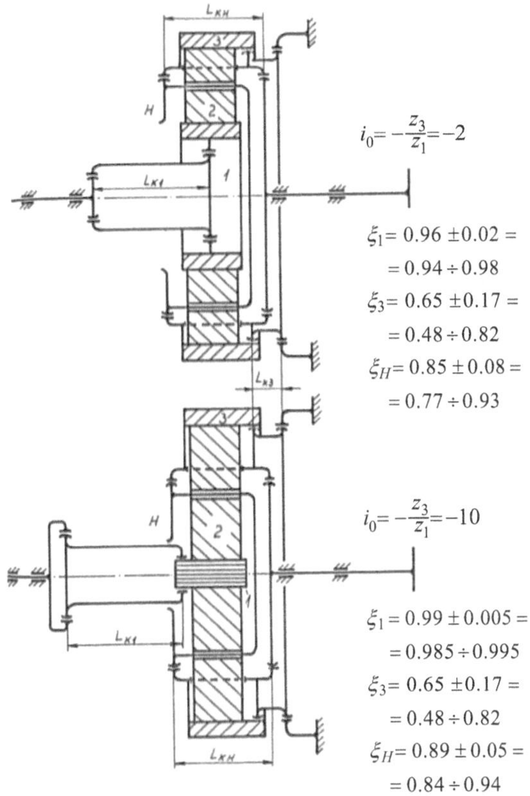

FIGURE 8.18
Sensitivity ξ_i of equalizing (distribution) devices on central elements of PGT with various basic speed ratios $i_0 = -z_3/z_1$ with equal other conditions [16].

dominant. In both cases, the compliance of gear elements as well as the clearances of bearings of the central elements—sun gear 1 and carrier H, of course, also plays a role, especially in the second case. These two cases refer to the so-called *static load equalizing* (see Figure 8.17) and *dynamic load equalizing*. For very slow-speed gear trains, the corresponding values for the mesh load factor $K_{\gamma\,max}$ depending on the ratio L/d, resp. the sensitivity of the equalizing device ξ, from the figure can be obtained. For example, for three planets,

$$K_{\gamma\,max} = 3 - 2\xi.$$

Such gears lack substantial dynamic effects, and the sensitivity of the equalizing device is dominant.

4. *Compensating ability* is a resultant characteristic and shows the extent to which the equalizing device can equalize load distribution and reduce the effect of various manufacturing anomalies (within the tolerances) of gear train elements.

   Moreover, in the said characteristics of the equalizing devices very clear appears the relationship between the three factors mentioned above (Section 8.3.4), namely, the mass–force–path (displacement), i.e., the mobile (floating or pliable) mass—the active and the resistive force—moving the movable element.

8.6.6 Complex Way

This is in fact a combination of the abovementioned ways and is expressed in the following:

- Use of a kinematic equalizing (distribution) device with a high sensitivity ξ on the element with a possible lower mass m_i
- High accuracy of gear train elements, especially of the carrier (above all the exact tangential position f_{tHj} of the planet pins)
- Equal thickness s_2 of the teeth of the planets
- Increased compliance α_i of gear train elements
- Equal radial clearance j_{L2j} of planet bearings
- Increased bearing clearance j_{L1} and j_{LH} of central elements of gear train (sun gear 1 and carrier H)
- *Sinphase montage* of planets, with radial orientated run-out (Figure 8.19)

The above can be expressed by the following formula:

High accuracy + floating element with high sensitivity + increased compliance + bearing clearance of central elements (1 and H or 1 and 3).

In this case, the value of mesh load factor $K_{\gamma\, max}$ (for $k = 3$ planets) is the lowest [9–11, 37]

$$K_{\gamma\, max} \approx 1.05 \div 1.10.$$

In the case without an equalizing device (floating element), but all other listed above actions are applied, the mesh load factor will be higher than the case of a floating element only:

$$K_{\gamma\, max} \approx 1.15 \div 1.3.$$

Finally, it is necessary to emphasize once again that the problem of the load imbalance and its equalization is extremely extensive, complicated, and intricate. Therefore, all the values of the mesh load factor $K_{\gamma\, max}$ mentioned here cannot be anything other than tentative! For more details, see specialized literature [7–12, 14–16, 19, 20, 29, 33, 34, 37–39, 42, 47, 48, 80, 90, 105, 111, 119, 122, 129, 139, 142–144, 147, 152, 165, 168, 181, 211, 222, 225, 233, 241, 245, 249, 252].

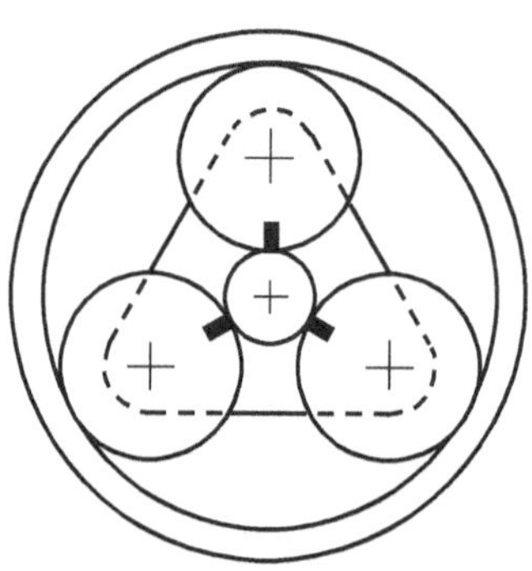

FIGURE 8.19
Sinphase montage of planets, with radial orientated run-out to the center of sun gear.

8.7 Ways of Teeth Loading Experimental Determination

As noted in Section 8.5, in the case of extremely responsive PGTs, or when the customer explicitly asks to establish the actual load imbalance, this may be done as illustrated in Figure 8.20.

If the PGT in question has, e.g., the number of teeth of the gears shown in the figure—$z_1 = 15$, $z_2 = 30$ and $z_3 = 75$, the kinematic cycle of the gear train, defined by the smallest total of the number of teeth, is $N_p = 150$ meshes. After this number of meshes, the loading of the gears repeats.

An example record of the load on the three planet pins I, II, and III is given in the figure.

For $k = 3$ planets, each tooth of the sun gear 1 has to rotate with respect to the carrier H at 120° or at $z_1/k = 5$ of its teeth to mesh the next planet. For the entire kinematic cycle $N_p = 150$, the concerned tooth will take part in $N_1 = N_p/5 = 30$ meshes at ten revolutions of sun gear 1 and, accordingly, in so many loadings. Thus, successively at every five teeth, from the individual planet pins I, II, and III, the loading of the concerned tooth is recorded. After arranging these loads in descending order, the load spectrum of the individual tooth of the sun gear is obtained.

Similar is the situation with the ring gear 3. Each of its teeth meshes with the next planet at every 120° toward the carrier or at every $z_3/k = 25$ of its teeth. For the kinematic cycle $N_p = 150$, there are $N_3 = N_p/25 = 6$ meshes in total for two revolutions of ring gear 3 and, accordingly, so many loadings of the tooth.

Per one revolution of the planets 2, their teeth are loaded twice, i.e., every $z_2/2 = 30$ teeth. For the kinematic cycle $N_p = 150$, there are totally $N_2 = N_p/15 = 10 = 5 + 5$ meshes for five revolutions of each planet 2 and so many loadings on the concerned tooth.

As shown in Figure 8.20, the loads of the individual tooth are arranged in descending order so that its equivalent load and equivalent stresses $\sigma_{F\,eq}$ and $\sigma_{H\,eq}$ can be determined and thus the respective safety factors S_F and S_H, what is ultimately sought.

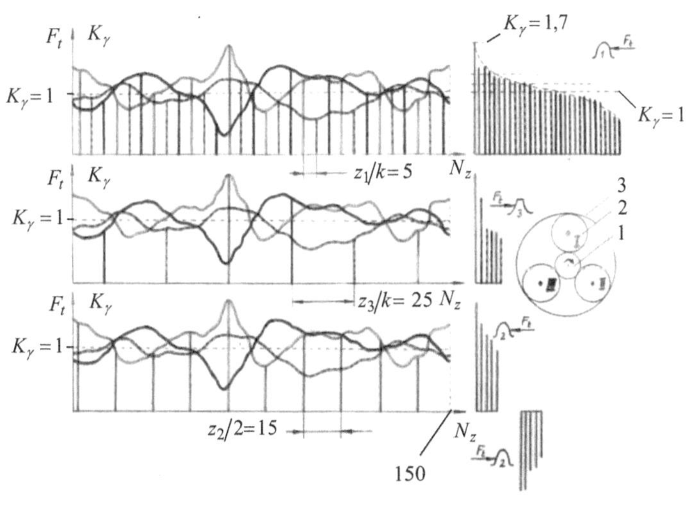

FIGURE 8.20
Determination of the load spectra of the individual teeth of sun gear 1 ($z_1 = 15$), ring gear 3 ($z_3 = 75$), and planets 2 ($z_2 = 30$).

All that has been said so far applies only to a particular individual tooth of a concrete gear wheel. Since most often the number of teeth z_1 of sun gear 1 is the smallest, and hence the number of cycles of its load N_1 is the highest, the determination of the safety factors S_F and S_H for all of its teeth according to the described procedure must be repeated $z_1 = 15$ times. Safety factors S_F and S_H for each tooth will show which of the teeth is most threatened.

If the sun gear is not the critical (weak) gear of the PGT, or this is not known with certainty, the above procedure must be applied to all the teeth of the other gears—planets and ring gear. Because the amount of this computing work is not small, the use of suitable software is advisable. This is particularly relevant for a long kinematic cycle, rather than as in the above example, where the number of teeth is specially selected in order to obtain a short cycle.

9

Loading on the Elements of $\overline{AI}$-Planetary Gear Train and its Peculiarities

Here are briefly noted the peculiarities of the loading of planetary gear train (PGT) elements, namely,

- Gear wheels
- Bearings
- Pins of the planets
- Shaft of the sun gear
- Gear couplings
- Carrier

These peculiarities need to be known to be taken into account in the strength calculations of the PGT elements given in the textbooks on machine elements [58, 120, 175] and reference literature.

9.1 Gear Wheels Loading

Peculiarities are as follows:

1. Load distribution (sharing) between the planets 2 and hence the loading of the other gears (sun gear 1 and ring gear 3) is uneven, which is taken into account in the calculations by the mesh load factor $K_{\gamma\,max}$ discussed in Chapter 8.
2. There is a difference in the magnitude of the contact stresses on the external (σ_{H12}) and internal (σ_{H23}) meshing (accepting that $F_{t12} = F_{t23}$, see Chapter 6), where always

$$\sigma_{H12} > \sigma_{H23}. \tag{9.1}$$

3. There is a difference in the type and magnitude of the bending loading of the teeth of the various gears as follows (Figure 9.1):
   - *Pulsating loading* (repeated, one-direction stress) on central gears (sun gear 1 and ring gear 3). The loading of the teeth of these gears can be completely reversed (zero-mean cyclic) in reverse PGTs, which along with the change in direction of rotation changes the direction of tooth load. This is not the case of hoisting mechanisms of cranes, hoists, etc. where in both lifting and lowering a load, the same sides of the teeth are always loaded and the load is pulsating.

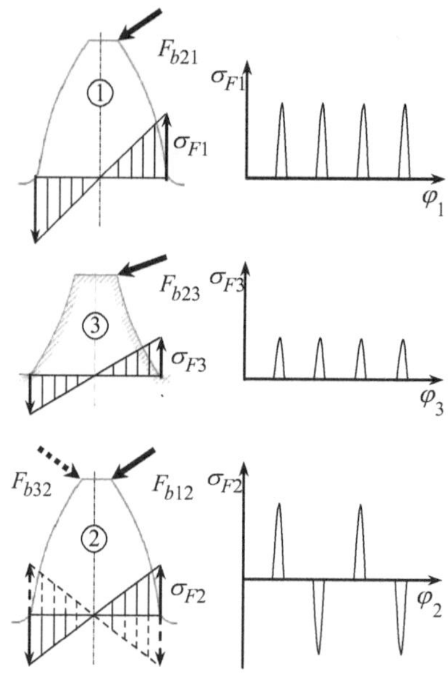

FIGURE 9.1
Loading of the gear teeth at bending.

- *Completely reversed loading* (zero-mean cyclic) of planets 2. As a result of the sequential mesh with the sun gear 1 and the ring gear 3, teeth of planets are loaded on different sides (repeated, reversed stress), and therefore, their permissible bending stress is reduced by 30% compared to the pulsating load [63, 112, 113, 215].
- Most often the bending stress σ_{F1} of sun gear 1 is greater than the bending stress σ_{F3} of ring gear 3, i.e.,

$$\sigma_{F1} > \sigma_{F3}. \tag{9.2}$$

4. The number of load cycles N is significantly different for individual gears as well as for their load type—N_F bending and N_H contact stress loading. Since the number of teeth of sun gear z_1 and planets z_2 is most commonly different, as a rule (at $|i_0| = z_3/z_1 > 3$) $z_1 < z_2$, then the number of load cycles N of the three gears (sun gear 1, planets 2, and ring gear 3) is arranged as follows:

$$N_1 > N_2 > N_3. \tag{9.3}$$

The number of load cycles for individual gears is calculated as follows:

$$N_{F1} = N_{H1} = N_1 = k \cdot 60(n_1 - n_H)L = k \cdot 60 n_{1rel} \cdot L$$

$$\left.\begin{aligned} N_{H2(1)} &= 60(n_1 - n_H)\frac{z_1}{z_2}L = 60 n_{1rel}\frac{z_1}{z_2}L \\ N_{F2} &= 2N_{H2(1)} \end{aligned}\right\} \text{— for external meshing} \tag{9.4}$$

$$\left.\begin{aligned} N_{H2(3)} &= 60(n_3 - n_H)\frac{z_3}{z_2}L = 60n_{3rel}\frac{z_3}{z_2}L = N_{H2(1)} \\ N_{F2} &= 2N_{H2(1)} \end{aligned}\right\} \text{— for internal meshing}$$

$$N_{F3} = N_{H3} = N_3 = k \cdot 60(n_3 - n_H)L = k \cdot 60n_{3rel} \cdot L,$$

where L is the operating time (life) of PGT, h, n is the rotation speed of the elements (gears and carrier) of PGT, min^{-1} (rev/min), and n_{rel} is the relative rotation speed of the respective gears to the carrier, min^{-1}.

5. When PGT operates a very long time L, the number of gear load cycles becomes high, and moreover, it is higher than the corresponding base number of cycles for both types of loading—$N_{F\,limb}$ and $N_{H\,limb}$ (Figure 9.2), i.e.,
   $N_F > N_{F\,limb}$ and $N_H > N_{H\,limb}$.
   In this case, the gears are calculated in terms of long (infinite) life fatigue (endurance strength). But at the last stage of multistage low-speed PGTs, the number of load cycles sometimes gets lower than the corresponding base number, i.e.,
   $N_F < N_{F\,limb}$ and $N_H < N_{H\,limb}$.
   In this case, the gears are calculated, as shown in Figure 9.2, with higher permissible stresses σ_{FP} and σ_{HP} due to the low number of load cycles N_F and N_H [limited (finite) life fatigue]. For this purpose, in the load capacity calculation, the life factors[1] $Y_N = f(N_F) > 1$ and $Z_N = f(N_H) > 1$ are used [63, 94, 112, 113, 215]. In the diagrams in Figure 17.10 for tooth root stress limit $\sigma_{FG} = f(N_F)$ and pitting stress limit $\sigma_{HG} = f(N_H)$, these factors are included in an implicit form. As the last stages of the multistage PGTs (when operating as reducers), on the one hand, are the most heavily loaded, and on the other hand, they have the lowest number of load cycles N, in this case, small gear wheels (and PGTs as a whole) are obtained. So a significant economic effect and a compact arrangement are achieved.
6. The smaller gear sizes (due to the torque split) lead to more precise machining (grinding), easier heat treatment, and lower peripheral velocity v in the meshing, resulting in lower dynamic loads, considered by the dynamic factor $K_v = f(v)$ in load capacity calculations (Section 17.4). This is beneficial not only for the load

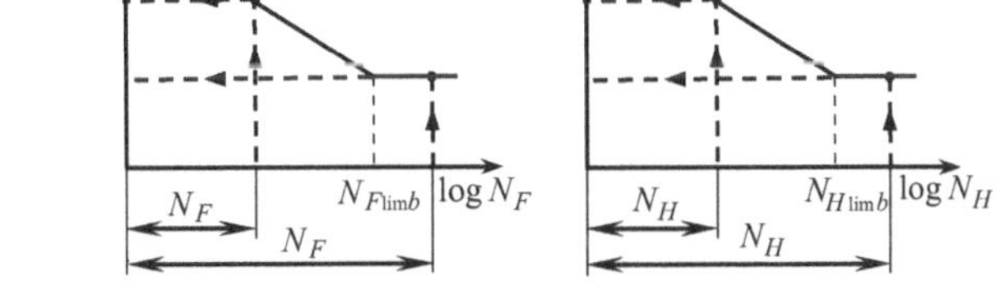

FIGURE 9.2
Wöhler diagrams for permissible bending (σ_{FP}) and contact (σ_{HP}) stresses.

[1] *Stress cycle factor* according Mott, R.L. *Machine elements in mechanical design*. 4th ed. Upper Stable River, NJ [USA]: Pearson, 2004. ISBN 0-13-061885-3.

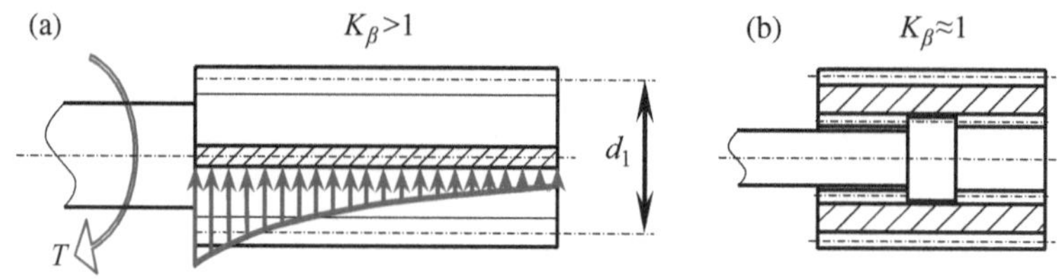

FIGURE 9.3
Nonuniform distribution of load over the sun gear face width (a) and a way to avoid it (b).

capacity of the gears and for the overall dimensions of the PGT, but also for its noise level.

7. In case of very wide gears and especially a very wide sun gear 1, i.e., $b_1/d_1 > 1$ (Figure 9.3), as a result of the considerable torsion of the gear, a very nonuniform distribution of the load over the gear face is obtained (Figure 9.3a), considered by the face load factor $K_\beta = f(b_1/d_1) > 1$ (see Section 17.4) in the load capacity calculations.

   For this reason, $b_1/d_1 < 0.8$ is recommended. This unevenness can be avoided if the torque is fed into the middle by a gear coupling (Figure 9.3b).

8. When the ring gear 3 is made pliable, i.e., the rim thickness under the tooth root is small, e.g., $(s_R/m < 3.5)$ (Figure 9.4), in this relatively rare case in the bending strength calculations, the rim thickness factor Y_B (which increases the computational load) is used [63, 94, 112, 135, 215].

$$Y_B = f\left(\frac{s_R}{m}\right) > 1.$$

Figure 9.5 shows the diagram of bending stresses in a ring gear caused by the loading of the teeth, which are cantilevers, so that a bending moment is created. These stresses cause deformation of the ring gear and depend on the way of its mounting [207]. Its local compliance is greatly increased with the use of axial slots (Figures 8.11 and 8.14c). Usually, the ring gears are made sufficiently thick $(s_R/m < 3.5)$, so most often the factor Y_B is not used.

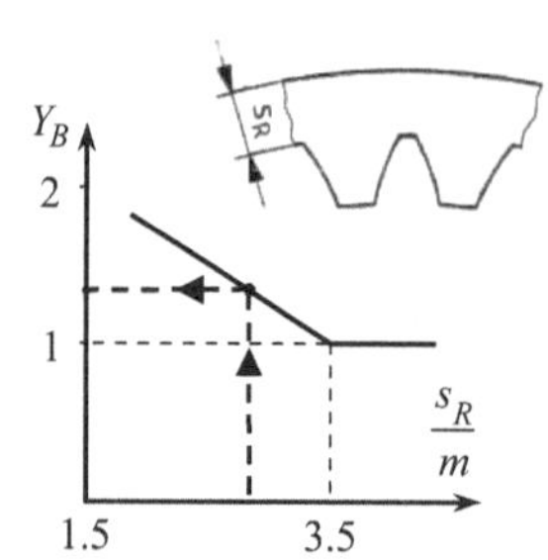

FIGURE 9.4
Rim thickness factor Y_B, considering the effect of rim thickness s_R under internal teeth on their bending strength. (Extracted from ISO 6336-3:2006(E) *Calculation of load capacity of spur and helical gears – Part 3: Calculation of tooth bending strength.*)

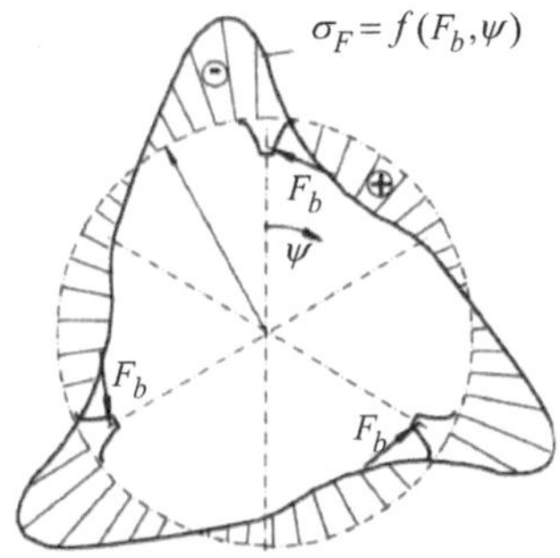

FIGURE 9.5
Bending stress in a ring gear.

9.2 Bearings Loading

1. Bearings of the central elements (sun gear 1 and carrier H) are theoretically unloaded (gravity forces are most often ignored), but their practical load depends on the uneven loading of the individual planets, considered by mesh load factor $K_{\gamma\, max}$. This load is not constant, neither in size nor in direction. It acts on impulses. In the calculations, it can be considered very approximate, only at discretion.

   Since the sun gear can be mounted either in the carrier or in the PGT housing, its relative speed $n_{1\,rel}$ is different. This must be taken into account when determining the bearing speed for a sun gear mounted in a carrier:

$$n_{1\,rel} = (n_1 - n_H) < n_1. \tag{9.5}$$

2. Planet bearings at low rotation speed of carrier n_H are assumed to be loaded with force:

$$F_{L2} \approx 2F_{tm} \cdot K_{\gamma\, max}. \tag{9.6}$$

Only at very high rotation speeds (e.g., in turbine PGTs), the centrifugal forces on the planets greatly increase the bearing load, and it can become critical. At relatively low-speed PGTs, such as industrial PGTs, this problem is practically nonexistent.

The type of inner race loading (rotating or stationary) must be taken into account depending on the way of bearing the planets:

- Bearings in the planet (most often)—Figure 9.6a
- Bearings in the carrier (less often)—Figure 9.6b

This, of course, matters for their life.

For planet bearings calculation, the relative rotation speed of the planets 2 with respect to the carrier H is significant:

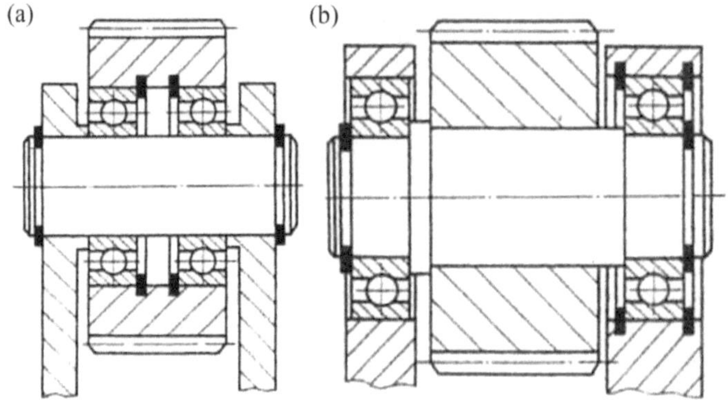

FIGURE 9.6
Types of inner bearing race loading depending on mounting: (a) Stationary loaded inner race (fixed pin); (b) rotating loaded inner race (rotating pin).

$$n_{2\,rel} = (n_1 - n_H)\frac{z_1}{z_2} = n_{1\,rel}\frac{z_1}{z_2}. \tag{9.7}$$

In [228], a high ratio gearbox with very low bearing load is proposed.

9.3 Planet Pins Loading

Two types of planet pins exist:

- Cantilever supported, in case of a single-wall carrier—Figure 9.7a
- At two ends supported, in case of a two-wall carrier—Figure 9.7b

In both cases, they are bending calculated.

9.4 Sun Gear Shaft Loading

Sun gear shaft is loaded by a combination of torsion and bending (Figure 9.8). It can be borne:

- With gear between bearings split (more favorable by strength and deformation)
- With gear outside of bearing split (cantilever borne in the housing)

When using a gear coupling for load distribution equalizing (Figure 20.5), the shaft is loaded with torsion only.

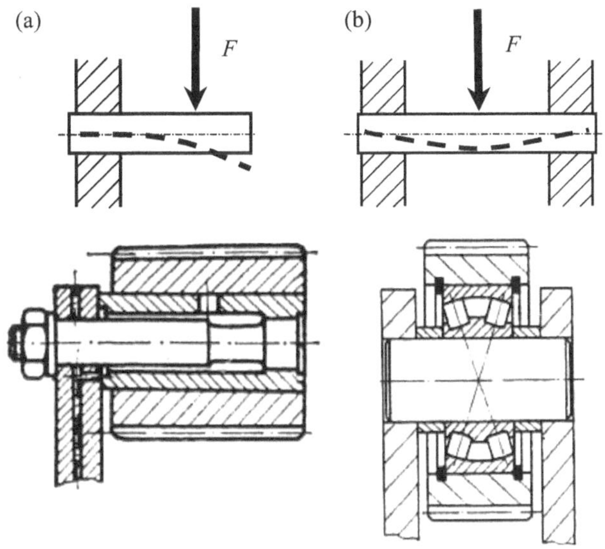

FIGURE 9.7
Planet pins loading and deformation: (a) Cantilever supported; (b) at two ends supported.

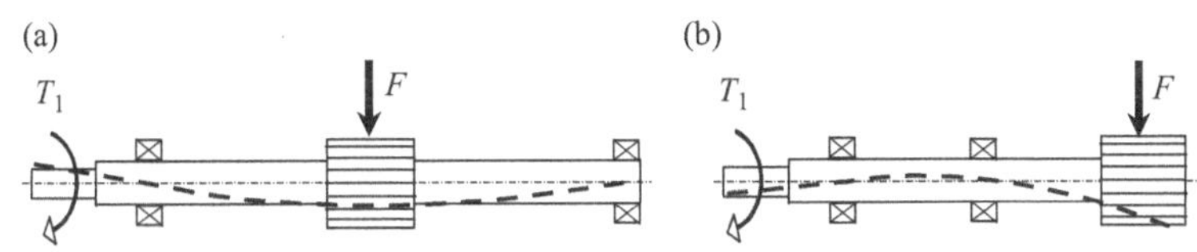

FIGURE 9.8 Loading and deformation of sun gear: (a) Between bearings split; (b) cantilever borne.

9.5 Gear Couplings Loading

Gear couplings are mainly used in PGTs with kinematic equalizing devices (Chapter 8). There is no load on their body (sleeve) on bending, as with the shafts, but only on torsion. Their teeth are calculated at the surface pressure [98, 175].

9.6 Carrier Loading

The load on the carrier and its deformations depend on its type. It can be single-wall (Figure 9.9a) or two-wall (Figures 9.9b and 9.9c). It is primarily required to be minimally deformable. Loaded by torque and the force on the planet pins, its calculation by the classical method of "strength of materials" is not always possible or expedient. Computer software (ANSYS®, MechSoft®) allows for the stresses and deformations of complicated details such as the carrier to be determined with a 3D model (SolidWorks®, Mechanikal Desktop®) using the finite element method (Figure 9.10). Practice shows that there is usually no

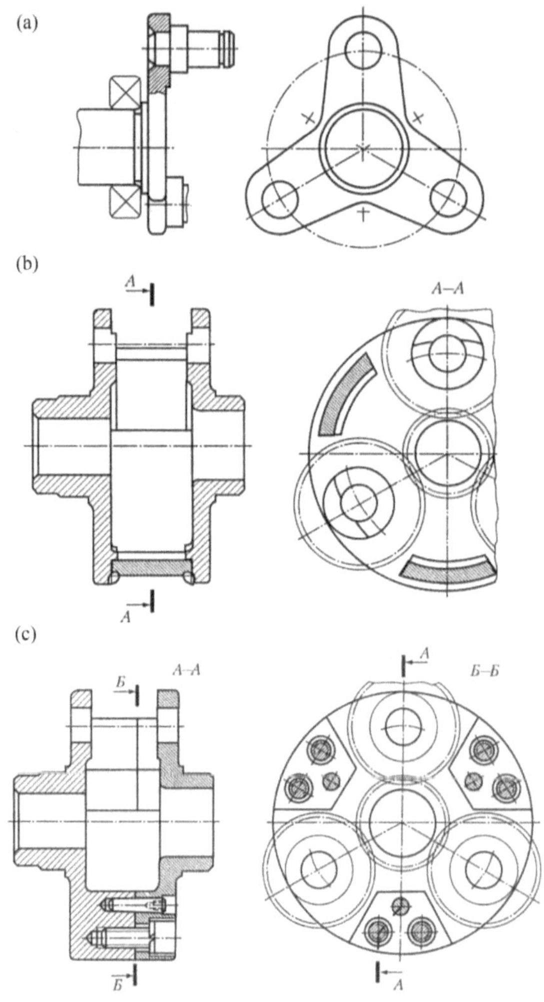

FIGURE 9.9
Carrier design: (a) Single-wall; (b) two-wall undemountable (welded); (c) two-wall demountable.

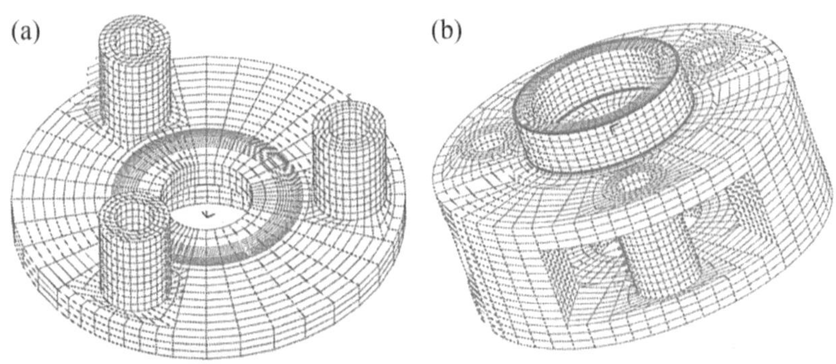

FIGURE 9.10
Carrier calculation by the finite elements method: (a) Single-wall carrier; (b) two-wall carrier. (Courtesy of Baumann, F. *Theoretische Untersuchungen zur Lastaufteilung und Lastverteilung in Planetengetrieben*. Dissertation. Dresden: Technische Universität, 2012.)

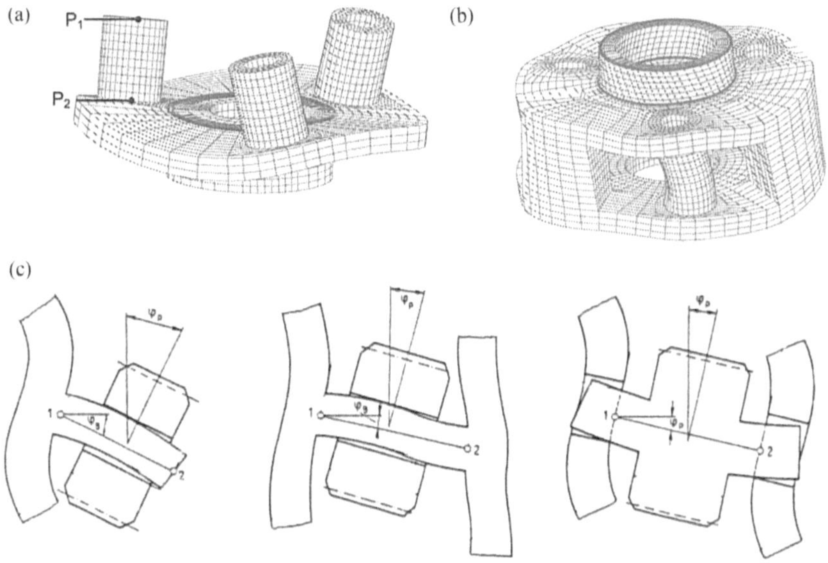

FIGURE 9.11
Carrier and planet deformations: (a) Single-wall carrier; (b) two-wall carrier; (c) planet tilling. (Courtesy of Baumann, F. *Theoretische Untersuchungen zur Lastaufteilung und Lastverteilung in Planetengetrieben*. Dissertation. Dresden: Technische Universität, 2012.)

problem with the strength of the carrier (except when it is intended to maximize lightening of the gear train) and its deformations are authoritative.

As shown in Figure 9.11, the deformations of the carrier lead to tilting of the planets and to nonuniform distribution of load over the tooth face, considered by face load factor $K_\beta = f\left(b_1/d_1\right) > 1$ in the load capacity calculations. For this reason, it is the best to use only one self-aligning planet bearing, whereby the influence of the carrier deformations is eliminated. Some self-alignment also allows for a single-row radial ball bearing as long as it has the necessary load capacity. The practice shows that in many cases, two bearings are also used (one needle roller bearing even). However, to avoid trouble $\left(K_\beta > 1\right)$, it is proper to select each planet bearing with equal radial clearance. Otherwise, the planets are tilted and the face load factor K_β can gain high values and it may lead to damage.

10

Types of Power in $\overline{AI}$-Planetary Gear Train

10.1 Absolute, Coupling, and Relative (Rolling) Power

When examining the question of power in planetary gear trains (PGTs), the notions derived from the superposition method, the Swamp method (Section 7.3), are all useful. Two partial movements (rotations) transmit their partial powers, e.g.,

Input (total, absolute, transmitting) power P_A in PGT consists of two types of power (Figure 10.1):

1. *Coupling power* P_{coup}, which transmits by coupling movement when PGT rotates as a coupling (as a whole) without relative rotation of gears to the carrier H. Consequently, this power is assumed to be transmitted without internal losses (the losses in the central element bearings are neglected).
2. *Relative (rolling) power* P_{rel} or the power in the mesh that is transmitted by relative movement of the gears with respect to the carrier. This is the power in PGT after inversion ("pseudo-PGT" with fixed carrier) that causes the meshing losses. These losses (lost power P_ψ) are considered by the *basic loss factor* ψ_0, respectively, and by the *basic efficiency* η_0.

Of the above follows

$$P_A = P_{coup} + P_{rel}. \tag{10.1}$$

In the most common case, the sun gear 1 is the input, the carrier H is the output, and ω_H is the coupling angular velocity. Then, the following dependencies are in effect:

$$P_A = T_A \cdot \omega_A = T_1 \cdot \omega_1, \tag{10.2}$$

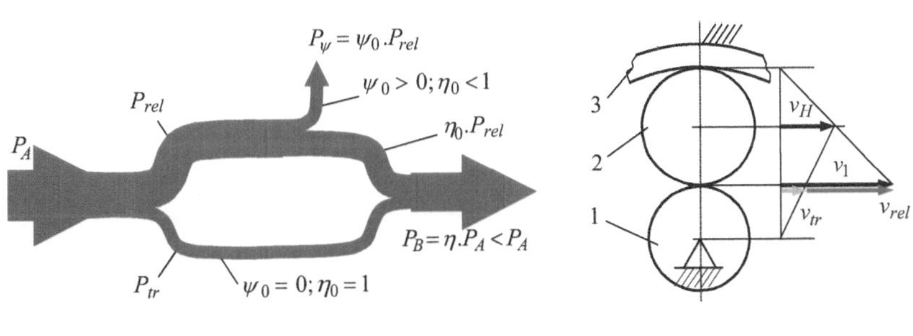

FIGURE 10.1
Types of power and peripheral velocities in $\overline{AI}$-PGT.

$$P_{coup} = T_A \cdot \omega_H = T_1 \cdot \omega_H, \tag{10.3}$$

$$P_{rel} = T_1(\omega_1 - \omega_H) = T_1 \cdot \omega_{1rel} = T_1 \cdot \omega_{1(H)} = T_1 \cdot \omega_1 \left(1 - \frac{\omega_H}{\omega_1}\right) = P_A \left(1 - \frac{1}{i_{1H(3)}}\right) < P_A, \tag{10.4}$$

respectively,

$$\frac{P_{rel}}{P_A} = 1 - \frac{1}{i_{1H(3)}} < 1. \tag{10.5}$$

It can be seen that in the $\overline{\text{AI}}$-PGT, the relative (rolling) power P_{rel} is smaller than the absolute power P_A, which results in smaller losses and higher efficiency of this type of PGT compared to other PGTs (AA, II, etc.). This is shown in Figure 10.1, where the peripheral velocity v_1 of sun gear 1 is represented as the sum of the coupling v_{coup} and the relative v_{rel} velocity:

$$v_1 = v_{coup} + v_{rel}, \tag{10.6}$$

and the coupling velocity is lower than the relative velocity:

$$v_{coup} < v_{rel}.$$

Figure 10.2 illustrates the increase of P_{rel}/P_A with i_0 ($i_{1H(3)}$, respectively) increasing. This increase leads to an increase in losses and, as a result, to a decrease in the efficiency $\eta_{1H(3)}$ (see Figure 12.1).

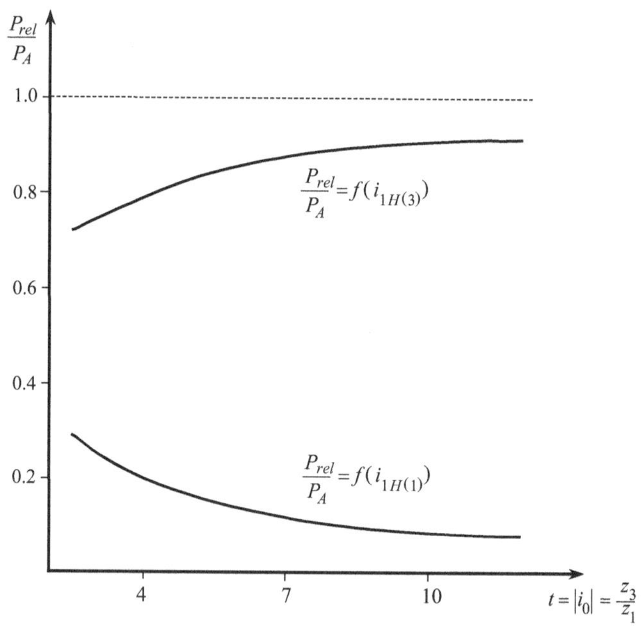

FIGURE 10.2
Changing the ratio P_{rel}/P_A of an $\overline{\text{AI}}$-PGT with $F = 1$ degree of freedom depending on the torque ratio t, respectively, on the basic speed ratio i_0.

The opposite is the case when the ring gear 3 is input and the carrier H—output. As shown in Figure 10.2, the ratio P_{rel}/P_A is much smaller than the previous case and also decreases with the increase of $i_{3H(1)}$. Losses also decrease, and the efficiency $\eta_{3H(1)}$ increases. The relationship between P_{rel}/P_A and $\eta_{1H(3)}$ and $\eta_{3H(1)}$ is clearly visible when comparing Figures 10.2 and 12.3.

For completeness, it will be noted that not only in $\overline{AI}$-PGT but in all PGTs working with unfixed carrier, the relative (rolling) power P_{rel} is never equal to the input (absolute) power P_A. It can be larger or smaller than the relative power:

$P_{rel} > P_A$ or $P_{rel} < P_A$.

This leads to a lower or higher efficiency η of the PGT compared to its basic efficiency η_0 (Section 11.5).

At the so-called negative-ratio PGTs ($i_0 < 0$), such as the $\overline{AI}$-PGT in question (Section 2.3), always

$P_{rel} < P_A$, so always $\eta > \eta_0$

The opposite is the case of the so-called positive-ratio PGTs ($i_0 > 0$), where

$P_{rel} > P_A$ and $\eta < \eta_0$

is obtained (see Chapters 22 and 23). It may even be possible to obtain $\eta < 0$; i.e., there is a self-locking, which in some cases may be desirable.

10.2 Real Torques

In PGT operation, regardless of whether it works with $F = 1$ or with $F = 2$ degrees of freedom and which of its elements is fixed at $F = 1$, as well as of the direction of input (absolute) power P_A, the possible directions of the relative (rolling) power P_{rel} are only two (Figure 10.3):

- From the sun gear 1 to the ring gear 3
- From the ring gear 3 to the sun gear 1

Criterion for determining the direction of relative (rolling) power is as follows: if in a converted PGT (after inversion) the direction of relative angular velocity $\omega_{1\,rel} = \omega_1 - \omega_H$ of sun gear 1 is the same as the direction of its torque T_1, the sun gear is driving (with respect to the carrier), and the relative (rolling) power P_{rel} is transmitted from it through the planets 2 to the ring gear 3 (driven gear) and vice versa. The criterion can be expressed by the following dependency:

$$T_1 \cdot \omega_{1\,rel} = T_1(\omega_1 - \omega_H)\begin{cases} > 0 - \text{sun gear 1 is driving} \\ < 0 - \text{sun gear 1 is driven} \end{cases}. \quad (10.7)$$

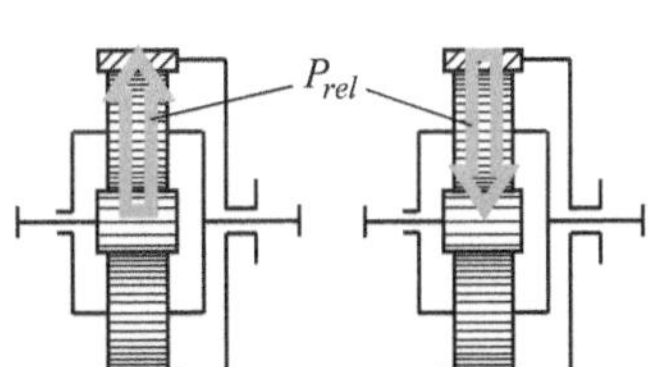

FIGURE 10.3
Possible directions of relative (rolling) power P_{rel}.

As a result of the existing internal losses in PGT (basic loss factor $\psi_0 > 0$ and basic efficiency $\eta_0 < 1$), the following dependencies for *real torques* T_1' and T_3' (on sun gear 1 and ring gear 3) are obtained:

In transmitting the relative (rolling) power P_{rel} from the sun gear 1 to the ring gear 3 (*driving* sun gear),

$$T_3' = \eta_0 \cdot t \cdot T_1 < t \cdot T_1 = T_3, \text{resp.} T_1' = \frac{1}{\eta_0} \cdot \frac{T_3}{t} > \frac{T_3}{t} = T_1. \tag{10.8}$$

In transmitting the relative power P_{rel} from the ring gear 3 to the sun gear 1 (*driven* sun gear),

$$T_1' = \eta_0 \frac{T_3}{t} < \frac{T_3}{t} = T_1, \text{resp.} T_3' = \frac{1}{\eta_0} \cdot t \cdot T_1 > t \cdot T_1 = T_3. \tag{10.9}$$

In the above formulae, T_1 and T_3 are the ideal, and T_1' and T_3'—real torques. Real torques are used to determine the efficiency of the gear train, and for them, an analogous formula such as formula (7.27) for the ideal torque is in force:

$$\boxed{\sum T_i' = T_1' + T_3' + T_H' = 0.} \tag{10.10}$$

This dependence is particularly useful in compound PGTs analysis (see Chapter 29 and in particular Figures 29.4, 29.6, 29.8–29.10, and 29.12).

The fact that one and the same gear (in this case the sun gear 1), with respect to one coordinate system (e.g., the housing), can be *driving* and, with respect to another coordinate system (e.g., the carrier H), can be *driven*, is in itself interesting and inherent in PGTs. This fact, however, is one that is not always easily understood.

11

Types of Losses and Basic Efficiency of AI-Planetary Gear Train

In gear trains in general, including planetary gear trains (PGTs), there are different types of losses during their work. Fundamentally, the losses in the gear trains are

- Load-dependent
- Non-load-dependent

Three types of losses are the main ones:

- Meshing losses
- Bearing losses
- Oil (hydraulic) losses

There are other losses (ventilation, in the seals), but they are either less common or small.

Losses in meshing and bearings depend on load, whereas other losses do not.

11.1 Meshing Losses

These losses depend on various factors, notably on tooth geometry, but to a large extent also on the coefficient of friction in teeth meshing μ_z. Various formulae and methodologies for their determination can be found in the specialized literature [102, 135, 150, 183, 188]. With sufficient accuracy for engineering practice, the loss factors in both meshings (external and internal) can be determined as follows [85, 135, 147, 148, 150, 175, 180]:

$$\boxed{\psi_{z12(H)} = \mu_z \frac{\pi}{z_1} \cdot \frac{u_{12}+1}{u_{12}} \left(1 - \varepsilon_{\alpha 12} + \varepsilon_{\alpha 1}^2 + \varepsilon_{\alpha 2(1)}^2\right) = f(\mu_z)}, \tag{11.1}$$

$$\boxed{\psi_{z23(H)} = \mu_z \frac{\pi}{z_2} \cdot \frac{u_{23}-1}{u_{23}} \left(1 - \varepsilon_{\alpha 23} + \varepsilon_{\alpha 2(3)}^2 + \varepsilon_{\alpha 3}^2\right) = f(\mu_z)}, \tag{11.2}$$

where μ_z is the coefficient of friction in meshing, z_1 and z_2 are the numbers of teeth of gears, $\varepsilon_{\alpha 12} = \varepsilon_{\alpha 1} + \varepsilon_{\alpha 2(1)}$ and $\varepsilon_{\alpha 23} = \varepsilon_{\alpha 2(3)} + \varepsilon_{\alpha 3}$ are the transverse contact ratios of both meshings—external and internal, $\varepsilon_{\alpha 2(1)}$ and $\varepsilon_{\alpha 1}$ are the transverse contact ratios of in-front-of-pitch-point and beyond-of-pitch-point meshing of external meshing, $\varepsilon_{\alpha 3}$ and $\varepsilon_{\alpha 2(3)}$ are the transverse contact ratios of in-front-of-pitch-point and beyond-of-pitch-point meshing of internal meshing, and $u_{12} = z_2/z_1 > 1$ and $u_{23} = z_3/z_2 > 1$ are teeth ratios of both meshings—external and internal.

In these formulae, the most uncertain value is the coefficient of friction in meshing μ_z, which determines the scattering of efficiency of each gear train. It depends on the type of used oil and its viscosity v at a given temperature θ_L, as well as on the intensity of the tooth load and on the peripheral velocity of gears. Figure 11.1 shows this dependence. As an average value is recommended $\mu_z = 0.05 \div 0.07$ [167]. More details about this coefficient can be found in the specialized literature [76, 78, 147, 167].

Transverse contact ratios of external meshing $\varepsilon_{\alpha 12}$ and of internal meshing $\varepsilon_{\alpha 23}$ can be determined by the following formulae in accordance with Figure 11.2:

$$\varepsilon_{\alpha 12} = \frac{\overline{CE} + \overline{AC}}{p_e} = \frac{g_{\alpha 1} + g_{\alpha 2(1)}}{p_e} = \frac{g_{\alpha 1}}{p_e} + \frac{g_{\alpha 2(1)}}{p_e} = \varepsilon_{\alpha 1} + \varepsilon_{\alpha 2(1)}$$

$$= \frac{z_1}{2\pi}(\tan\alpha_{a1} - \tan\alpha_{w12}) + \frac{z_2}{2\pi}(\tan\alpha_{a2} - \tan\alpha_{w12}) \quad (11.3)$$

$$\varepsilon_{\alpha 23} = \frac{\overline{CE} + \overline{AC}}{p_e} = \frac{g_{\alpha 2(3)} + g_{\alpha 3}}{p_e} = \frac{g_{\alpha 2(3)}}{p_e} + \frac{g_{\alpha 3}}{p_e} = \varepsilon_{\alpha 2(3)} + \varepsilon_{\alpha 3}$$

$$= \frac{z_2}{2\pi}(\tan\alpha_{a2} - \tan\alpha_{w23}) + \frac{z_3}{2\pi}(\tan\alpha_{w23} - \tan\alpha_{a3}), \quad (11.4)$$

where p_e is the meshing pitch, equal to base circle pitch $p_b = \pi \cdot m \cdot \cos\alpha$, $\varepsilon_{\alpha 2(1)}$ and $\varepsilon_{\alpha 1}$ are the transverse contact ratios of in-front-of-pitch-point and beyond-of-pitch-point meshing of external meshing, $\varepsilon_{\alpha 3}$ and $\varepsilon_{\alpha 2(3)}$ are the transverse contact ratios of in-front-of-pitch-point and beyond-of-pitch-point meshing of internal meshing, α_{w12} and α_{w23} are the operating pressure angles of external and internal meshing, and α_{a1}, α_{a2}, and α_{a3} are the profile angles of tip circles with diameters d_{a1}, d_{a2}, and d_{a3} of gears, determined as follow:

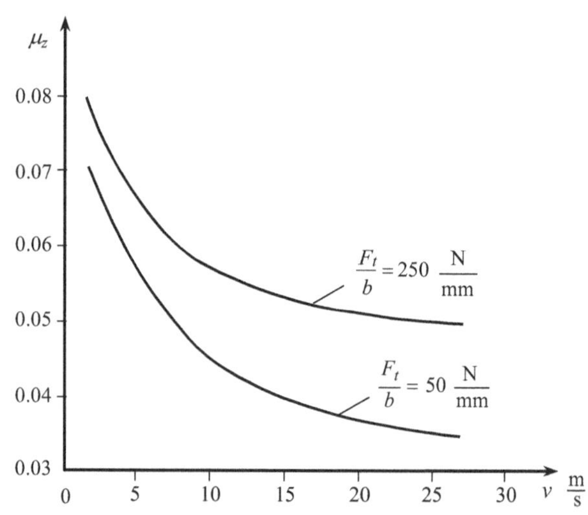

FIGURE 11.1
Coefficient of friction in meshing μ_z depending on the intensity tooth load F_t/b and on peripheral velocity v. (Redrawn from Müller, H.W. *Die Umlaufgetriebe – Auslegung und vielseitige Anwendungen*. 2. Auflage. Berlin: Springer-Verlag, 1998.)

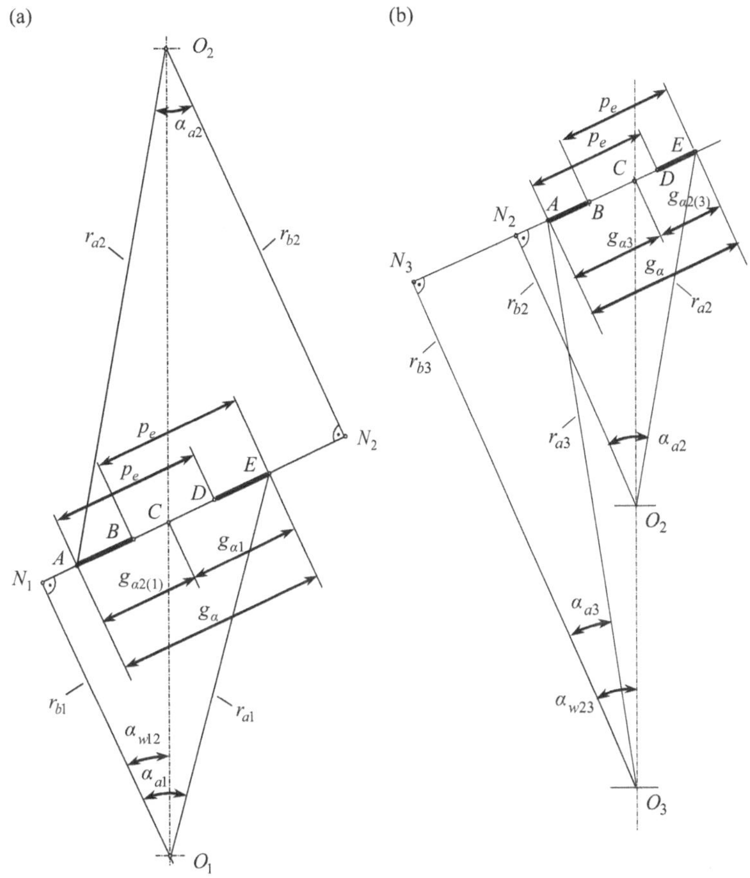

FIGURE 11.2
Transverse contact ratio determination: (a) External meshing; (b) internal meshing.

$$\alpha_{a1} = \arccos\frac{d_{b1}}{d_{a1}};\ \alpha_{a2} = \arccos\frac{d_{b2}}{d_{a2}};\ \alpha_{a3} = \arccos\frac{d_{b3}}{d_{a3}}; \tag{11.5}$$

where d_{b1}, d_{b2}, and d_{b3}—diameters of basic circles of gears.

Recommended values for transverse contact ratios are [175]

$$\varepsilon_{\alpha 12} \geq 1.2 \text{ and } \varepsilon_{\alpha 23} \geq 1.4.$$

In a PGT with its invariable geometry mentioned above, only its load F_t and peripheral velocity v can be changed during operation, so the lost power P_ψ is only a function of these two dimensions, i.e.,

$$P_\psi = f(F_t, v), \text{ resp. } P_\psi = f(T_B, \omega_A).$$

Losses in meshing typically account for about 85% of total gear train loss [60, 206]. According to [78], the losses in each meshing are on average 0.1% of the transmitted power. An interesting investigation of the dependence of mesh loss factor ψ_z on h_a^*, x, and α is given in [95].

11.2 Bearing Losses

Losses in bearings are very different in size depending on the types of bearings—rolling or sliding. With rolling bearings, they are much smaller than meshing losses, so they are either ignored $(\psi_B = 0)$ or estimated by a small addition to meshing losses (approximately 5%–6% of meshing losses). If, however, their calculation is necessary, specialist literature can be used [83, 175].

It is quite another situation with high-speed powerful gear trains (e.g., turbine gear trains) with sliding bearings, in which the bearing losses greatly increase, and they become dominant and bigger than the meshing losses and cannot be ignored. In these gear trains, the bearing loss factor is $\psi_B > 0$. This very special type of PGT is not an object of this book.

For power transmission, design solutions to reduce bearing loads are sought. For example, [228] proposes a gear train with a high speed ratio, whereby the reduced bearing load is highlighted as a major advantage.

11.3 Oil (Hydraulic) Losses

Only the oil losses due to churning and crushing (oil bath lubrication), as used in the most common manufactured small and medium power PGT, will be considered here.

As shown in Figure 11.3, during operation of PGT, the planets sequentially (one by one) splash (crush) into the oil, which of course is connected with losses and the corresponding loss factor ψ_L. These losses are further increased by the hydraulic resistance of the oil ejection from the tooth spaces of the meshed gears. Unfortunately, these losses cannot be reliably determined in a theoretical way, so that in each case, only the results of the experimental research are relevant.

In design calculations, the results of some theoretical [60] or experimental [59, 86] studies can be used. For example, in the theoretical–experimental study [190] of the two-stage PGT are obtained $\psi_L = 0.1\psi$ for fast-speed stage $\left(n_A = 1486\ \text{min}^{-1}\right)$ and $\psi_L = 0.017\psi$ for slow-speed stage $\left(n_A = 173\ \text{min}^{-1}\right)$—both with fixed ring gears.

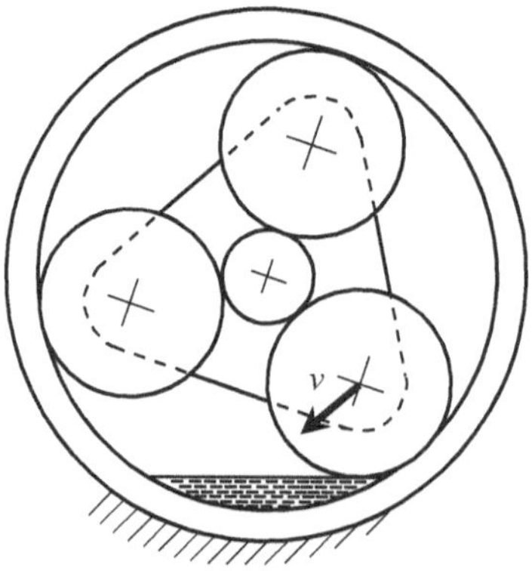

FIGURE 11.3
Planets splashing in the oil.

11.4 Other Losses

Of the three types of losses listed above, the meshing losses are the most accurately quantifiable, and the hydraulic losses are the most inaccurately determinable and requiring experimental tests. However, for the most common low-power PGTs of general industrial use, these losses are the smallest.

Generally, PGTs also have other losses, e.g., losses in seals, that are only relevant for very small-sized gears, or ventilation (windage) losses only relevant to high-speed, large-sized turbine gear trains. Theoretical studies for their determination exist [59]. Experiments also provide data that can be used in design calculations [86]. For example, according to [190], the losses in the fast-speed stage seals are below 7% of all losses, and in slow-speed stage—negligible (0.7%).

11.5 Basic Efficiency

As can be seen from the above, the losses in PGTs depend on a number of factors, and most often, the meshing losses dominate. For this reason, for simplicity, the calculation of efficiency is based only on the meshing losses, and the others losses are evaluated with an addition of $\Delta\psi$. In the most common PGTs (with rolling bearings and low peripheral velocities, most often <10 m/s), at 100% loading (or load close to nominal), only considering the meshing losses (with loss factors $\psi_{z12(H)}$ and $\psi_{z23(H)}$ and addition $\Delta\psi$) gives satisfactory results.

Under this stipulation, the basic loss factor ψ_0 of converted PGT (with fixed carrier, see Section 7.1.1) can be expressed as follows:

$$\psi_0 = \psi_{z12(H)} + \psi_{z23(H)} + \psi_B + \psi_L = \psi_z + \psi_B + \psi_L = \psi_z + \Delta\psi \approx \psi_{z0} \tag{11.6}$$

There is a simplified formula for the basic loss factor, which takes into account the dominant influence of the number of teeth of the gears [85, 155]

$$\boxed{\psi_0 \approx \underbrace{0.15\left(\frac{1}{z_1}+\frac{1}{z_2}\right)+0.2\left(\frac{1}{z_2}-\frac{1}{z_3}\right)}_{\psi_z}+\Delta\psi \approx \psi_{z0}}, \tag{11.7}$$

where

$\Delta\psi = 15 \div (30)\%$ of ψ_z—additional accounting of the other losses in the gear train.

On the basis of experimental studies [85, 167], it was found that in practice, there is an equality of the meshing losses in $\overline{AI}$-PGT (i.e., deviation of less than 5%), in the transmission of relative power P_{rel} from sun gear 1 to ring gear 3 and vice versa. This means

$$\psi_{z13(H)} \approx \psi_{z31(H)} = \psi_{z0} = \psi_0, \tag{11.8}$$

respectively, for *basic* efficiency η_0, it is obtained:

$$\boxed{\eta_0 = 1-\psi_0 = \eta_{13(H)} = 1-\psi_{z13(H)} \approx \eta_{31(H)} = 1-\psi_{z31(H)}}. \tag{11.9}$$

Because the basic loss factor ψ_0, resp. basic efficiency η_0, has a scattering due to the scattering of the coefficient of friction in the meshing μ_z, the efficiency of PGT η will ultimately be scattered.

$$\eta = f(\eta_0) = f(\mu_z).$$

This is because, at a particular PGT with its unchanging geometric parameters, μ_z is the only parameter that can affect ψ_0 and η_0, and hence affect the η. The expression of efficiency η as a function of the coefficient of friction μ_z has this important advantage that it outlines the limits within which its actual numerical value is located and does not present it as a fixed but unreliable numerical value.

For PGTs, losses are more important than for other types of gears, because the compactness (one of their advantages) of PGTs results in a smaller cooling surface and higher heating. In some cases, this may require circulating lubrication, which significantly complicates the structure and raises the cost of the gear train.

The formula (11.7) can also be converted, and the meshing loss factor ψ_z and the corresponding meshing efficiency η_z, respectively, can be expressed as a function of the torque ratio t and the number of teeth of sun gear z_1 [230]:

$$\eta_z = f(t, z_1) = 1 - \frac{1}{z_1} \cdot \frac{t+1}{t(t-1)}(0.15t + 0.2). \tag{11.10}$$

This dependence is shown in Figure 11.4.

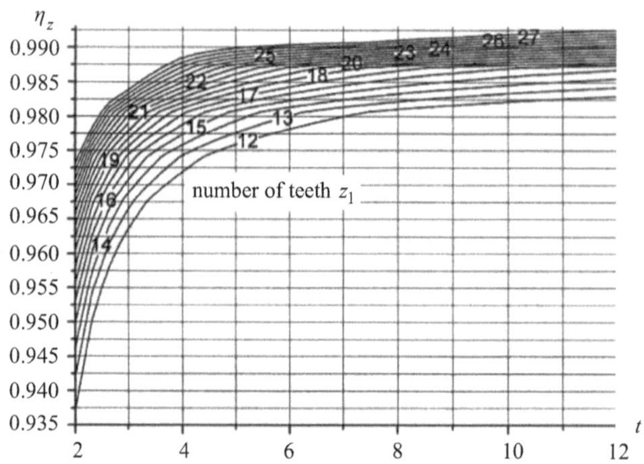

FIGURE 11.4
Determining of meshing efficiency η_z (11.10) depending on the torque ratio t and the number of teeth z_1 of sun gear 1. (Courtesy of Troha, S. *Analysis of a planetary change gear train's variants.* Dissertation. Rijeka: Engineering Faculty, University of Rijeka [Croatia], 2011 (in Croatian).)

12

Efficiency of $\overline{AI}$-Planetary Gear Train

12.1 General

Basic efficiency η_0 of converted by inversion (see Section 7.1.1) planetary gear train (PGT) (pseudo-PGT) gives possibility to calculate efficiency for six possible modes of work of PGT with $F = 1$ degree of freedom (as reducer or multiplier) as well as with $F = 2$ degrees of freedom (as summing or devising differential) (Figure 3.2). The dependencies are as follows:

$$\eta = -\frac{P_B}{P_A} = -\frac{-(P_A + P_\psi)}{P_A} = 1 + \frac{P_\psi}{P_A} = 1 - \psi < 1, \tag{12.1}$$

$$\psi = \left|\frac{P_\psi}{P_A}\right| = \frac{\psi_0 \cdot P_{rel}}{P_A} < \psi_0, \tag{12.2}$$

where P_A is the input power $(P_A > 0)$, P_B is the output power $(P_B < 0)$, P_ψ is the lost power $(P_\psi < 0)$, P_{rel} is the relative (rolling) power $(P_{rel} > 0)$, ψ is the loss factor of PGT working as planetary $(\psi > 0)$, and ψ_0 is the basic loss factor of converted PGT working with fixed carrier (as pseudo-planetary) $(\psi_0 > 0)$.

Formula (12.1) is derived from the sum of the powers in the gear train:

$$\sum P_i = P_A + P_B + P_\psi = 0, \tag{12.3}$$

respectively,

$$\sum P_i = \eta \cdot P_A + P_B = 0, \tag{12.4}$$

from where the above expression follows $\eta = -\frac{P_B}{P_A}$.

As a result of the lower relative (rolling) power P_{rel} compared to the input power P_A (see Chapter 10 and Figure 10.1)

$$P_{rel} < P_A \tag{12.5}$$

Loss factor of PGT ψ working as a planetary one is smaller than basic loss factor ψ_0 in converted (pseudo-planetary) PGT:

$$\psi < \psi_0, \tag{12.6}$$

and the efficiency η accordingly increases and becomes larger than η_0:

$$\eta > \eta_0. \tag{12.7}$$

12.2 Theoretical Determination of Efficiency

There are different ways for the theoretical determination of PGT efficiency η known from the literature [85, 106, 147–150, 155, 175, 217]. The torque method is very appropriate because, unlike other methods, it is primarily very visual, and also simple and easy to use (see Section 7.4).

Starting from formulae (12.4) and (7.33), respectively,

$$\sum P_i = \eta \cdot P_A + P_B = \eta \cdot T_A' \cdot \omega_A + T_B' \cdot \omega_B = 0, \tag{12.8}$$

where P_A and P_B are the input and output powers $(P_A > 0; P_B < 0!)$, T_A' and T_B' are the real input and output torques, and ω_A and ω_B are the input and output angular velocities.

From formula (12.9), efficiency is [see Section 7.4.1 formula (7.32)]

$$\boxed{\eta = -\frac{P_B}{P_A} = -\frac{T_B' \cdot \omega_B}{T_A' \cdot \omega_A} = -\frac{T_B'/T_A'}{\omega_A/\omega_B} = -\frac{i_T}{i_k}}, \tag{12.9}$$

where i_T is the torque transmit ratio (torque transformation) [see Section 7.4.1 formula (7.31)] and $i_k \equiv i$ is the speed ratio (kinematic transmit ratio).

Except through formula (12.9), efficiency η also may be determined by the following formula (Figure 10.1), the result, of course, being the same:

$$\boxed{\eta = -\frac{P_B}{P_A} = \frac{P_{tr} + \eta_0 \cdot P_{rel}}{P_{tr} + P_{rel}} = \frac{T_1 \cdot \omega_H + \eta_0 \cdot T_1 \cdot \omega_{1(H)}}{T_1 \cdot \omega_H + T_1 \cdot \omega_{1(H)}}}, \tag{12.10}$$

where P_{coup} and P_{rel} are the coupling and relative (rolling) powers determined by formula (10.2), ω_H is the angular velocity of carrier H, and $\omega_{1(H)} \equiv \omega_{1\,rel}$ is the relative angular velocity of sun gear to the carrier H.

In formula (12.9) to obtain $\eta < 1$, the torque ratio *must necessarily* be smaller than the speed ratio, i.e., $i_T < i_k$! In view of this requirement, it is easy to obtain efficiency for the six working modes of PGT with $F = 1$ degree of freedom (such as reducer or multiplier).

In the formulae for torque transformation i_T (Figure 7.5), it is taken either $\eta_0 \cdot t$ or t/η_0 so as to obtain $i_T < i_k$.

Determination of PGT efficiency is done through real torques. The two possible directions of relative (rolling) power P_{rel}—from sun gear 1 to the ring gear 3 or vice versa, shall be determined according to the criterion set out in Section 10.2—matching or non-matching of the torque T_1 direction with the direction of the relative angular velocity $\omega_{1\,rel} = \omega_1 - \omega_H$ of the sun gear 1 (10.4).

As already noted in Section 11.5, efficiency of the gear train η is ultimately a function of the friction coefficient μ_z in meshing and this coefficient determines its scattering:

$$\eta = f(\mu_z). \tag{12.11}$$

This should be taken into account when using the formulae given below for η, depending on the basic efficiency η_0, which in turn depends on μ_z, i.e.,

$$\eta = f(\eta_0) = f(\mu_z). \tag{12.12}$$

For this reason, it is right and more credible efficiency not to be given as a fixed numerical value but with its scattering.

Naturally, efficiency is with scattering when the PGT operates with both $F = 1$ and $F = 2$ degrees of freedom (as summation or division differential).

12.3 Efficiency at Work with $F = 1$ Degree of Freedom

Three modes of work with $F = 1$ degree of freedom are possible, depending on which of the central elements is fixed.

$\underline{\omega_H = 0}$

In this case, PGT works as non-planetary, and it is known from Section 11.5 that the efficiencies in the two possible directions—from sun gear 1 to the ring gear 3 or vice versa—are practically the same [76, 167]:

$$\eta_{13(H)} \approx \eta_{31(H)} = \eta_0 < 1. \tag{12.13}$$

The difference $\Delta\eta_0 = \left|\eta_{13(H)} - \eta_{31(H)}\right|$ is very small and is of the order of error in the experimental studies $(\Delta\eta_0 < 5\%)$ [76]. This difference is also smaller than the change of efficiency in operating (caused by temperature fluctuations, hence, oil viscosity, angular velocity, load), so the above acceptance (12.8) is fully justified for practice.

In the following pictures, the direction of transmitted power is indicated by solid arrows and of the relative power—with dotted arrows.

$\underline{\omega_3 = 0}$—the most common case (Figure 12.1)

Work as reducer $i > 1$

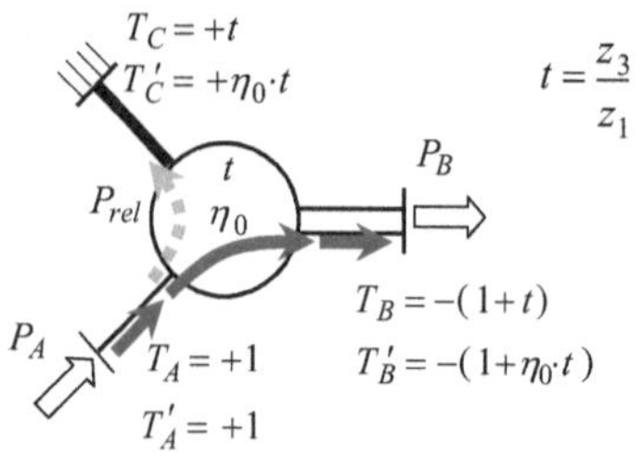

FIGURE 12.1A
AI-PGT with $F = 1$ at fixed ring gear 3 as a reducer.

$$i_k = i_{1H(3)} = -\frac{T_B}{T_A} = -\frac{-(1+t)}{+1} = 1 + \frac{z_3}{z_1}$$

$$i_T = \frac{T_B'}{T_A'} = \frac{-(1+\eta_0 \cdot t)}{+1} = -\left(1 + \eta_0 \frac{z_3}{z_1}\right)$$

$$\eta_{1H(3)} = -\frac{i_T}{i_k} = -\frac{-\left(1+\eta_0 \frac{z_3}{z_1}\right)}{1+\frac{z_3}{z_1}} = \frac{1+\eta_0 \frac{z_3}{z_1}}{1+\frac{z_3}{z_1}} < 1$$

Because of equality $T_A' = T_A = +1$ (value + 1 is not obligatory!), it is possible to define efficiency and in the following second way, using the output torques—ideal T_B and real T_B'

$$\eta_{1H(3)} = \frac{T_B'}{T_B} = \frac{-(1+\eta_0 \cdot t)}{-(1+t)} = \frac{1+\eta_0 \frac{z_3}{z_1}}{1+\frac{z_3}{z_1}} < 1,$$

i.e., the same result is obtained as above.

Work as multiplier $i < 1$

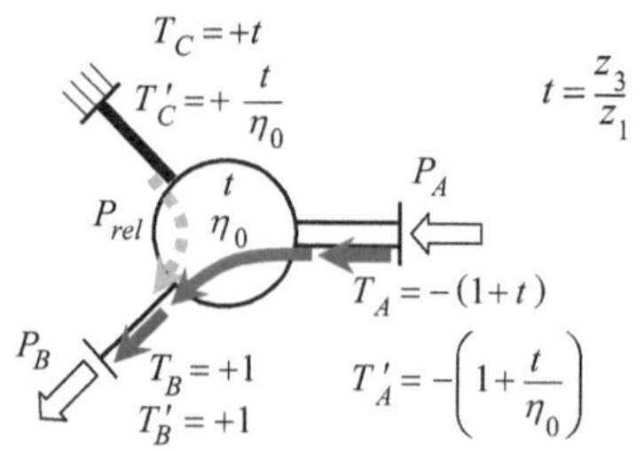

FIGURE 12.1B
AI-PGT with $F = 1$ at fixed ring gear 3 as a multiplier.

$$i_k = i_{H1(3)} = -\frac{T_B}{T_A} = \frac{+1}{-(1+t)} = \frac{1}{1+\frac{z_3}{z_1}}$$

$$i_T = \frac{T_B'}{T_A'} = \frac{+1}{-\left(1+\frac{t}{\eta_0}\right)} = -\frac{1}{1+\frac{1}{\eta_0}\cdot\frac{z_3}{z_1}}$$

$$\eta_{H1(3)} = -\frac{i_T}{i_k} = -\frac{-\frac{1}{1+\frac{1}{\eta_0}\cdot\frac{z_3}{z_1}}}{\frac{1}{1+\frac{z_3}{z_1}}} = \frac{1+\frac{z_3}{z_1}}{1+\frac{1}{\eta_0}\cdot\frac{z_3}{z_1}} < 1$$

And here because of $T_B' = T_B = +1$, efficiency can be defined in the same second way as in the work as a reducer

$$\eta_{H1(3)} = \frac{T_A}{T_A'} = \frac{-(1+t)}{-\left(1+\frac{t}{\eta_0}\right)} = \frac{1+\frac{z_3}{z_1}}{1+\frac{1}{\eta_0}\cdot\frac{z_3}{z_1}} < 1$$

The same result as above is obtained, too.

$\omega_1 = 0$ (Figure 12.2)

Work as reducer $i > 1$

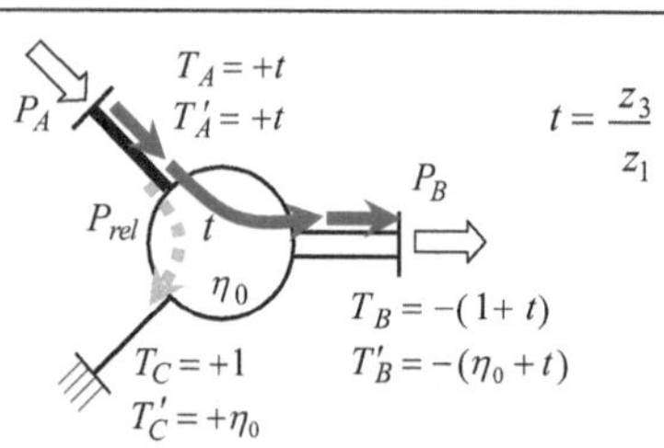

FIGURE 12.2A
AI-PGT with $F = 1$ at fixed sun gear 1 as a reducer.

Work as multiplier $i < 1$

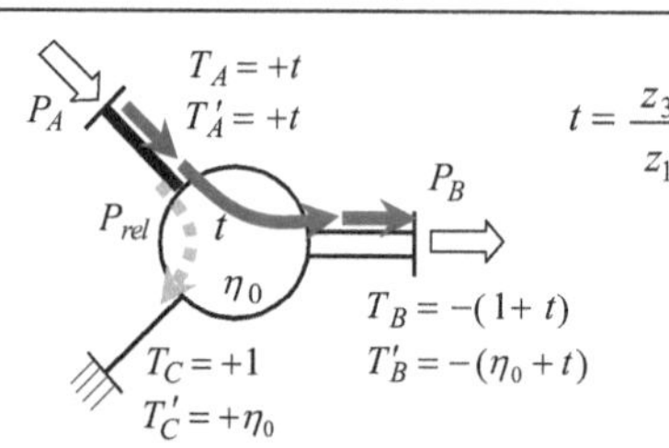

FIGURE 12.2B
AI-PGT with $F = 1$ at fixed sun gear 1 as a multiplier.

(Continued)

Work as reducer $i > 1$	**Work as multiplier $i < 1$**
$i_k = i_{3H(1)} = -\frac{T_B}{T_A} = -\frac{-(1+t)}{+t} = 1+\frac{1}{t} = 1+\frac{z_1}{z_3}$	$i_k = i_{H3(1)} = -\frac{T_B}{T_A} = -\frac{+t}{-(1+t)} = \frac{1}{1+\frac{1}{t}} = \frac{1}{1+\frac{z_1}{z_3}}$
$i_T = \frac{T'_B}{T'_A} = \frac{-(\eta_0+t)}{t} = -\left(\frac{\eta_0}{t}+1\right) = -\left(1+\eta_0\frac{z_1}{z_3}\right)$	$i_T = \frac{T'_B}{T'_A} = \frac{\eta_0 \cdot t}{-(1+\eta_0 \cdot t)} = -\frac{1}{1+\frac{1}{\eta_0 \cdot t}} = -\frac{1}{1+\frac{1}{\eta_0}\cdot\frac{z_1}{z_3}}$
$\eta_{3H(1)} = -\frac{i_T}{i_k} = -\frac{-\left(1+\frac{\eta_0}{t}\right)}{1+\frac{1}{t}} = \frac{1+\eta_0\frac{z_1}{z_3}}{1+\frac{z_1}{z_3}} < 1$	$\eta_{H3(1)} = -\frac{i_T}{i_k} = \frac{-\frac{1}{1+\frac{1}{\eta_0 \cdot t}}}{\frac{1}{1+\frac{1}{t}}} = \frac{1+\frac{1}{t}}{1+\frac{1}{\eta_0 \cdot t}} = \frac{1+\frac{z_1}{z_3}}{1+\frac{1}{\eta_0}\cdot\frac{z_1}{z_3}} < 1$
And here because of $T'_A = T_A = +t$, efficiency can be defined in the second way $\eta_{3H(1)} = \frac{T'_B}{T_B} = \frac{-(\eta_0+t)}{-(1+t)} = \frac{\frac{\eta_0}{t}+1}{\frac{1}{t}+1} = \frac{1+\eta_0\frac{z_1}{z_3}}{1+\frac{z_1}{z_3}} < 1$ That is, the same result as above is obtained, too.	Because neither $T'_A \equiv T_A$ nor $T'_B \equiv T_B$, the second way to define efficiency cannot be applied here. However, this can happen if it is started from $T'_B \equiv T_B = +t$ instead of $T'_C \equiv T_C = +1$, where the other torques will be $T'_C = +\frac{1}{\eta_0}$ and $T'_A = \left(\frac{1}{\eta_0}+t\right)$.

From the above, it can be seen that as a third possibility, in individual cases, it can be taken directly $\eta_0 \cdot t$ or t/η_0 in the formulae for the kinematic (speed) ratio i_k, so as to obtain the condition $i_T < i_k$, as outlined above. In fact, it has been shown that efficiency of PGT can be defined in three different ways, which is advantageous because it gives more freedom to the designer and allows for the result to be checked. The third, simplest way—the sampling method—can be used, however, primarily for simple, single-carrier PGTs, but not always for compound ones, where things are no longer so obvious to achieve the condition $i_l < i_k$.

The aforementioned dependencies on the work of PGT with $F = 1$ degree of freedom can be illustrated as follows (Figure 12.3) and the following *comment* can be made:

1. Efficiency of an $\overline{AI}$-PGT, working as reducer or as multiplier at established mode, practically is the same. Efficiency η of an $\overline{AI}$-PGT, working as planetary, is always higher than the basic efficiency η_0, when it works as non-planetary (with a fixed carrier), i.e.,

$$\eta > \eta_0. \tag{12.14}$$

As noted above, the reason for this is in the lower relative power P_{rel} compared to the input power P_A

$$P_{rel} < P_A, \tag{12.15}$$

which power P_{rel} causes internal losses.

Although in the two cases of work (as reducer or multiplier) with the same fixed element, different formulae for efficiency are obtained, in the calculation with concrete values of t and η_0, practically the same results are obtained (as it is confirmed by experiments).

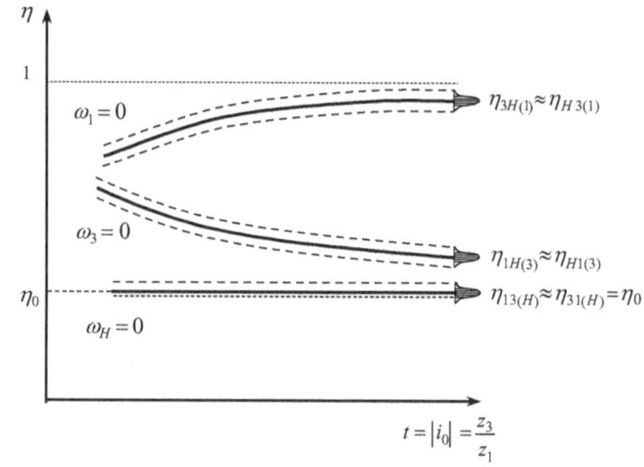

FIGURE 12.3
Efficiency η of an $\overline{\text{AI}}$-PGT working with $F = 1$ degree of freedom as a function of torque ratio t, resp. of basic speed ratio i_0, in an established mode (T_B = const), taking into account the scattering of the base efficiency η_0, resp. the coefficient of friction in meshing μ_z.

2. Trends of the change of efficiency $\eta_{1H(3)}$ and $\eta_{3H(1)}$ are different. With the increase of $t = |i_0| = z_3/z_1$, $\eta_{1H(3)}$ decreases and $\eta_{3H(1)}$ increases.
3. In the whole diapason, $\eta_{3H(1)} > \eta_{1H(3)}$, whereby this difference increases with the increase of $t = |i_0| = z_3/z_1$.
4. Self-locking $(\eta < 0)$ at $\overline{\text{AI}}$-PGT is generally not possible unlike other PGTs. This danger exists in the so-called positive-ratio PGTs (Section 2.3), in which $P_{rel} > P_A$, while the $\overline{\text{AI}}$-PGT under consideration is in the negative-ratio category.

12.4 Efficiency at Work with F = 2 Degrees of Freedom

When PGT operates at $F = 2$ degrees of freedom (as a differential), it can only work in two ways (as already explained in Chapter 3):

- As a summation PGT (Figure 12.4a)
- As a division PGT (Figure 12.4b)

Using the real external torques T_1', T_3', and T_H' defined, and taking into account the direction of the relative power P_{rel}, the efficiency formulae for both cases are obtained:

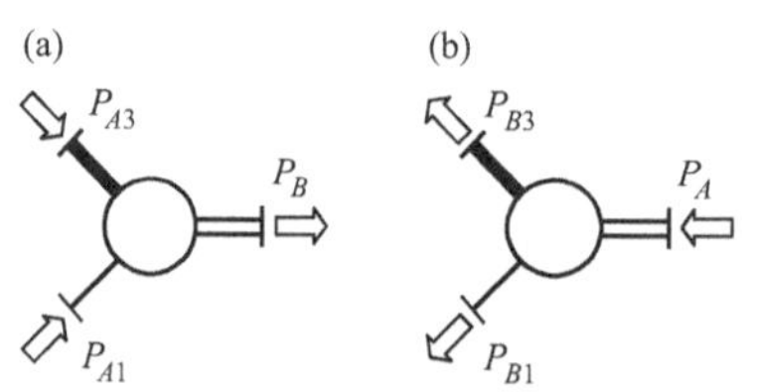

FIGURE 12.4
Work of a PGT as a summation (a) and as a division one (b).

Summation PGT

In twin-motor drive, the shafts of sun gear 1 and ring gear 3 are the most common input, and the carrier H shaft—output (Figure 12.4a). In this case, efficiency is determined by the following formula:

$$\boxed{\eta = -\frac{P_B}{\sum P_{Ai}} = -\frac{T'_B \cdot \omega_B}{T'_{A1} \cdot \omega_{A1} + T'_{A3} \cdot \omega_{A3}} = -\frac{T'_H \cdot \omega_H}{T'_1 \cdot \omega_1 + T'_3 \cdot \omega_3}}. \tag{12.16}$$

This formula is derived analogously to formula (12.8), starting from formulae (12.3) and (12.4).

Division PGT

If the carrier shaft H is input, then the following formula is obtained:

$$\boxed{\eta = -\frac{T'_1 \cdot \omega_1 + T'_3 \cdot \omega_3}{T'_H \cdot \omega_H}}. \tag{12.17}$$

Unlike the first case of the twin-motor drive, when the angular velocities are set $\omega_1 \approx$ const and $\omega_3 \approx$ const (e.g., electric motors!), in the second case, the output angular velocities ω_1 and ω_3 depend on the working resistances on the output shafts (see Sections 7.1.3 and 7.4.3 and Figures 7.1 and 7.9).

Due to internal idle losses, some of which are permanent and not load-dependent, whereas others are subject to some law, then the internal losses of a PGT gear are not proportional to the load F_t and the lost power P_ψ.

$P_\psi \neq \text{const} \cdot F_t$, resp. $P_\psi \neq \text{const} \cdot P_A$, and from there, $\psi \neq$ const.

Therefore, the gear train efficiency η will not be constant, but will vary depending on the authoritative factors. For a given PGT with its tooth geometry (z_1, z_2, z_3, x_1, x_2, x_3) and lubricating oil with its viscosity ν at a stationary oil temperature θ_L, the efficiency of the gear train depends only on two other factors—a load on the train T_B (resp. on the teeth F_t) and an angular velocity, e.g., ω_A, resp. peripheral velocities of gears v (Section 11.1), i.e.,

$$\eta = f(T_B, \omega_A) = f(F_t, v). \tag{12.18}$$

Figure 12.5 shows the principal change of efficiency η over time t under constant load T_B = const and at various input angular velocities ω_A.

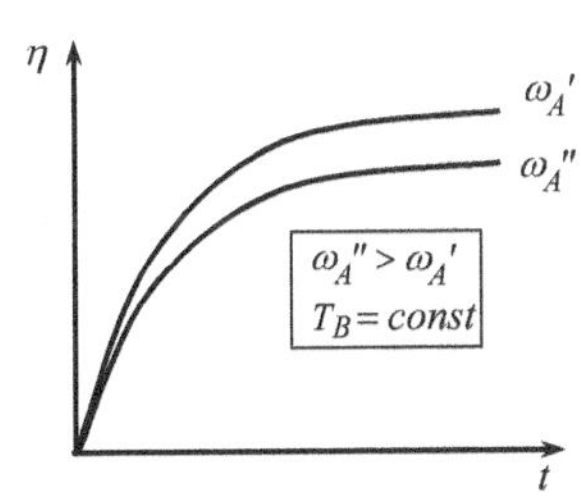

FIGURE 12.5
Efficiency η of a PGT as a function of time t under constant load T_B = const and various angular velocities ω_A.

Since the calculation of efficiency is based on different assumptions, it is unrealistic to express the result with an accuracy of more than 0.1%. For example, it is unrealistic to give $\eta = 97.385\%$ instead of $\eta = 97.4\%$.

12.5 Methods for Experimental Determination of Efficiency

Up to this point, the theoretical determination of efficiency has been exposed, which is only possible in the design. Different methods for its experimental determination exist [150, 221, 226]. One such method is the method by which the torques T'_A and T'_B on the input and output shaft of a PGT are measured and replaced in (12.6):

$$\eta = -\frac{T'_B/T'_A}{i_k} = -\frac{i_T}{i_k}. \tag{12.19}$$

Since certain inaccuracy is inherent in each experimental method, and because of the very high PGT efficiency (unlike other types of gears, especially worm gears), it is possible to obtain an absurd result in this method, e.g.,

$$\eta = 98 \pm 3\% = 95 \div 101\%.$$

Such an absurdity cannot be obtained with the thermal method for efficiency determining by the loss factor ψ. This method, described in [17], is characterized by the following.

As shown in Figure 12.6, the test gear train is heat insulated from the input and output couplings, as well as from the seating. In this way, its heat can be dissipated in the surrounding space only from its surface and not transmitted through the shafts to the engine, brake, and seating.

Efficiency is determined at uniform load $T_B = \text{const}$ and constant angular velocity $\omega_A = \text{const}$. This usually takes a few hours (usually $= 4 \div 6\,\text{h}$ and theoretically $t = \infty$) until a constant temperature $\theta_L = \text{const}$ of the oil in the housing is obtained (Figure 12.7). At the same time, the ambient (surrounding) air temperature θ_0 is measured with a *second thermometer*, and the temperature rise (difference) is determined:

$$\Delta\theta = \theta_L - \theta_0. \tag{12.20}$$

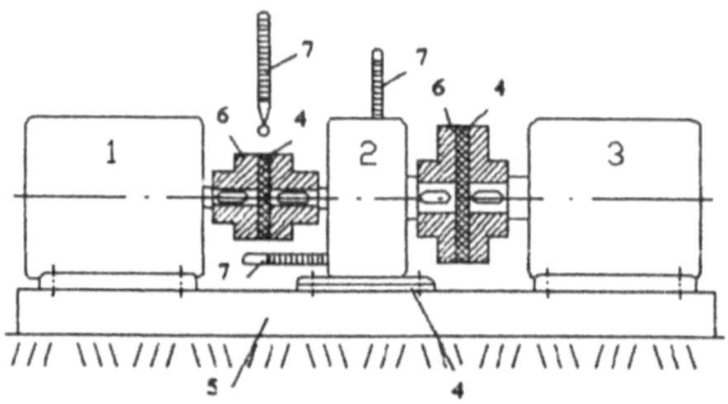

FIGURE 12.6
Test rig for efficiency determination by thermal method [17]: 1—motor; 2—tested PGT; 3—break; 4—insulation; 5—seating; 6—couplings; 7—thermometers.

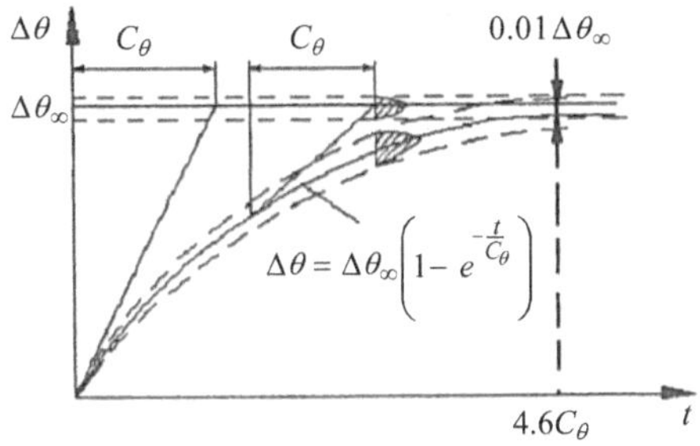

FIGURE 12.7
Rise of oil temperature in tested gear train.

This temperature rise (difference) changes over time with the following law:

$$\boxed{\Delta\theta = \Delta\theta_\infty\left(1 - e^{\frac{t}{C_\theta}}\right) = f(t)}, \tag{12.21}$$

where $\Delta\theta_\infty$ is the equilibrium temperature rise (difference) that is asymptotically reached after a long enough time—approximately $t = 4.6\ C_\theta$, and C_θ is the temperature–time constant, characteristic and different for different gear trains.

Lost power P_ψ dissipated as a heat from the outer surface A of PGT is determined by the following formula:

$$P_\psi = \psi \cdot P_A = (1-\eta)P_A = \alpha \cdot A \cdot \Delta\theta_\infty, \tag{12.22}$$

where α is the heat transfer coefficient and has the following values [175]:

$$\alpha = 15 \div 25 \frac{\text{W}}{\text{m}^2 \cdot \text{K}^\circ}.$$

For the loss factor is obtained:

$$\boxed{\psi = \frac{\alpha \cdot A \cdot \Delta\theta_\infty}{P_A}}, \tag{12.23}$$

respectively, for efficiency

$$\eta = 1 - \psi. \tag{12.24}$$

The analysis carried out in [17] at the maximum possible deliberately increased scattering of the individual variables α, A, $\Delta\theta_\infty$, and P_A ($\Delta\alpha/\alpha \approx 25\%$; $\Delta A/A \approx 15\%$; $\Delta(\Delta\theta_\infty)/\theta_\infty \approx 8\%$; and $\Delta P_A/P_A \approx 4\%$) shows that in the thermal method under no circumstances, an absurd result ($\eta > 1$) can be obtained, such as in the torque method. This is a very significant advantage of the thermal method.

In [135], a rig[1] for an experimental study of losses in PGTs, both by the torque method and the thermal method, is presented (Figure 12.8).

The rig operates with open-loop (contour) loading. The loading torque is formed by the block-drum brake and measured by means of a tensobeam. The input torque is measured by a measuring shaft (torquemeter) and the output—by a lever attached to the brake, and electronic scales (Figure 12.8a).

The rig enables the losses by the thermal method to be determined too, by measuring of the oil temperature and the surface temperature of PGT housing (Figure 12.8c). Figure 12.8b shows the temperature measurement with ATEL Electronics® digital multimeter MS 8229. The same instrument allows for the measurement of noise level of the gear train.

The rig enables the determination of the train's losses under static and quasi-static loads. The static load is one of the ways to determine the loss in gear trains [1], which largely eliminates permanent (non-load-bearing) losses. This method can be used to control the quality of serially produced gears [230]. The "static efficiency" can be used to benchmark the energy efficiency of different gears of planetary change-gears.

In the manner shown in Figure 12.9, the gear train is loaded with static torque (by weight 5, hinged to pulley 4 attached to the input shaft of the gear train).

The loading pulley 4 has a working diameter 92 mm.

The load is gradually increased from 300 g (0.135 Nm) to 4.8 kg (2.166 Nm).

The output torque is measured by using a lever 1 with a load arm of 0.9 m and electronic scales.

The torque transmit ratio (torque transformation) i_T of the gear train and the bearings of the shaft of the braking drum is equal to the ratio of the output torque T_B' (measured on the braking drum) and the input torque T_A' (applied on the loading pulley):

$$i_T = \frac{T_B'}{T_A'}. \tag{12.25}$$

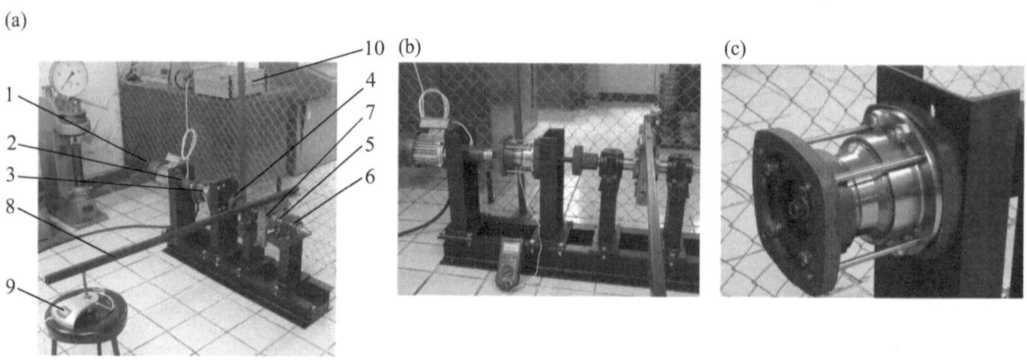

FIGURE 12.8
Rig for dynamic (in motion) testing of PGTs [135] (1—electric motor; 2—measuring shaft (torquemeter); 3—tested PGT; 4—gear coupling; 5—brake drum; 6—bearings; 7—brake; 8—lever; 9—scales; 10—tensometric equipment): (a) The rig during torques measuring; (b) the rig during temperature and noise measuring; (c) tested PGT. (Photographer: D. Karaivanov.)

[1] Developed by author D. Karaivanov. Some of the funds are provided by the University of Chemical Technology and Metallurgy, Sofia. Bearing set is a gift from the SKF® office in Bulgaria.

FIGURE 12.9
Static loading of a PGT [135]: 1—brake; 2—tested gear train; 3—bearing; 4—loading pulley; 5—loading weight. (Photographer: D. Karaivanov.)

In [135], a comparison between the results for efficiency, obtained with dynamic and static loading, is made.

13

Lubrication of $\overline{AI}$-Planetary Gear Train

Gear trains lubrication is very important. Their reliability depends to a great extent on it.

13.1 Purpose of the Lubrication

Lubrication purpose is

1. Creating, if possible, a load-bearing oil film between the contacting and friction surfaces (teeth, bearings) which reduces wear and friction losses
2. Heat removal and thus cooling the gear train
3. Removal of wearing products
4. Reducing dynamic loads in the gear mesh, which reduces the noise of the gear train

13.2 Ways of Lubricating

There are different ways to lubricate the gear trains:

1. Splash lubrication (oil bath lubrication) is the most commonly used lubrication method and is recommended at a lower peripheral velocity within the range

   $$v = 1 \div 10(15)\text{m/s}.$$

2. Forced oil circulation lubrication is practiced when the peripheral velocity is higher, i.e.,

   $$v > 10(15)\ \text{m/s},$$

   as well as when the cooling surface of the gear train is not sufficient for the required heat transfer and the train is overheated, or in the case of planetary gears with vertical shafts when splash lubrication is not applicable.
3. Grease lubrication is used for nonresponsive, open gears with very low peripheral velocities $v < 1\ \text{m/s}$.

   Colloidal graphite or molybdenum bisulfide MoS_2 may be added to the grease. More about lubricant additives one can find in [201].

TABLE 13.1

Recommended Kinematic Viscosity ν of the Oil (at 50°C) Depending on the Peripheral Velocity v of the Gears

Peripheral Velocity v (m/s)	Viscosity ν	
	cSt	°E
Over 0–0.5	150	20
0.5–1.0	110	15
1.0–2.5	70	12
2.5–5.0	50	8
5.0–12.5	40	5.5
12.5	30	4

The most appropriate is a viscosity of 7°E–9°E.

13.3 Oil-Type and Quantity Choice

13.3.1 Oil-Type Choice

Lubricating oil is mainly used in enclosed gear trains. An important characteristic of it, though not the only one, is the kinematic viscosity ν, by which it is chosen primarily according to the peripheral velocity of gears. The *guiding principle* for choosing lubricating oil is as follows: the lower peripheral velocity v, the higher load F_t, and the higher roughness demand higher viscosity ν. Table 13.1 gives an approximate indication of the viscosity of lubricant oil [66, 78, 148, 162]. For convenience, the table gives not only the dimension (in the CGS[1] system) cSt[2] at 50°C but also the older but practical measure in Engler degrees (°E).

13.3.2 Oil Quantity Choice

The required quantity of lubricating oil in splash (oil bath) lubrication can be determined using the following formula [147, 162]:

$$Q = (0.35 \div 0.7) P_A$$

or

$$Q = (5 \div 10) P_\psi, \tag{13.1}$$

where Q is the required quantity (volume) of oil, l, P_A is the input power, kW, and P_ψ is the lost power, kW.

13.4 Depth of Gears Immersion

This depth (oil level) is important not only for lubrication but also for hydraulic losses, which can increase at a high peripheral velocity of gears. They affect the efficiency and

[1] Centimetre–gram–second system of units.
[2] 1 St = 10^{-4} m^2/s.

hence the heating of the gear train. During PGT operation, the planets splash (crush) into the oil and spray it, giving ubiquitous lubrication. In addition to spraying, lubrication can be facilitated by additional attachments, such as spreading rings or rockers.

The following depth of gears immersion in oil bath is recommended:

$$h_M = (1 \div 6)m, \tag{13.2}$$

where m is the module of the gear.

The most recommended value is

$$h_M = 2m. \tag{13.3}$$

At PGTs with cylindrical housing on which the internal teeth of the ring gear (e.g., the hoists) are cut, deeper immersion of planets may be necessary to achieve the required quantity of oil.

13.5 Lubricating Oil Changeover

There are quite different values in the literature for the time of changing the lubricating oil. It should be borne in mind that the oil ages very fast as the temperature increases, especially above 60°C.

Any exceeding of this temperature by 10°C reduces the oil life (resource) by half [175]. For this reason, the following recommendation can be accepted for rough orientation:

I change—after L = 100 ÷ 200 hours of work
II change—after L = 500 ÷ 700 hours of work
III change—after L = 1000 ÷ 2000 (2500) hours of work

Since gear trains lubrication issues are very specific (lubrication is science in itself!), it is the best to use oil-producer data [78] as well as specialized literature [78, 147–149, 201]. It is also necessary to work in close contact with the lubricant manufacturing companies.

When designing a new PGT, the best thing to do is to determine the type of oil, its quantity and changeover time, depth of immersion of gears, and gear train heating experimentally. All these questions are interrelated, and therefore, the best approach is experimental.

It should be noted that, for economic and environmental reasons, ways to reduce the amount of oil or its complete elimination are always sought (e.g., with special coatings on the tooth flank[3]). The article [77] shows the results of a study of working behavior of self-lubricating gears based on oil-impregnated sintered material.

[3] Bozzolo, A. and D. Zangani. Development of gear drive-trains based on new materials and novel gear systems. *Proceedings of the 3rd International Conference "Power Transmissions'09"*, edited by A. Mihailidis, Kallithea [Greece], Oct. 1–2, 2009, pp. 249–253. ISBN 978-960-243-662-2.

14

Heating and Cooling of $\overline{AI}$-Planetary Gear Train

Losses in a planetary gear train (PGT), of course, lead to its warming up. $\overline{AI}$-PGT in question has really the highest efficiency of all types of gears (with proper design and quality fabrication!) but also has a small cooling surface (as a result of its compactness), and in some cases, because of this, the heating can cause problems. In such cases, forced circulating lubrication or artificial cooling (blowing or water coil) is required. In case of insufficient heat dissipation, the gear train is overheated, and the viscosity of the oil decreases and thus the oil film load bearing capacity. In addition, at a higher temperature, the oil ages (degenerates) more rapidly and needs to be changed more often. Therefore, experience has shown that the oil temperature should be within the range $\theta_{max} \approx 60 \div 70$°C and only in rare $\theta_{max} \approx 80 \div 90$°C is allowed [175]. In any case, these limits must be consistent with the particular type of oil and the manufacturer's prescriptions [78].

14.1 Cases of Work of PGT

The following three main working cases (modes) exist (Figure 14.1):

1. Working at constant full (nominal, maximum) load and uninterrupted rotation indefinitely (permanent operation—intermittent duty rating ED[1] = 100%), whereby

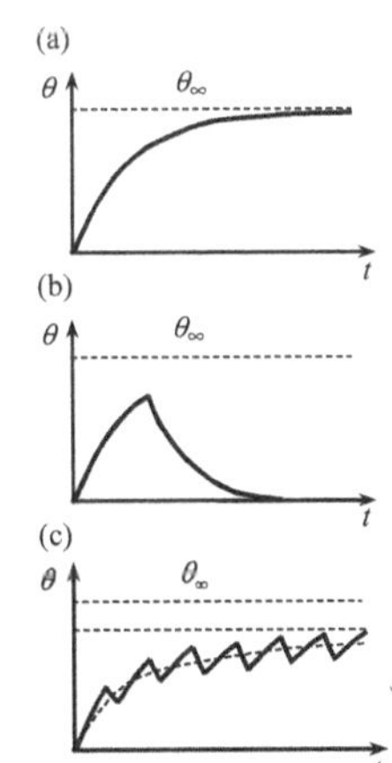

FIGURE 14.1
Temperature difference between oil temperature θ_L in the gearbox and ambient air temperature θ_0 at different operating modes: (a) Permanent (uninterrupted) operation; (b) short-time operation with full cooling; (c) short-time operation without full cooling.

[1] From Einschaltdauer (in German) – duty cycle, operating time.

equilibrium temperature θ_∞ is asymptotically raised (Figure 14.1a). This case is the most unfavorable in terms of gear train heating.

2. Working at full or partial load, at which a certain temperature is raised, after which a sufficiently long pause occurs when the gear train is fully cooled (Figure 14.1b).
3. Working at full or partial load with rhythmically repetitive operating and rest periods (the so-called short-time rating), where the gear train fails to cool completely (Figure 14.1c). The temperature that is raised asymptotically is lower than in the first case (permanent operation).

14.2 Heating Temperature

The equilibrium oil temperature in the gear train for the most unfavorable first case (Figure 14.1a) is calculated from the heat balance of the system (Section 12.5)—the lost power P_ψ is equal to the amount of dissipated (by convection) heat (formula 12.13):

$$P_\psi = \psi \cdot P_A = (1-\eta) P_A = \alpha \cdot A(\theta_L - \theta_0), \tag{14.1}$$

where P_A is the input power, W, η is the efficiency of PGT, ψ is the loss factor of PGT, α is the heat transfer coefficient of cooling (housing) surface, $\frac{\text{W}}{\text{m}^2 \cdot °\text{C}}$, A is the cooling surface of housing, m^2, θ_L is the oil temperature ([175] recommends using the oil temperature instead of the surface temperature), °C, and θ_0 is the ambient air temperature, °C.

Heat transfer coefficient α depends on the shape of the cooling surface and air circulation around it, and for a smooth surface can be accepted [175] $\alpha = (15 \div 25)\frac{\text{kW}}{\text{m}^2 \cdot °\text{C}}$. Lower values should be chosen for a dirty surface and higher ones if air currents exist. Ambient air temperature is usually taken $\theta_0 = 20°\text{C}$, but another value may be taken, if necessary.

Oil temperature from (14.1) is obtained:

$$\boxed{\theta_L = \frac{(1-\eta) P_A}{\alpha \cdot A} + \theta_0}. \tag{14.2}$$

The heating temperature for the other two cases is lower than the calculated for the first case.

When the oil temperature is unacceptably high, one of the following measures is used to improve cooling:

- Adding gills to the housing surface
- Blowing the housing
- Forced oil circulation lubrication
- Cooling coils

More details regarding the heating and cooling of PGTs can be found in the specialized literature [78, 147–149, 175, 251].

15

Design, Manufacturing, and Measurement of the Elements of $\overline{AI}$-Planetary Gear Train

The main elements of planetary gear trains (PGTs) are

- Gears
- Shafts and pins
- Carrier
- Housing – and in cases with equalizing (distribution) device, most often
- Gear couplings

The design and strength calculation of such basic machine elements, such as gears, shafts, and pins as well as gear couplings, are studied in detail in textbooks and reference literature on machine elements [58, 175, 192, 208]; therefore, they are not considered here. Only the specific elements of PGTs that are not encountered in other non-PGTs are treated, namely,

- Planets
- Ring gear
- Carrier
- Housing

The manufacturing of PGT elements must be consistent not only with the requirements of the gear train but with the capabilities of technology. The accuracy must be economically feasible and technically easy to achieve.

15.1 Sun Gear and Planets Design and Manufacturing

The design of a sun gear largely depends on how the load distribution between the planets is equalized (see Section 8.6). Some design solutions are shown in Figure 8.15 as well as in Chapter 20. Most often, the sun gear is made of carbonized steel (16MnCr5, 15CrNi6, 17CrNiMo6, etc.). Carbonization and case hardening till 56 ÷ 62 HRC[1] are made.

Planets are intermediate wheels, which, in a certain sense, act as parasitic (auxiliary) wheels as they do not affect the speed ratio. Most often planets are made of the same steels as the sun gear and are also carbonized and case hardened to 56 ÷ 62 HRC. It should be remembered that in the same gear train, it is desirable that the matched gears are made of a different material (different grades of steel) in order to reduce the risk of scuffing.

[1] Rockwell scale hardness.

In order to obtain a minimal inaccuracy in the direction of planet teeth, which would result in uneven load on their face width (i.e., face load factor $K_\beta > 1$), which will lead to an uneven load, the following applies to particularly precise gear trains. On a surface grinding machine, the faces of all planets of the gear train are simultaneously ground to be parallel. When toothing (cutting or grinding) a package, the orientation of planets can be marked with paint, so that the so-called *sinphase montage* can be carried out (Figure 8.19). Even greater accuracy is obtained by measuring the radial run-out of each planet (after its finishing), marking its orientation and performing a sinphase montage. Sinphase montage helps reduce the unevenness of the planets load (see Chapter 8, mesh load factor $K_{\gamma\, max}$).

15.2 Ring Gear Design and Manufacturing

Ring gear is a specific gear of $\overline{\text{AI}}$-PGT. Unlike the sun gear and planets, the ring gear does not usually harden and very rarely grind. It is made of hardened (tempered), alloy or non-alloy steel (C45N, 42CrMo4, 30CrNiMo8, etc.), through hardening (tempering) within 250 ÷ 280 HB[2], for easy machining. As a rule, it is done on a gear-cutting machine. Only in extremely rare cases of very rough and not very responsive gears, as well as in a very large module for which there is no suitable gear cutter (and a gear-cutting machine), this is done on a slotting machine by a shaped tool or on a milling machine with the corresponding equipment and a profile mill. There are other ways of machining (e.g., with a non-cylindrical worm mill, wire-cut EDM[3], laser cutting, etc. [123]) that have not been widely used yet.

Since the ring gear is a thin-walled and easily deformable part, it is necessary that when tightened on the gear-cutting machine, there are no deformations which will subsequently occur in the performance of the gear train in a very unfavorable manner.

There is another feature of the ring gear cutting—the danger of a radius of curvature of the fillet curve $\rho_{f3} = 0$ (see Section 5.5).

Figure 15.1 shows some examples of stiff, pliable, and floating ring gear (see also Figure 8.14c).

15.3 Gears Thermal and Chemico-thermal Treatment

With respect to the thermal (heat) and chemico-thermal treatment of gears, it is imperative that the designer works in close contact with the metallurgical specialist in the plant. Thermal and chemico-thermal treatment is a very specific, extremely complex, and risky area, so that there is a lot of personal technological experience not only generally but at the specific production conditions.

For PGTs gears, the following thermal and chemico-thermal treatments are the most common:

- Carburizing and case hardening;

[2] Brinell scale hardness.

[3] Electrical discharge machining.

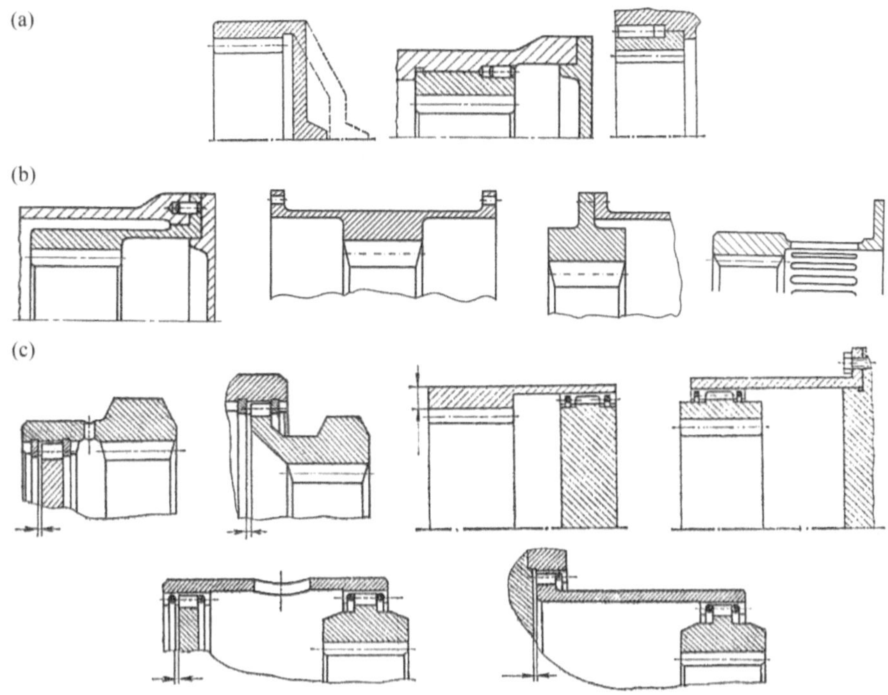

FIGURE 15.1
Some ring gear design solutions: (a) Stiff (non-pliable); (b) pliable; (c) floating.

- Through hardening (tempering);
- Nitriding, respectively, ion-nitriding.

Sun gears and planets are generally carburized and case-hardened and ring gears—hardened (tempered). To nitriding, and resp. ion-nitriding, all gears may be subjected, including ring gear.

15.4 Gears Accuracy Measurement

From every point of view, gears require special attention to design, high-quality machining, and reliable quality control. Quality of PGT (its reliability, durability, heating, noise and vibration, etc.) first depends on its gears. The type and number of accuracy measurements of gears depend on the requirements and responsibility of the PGT (on the estimations of the designer, manufacturer, and user), on the level of equipment of the manufacturer, as well as the training of its staff [120, 162].

The gear accuracy measurements are as follows:

On the cutting machine, in all cases, the base tangential length (span measurement, measurement of Wildhaber) of gears with external teeth is measured (Figure 15.2a), and in some cases (a large module), this is also possible for gears with internal teeth. Span length is determined as follows:

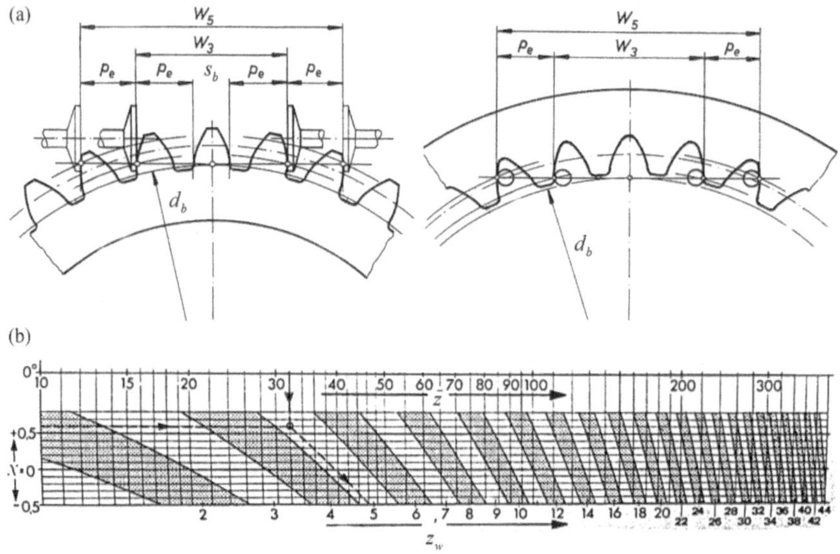

FIGURE 15.2
Base tangential length (span) W measurement of external and internal teeth: (a) Ways of measuring; (b) determining of the number of measured teeth z_w.

Over $z_w = 3$ teeth

$$W_3 = 2p_b + s_b = (z_w - 1)p_b + s_b;$$

Over $z_w = 5$ teeth

$$W_5 = 4p_b + s_b = (z_w - 1)p_b + s_b,$$

where $p_b = p_e = \pi \cdot m \cdot \cos\alpha$ are the base and meshing pitch, respectively, $s_b = d_b\left(\frac{s}{d} + \text{inv}\alpha\right)$ is the external tooth thickness at the base circle with diameter d_b, and $s = \frac{\pi \cdot m}{2} + 2m \cdot x \cdot \tan\alpha$ is the external tooth thickness at the reference circle with diameter d.

The number of measured teeth z_w can be determined by the nomogram in Figure 15.2b depending on the number of teeth z and profile shift coefficient x of the measured gear.

Internal teeth are generally measured with pins or balls [123, 162] (Figure 15.3a), which is a very accurate indirect way to inspect tooth thickness at a given reference diameter. Although very rare, it is also practiced for external teeth (Figure 15.3b). Moreover, measurement distance between M_i or over M_a pins or balls depends not only on the module m, the number of teeth z, and the shift coefficient x, but also on whether z is an even or odd number, as shown in the figure.

Determination of the distance between pins (Figure 15.4) starts with the following data: m, z, x, $\alpha = 20°$, resp. $d = m\,z$, and $d_b = d \cdot \cos\alpha$, as well as the pins (rolls) diameter d_R that is recommended to be

$$d_R = 1.44m \text{ or } d_R = 1.65m.$$

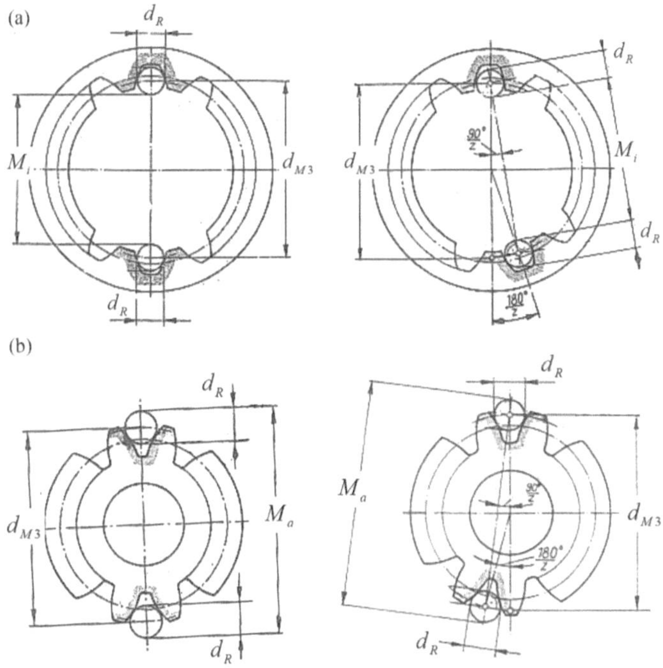

FIGURE 15.3
Measuring with pins or balls of internal (a) and external (b) teeth.

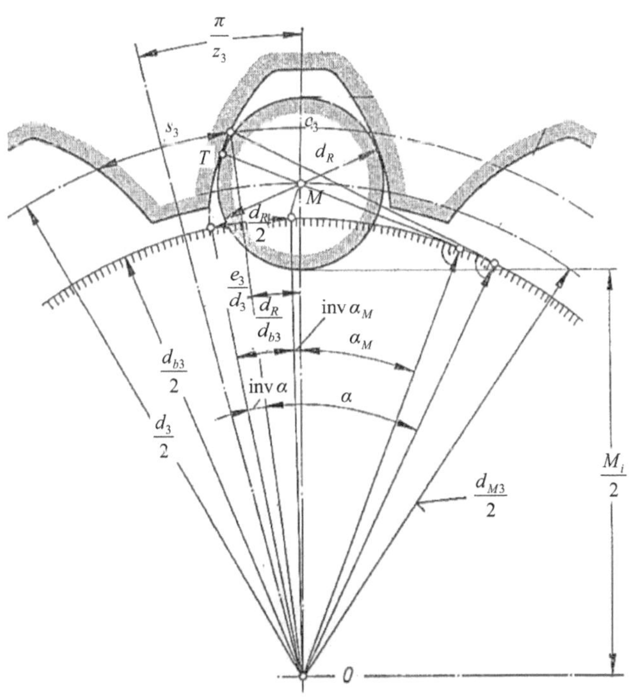

FIGURE 15.4
Determination of the distance between pins or balls.

This is done with the aim of points of pins touching with the teeth to be within the reference circle.

The sequence of calculations is as follows:

1. Tooth space at reference circle

$$e_3 = \frac{\pi \cdot m}{2} + 2m \cdot x_3 \cdot \tan\alpha. \tag{15.1}$$

2. Profile angle of the pins center M

$$\operatorname{inv}\alpha_M = \frac{e_3}{d_3} - \frac{d_R}{d_{b3}} + \operatorname{inv}\alpha, \tag{15.2}$$

from where the angle α_M is determined.

3. Diameter of the circle passing through the pins center M

$$d_{M3} = d_3 \frac{\cos\alpha}{\cos\alpha_M} = \frac{d_{b3}}{\cos\alpha_M}. \tag{15.3}$$

4. Distance between pins

$$M_i = \begin{cases} d_{M3} - d_R & \text{– an even teeth number} \\ \cos\dfrac{90^\circ}{z_3} d_{M3} - d_R & \text{– an odd teeth number} \end{cases} \tag{15.4}$$

These are the nominal values. It is also necessary to take into account the tolerances of the gear.

Detailed numerical and tabular data required to calculate the base tangential length W and distance between M_i or over M_a pins (balls) one can find in the specialized literature [46, 156, 162].

After toothing, depending on the requirements to the gear train, the following measurements are also made:

- In all cases radial run-out F_r (Figure 15.5), from which eccentricity of the toothing f_e is determined

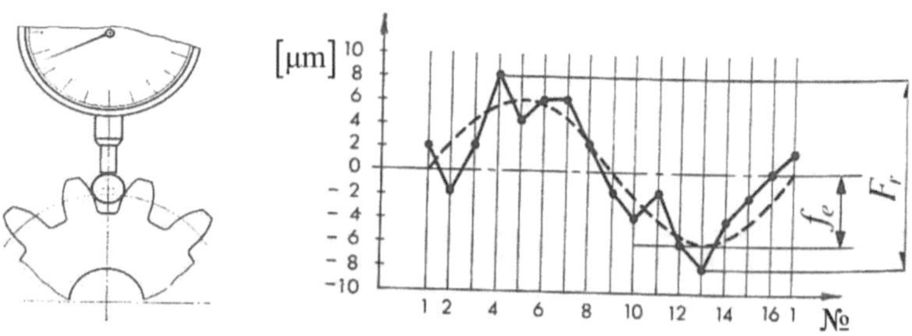

FIGURE 15.5
Radial run-out F_r measuring and toothing eccentricity f_e determination from the diagram.

Although not all gears are measured as needed:

- Reference pitch p and base pitch p_b (Figure 15.6), by plotting the accumulated deviations of the pitch from the individual measurements
- Involute curve (Figure 15.7)
- Tooth flank line

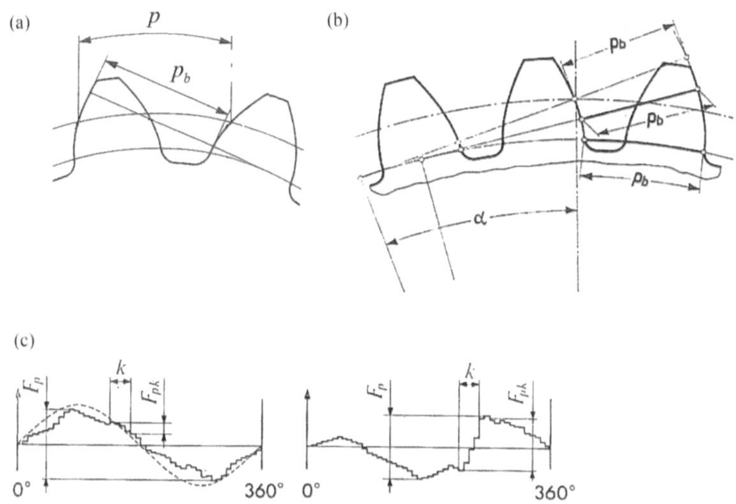

FIGURE 15.6
Pitch measuring: (a) Reference circle pitch p; (b) base circle pitch p_b, equal to the mesh pitch p_e; (c) diagrams of accumulated pitch deviations.

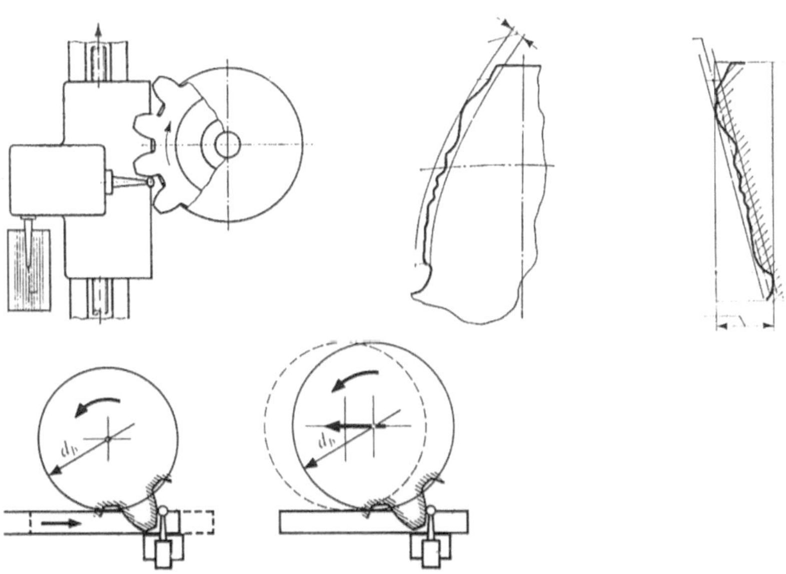

FIGURE 15.7
Involute profile measuring and diagram creating.

It is very important that the variation of accumulated pitch deviations should be smooth as in the left diagram in Figure 15.6c.

A tooth thickness measurement should be strictly avoided (Figure 15.8), because in this measurement, the instrument is based on a not very accurate tip circle (which diameter has a wide tolerance) so the measurement result is not accurate.

Since the radial run-out F_r, respectively, eccentricity f_e, and especially those of the planets adversely affect the load distribution between them (considered by mesh load factor $K_{\gamma\, max}$ in load capacity calculations), the measurements shown in Figure 15.9 are also possible [162]. They simultaneously measure the eccentricity of the toothing f_e and the deviation from the theoretically correct position of the flank (working) profile of the tooth. For gears that only work on one side of the teeth (the sun gear), the measurement is made on that side of the teeth (Figure 15.9a). For gears loaded on both sides of the teeth (planets), both sides of the teeth must be measured (Figure 15.9b).

Measurement of the ring gear of particularly accurate and responsible PGTs is done on a 3D CMM [coordinate-measuring machine (CNC[4] gear measuring center [120])].

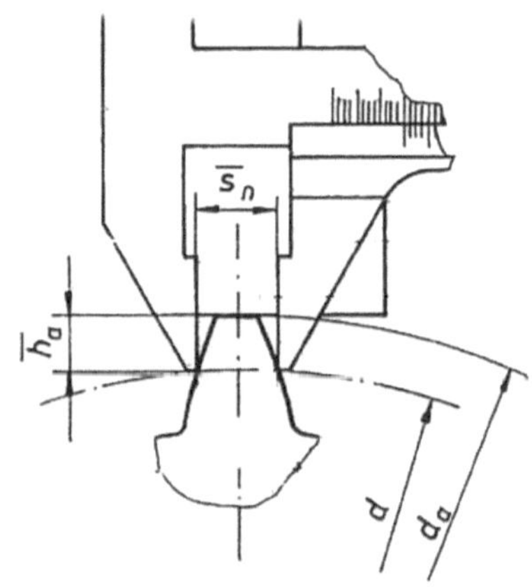

FIGURE 15.8
Tooth thickness measuring.

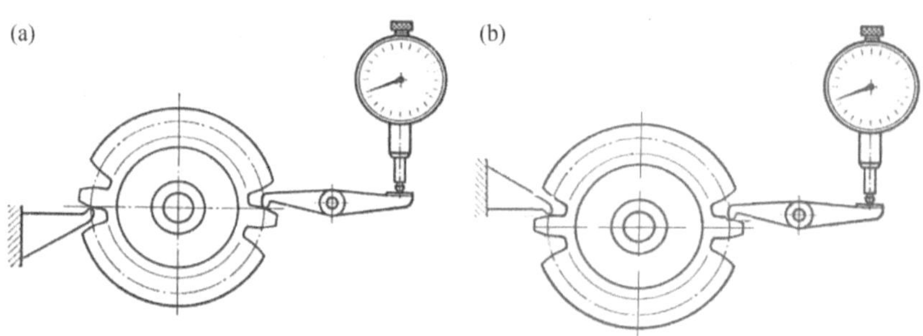

FIGURE 15.9
A complex way of measuring both the toothing eccentricity f_e and the flank deviation of the teeth of the same (a) or different (b) sides.

[4] Computer numerical control.

For the measurement and control of gears, it is best for the designer to be in close contact with the technology and control specialists.

15.5 Carrier Design and Manufacturing

The carrier is the structural element on which the planet pins are attached.

This is the most specific part of PGTs, with a complex shape, which part is missing in the other gear trains. The carriers are

- Undemountable (monolithic) or demountable
- Single- or two-wall
- Cast, forged or welded
- Of steel or iron

Some of the most common constructions are shown in Figure 15.10.

Requirements to the carrier are as follows:

- Stiffness (low deformations) in firs
- High accuracy
- Strength

Generally, the strength of a carrier is ample, and therefore, first of all, it is its deformation, in which the planet pins till (see Section 9.6 and Figure 9.11). In case of inadequate choice of planet bearings, this may cause uneven loads over the tooth face (face load factor $K_\beta > 1$), which in some cases may cause damage. This is especially true when using two bearings of a planet (or one needle bearing). When using a planet bearing that allows for some self-alignment, such as a single-row radial ball bearing or a double-row spherical self-aligning bearing, the carrier deformation has no detrimental effect. That is why it is advisable, if possible, one bearing to allow for some self-alignment. When it is necessary to use two planet bearings, the rule is that the bearings of the individual planet should be selected with equal radial clearance.

The uniformity of planets load (mesh load factor $K_{\gamma\,max}$) depends largely on the carrier accuracy. It is particularly important that the central angle between the holes for planet pins should be the same, i.e.,

$$\psi = \frac{360^\circ}{k} = \text{const.} \tag{15.5}$$

Tangential misalignments f_{tHj} have a very negative effect on the load distribution between planets (see Section 8.3.1 and Figure 8.2). Therefore, the holes for planet pins in the carrier have to be made very precisely by a jig boring machine. Similar to the ring gear, the measurement of the carrier of the particularly accurate and responsible PGTs is done on a three-coordinate measuring machine.

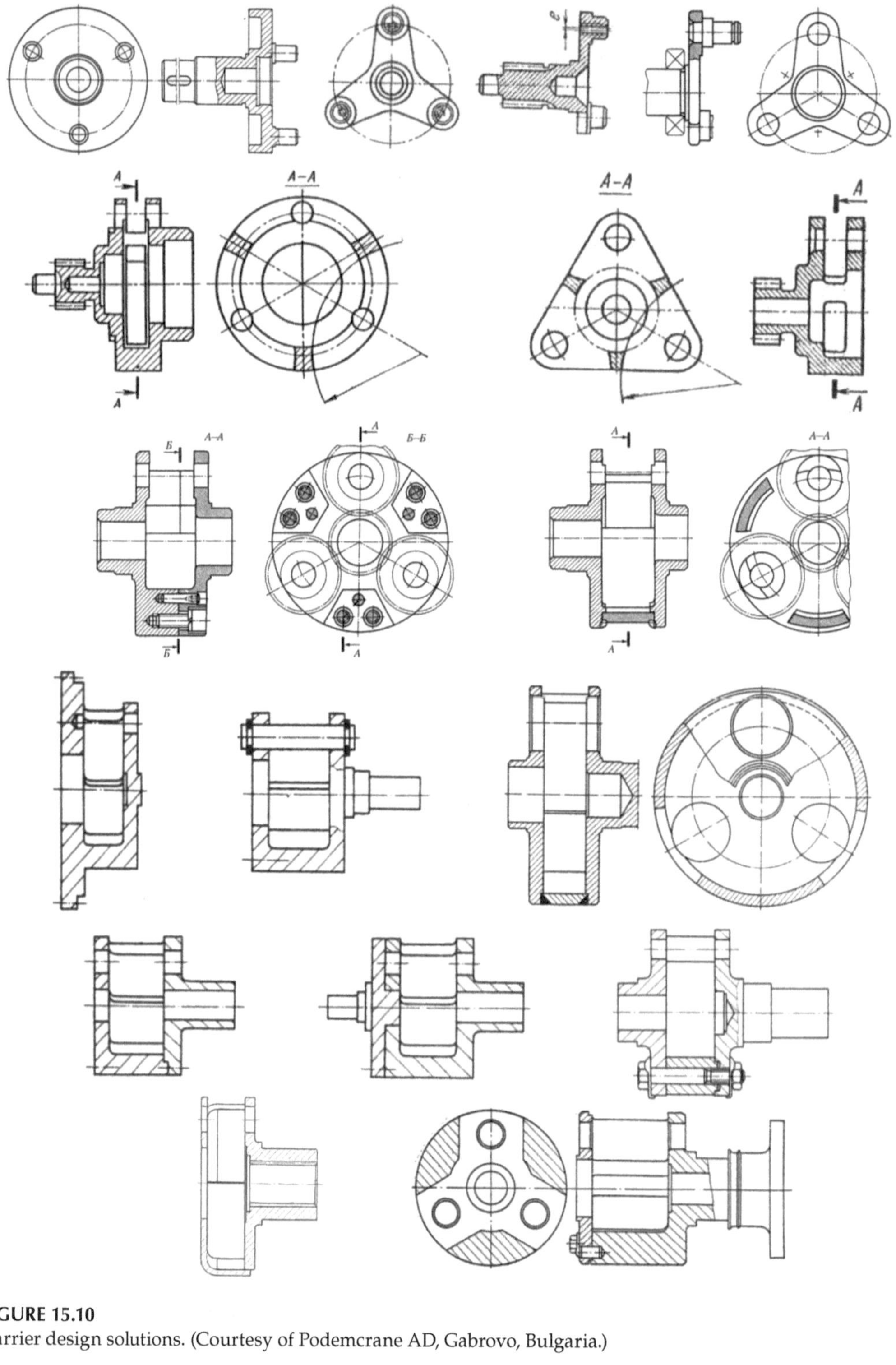

FIGURE 15.10
Carrier design solutions. (Courtesy of Podemcrane AD, Gabrovo, Bulgaria.)

15.6 Housing Design and Manufacturing

The purpose of the housing is to close the inner space and, when either the ring gear or the carrier is fixed, take up the reactive torque of the seating. Its structure is shown in Figure 15.11, as well as in the many examples of existing PGTs given in Chapter 20.

Except for closing the inner space and for taking the reactive torque, the housing naturally also serves for heat transfer. In cases where its surface is insufficient for heat transfer, the housing is blown or gills are added. The gills are also beneficial in terms of the noise behavior of the gear train.

The housing can be made of different materials—cast iron or steel, and in different ways—by casting or welding. For single units or for lightening purposes, a well-gilled welded structure is used. With a larger number of gear trains, it is suitable to produce a casting mold (matrix) and the housing to be a cast structure.

Typically, the housing feet are located beneath the central geometric axis of the gear train. In case of powerful, large-scale gears with significant heat dissipation and hence significant thermal expansion, the feet are positioned at the center axis level (Figure 15.12). In this way, temperature deformations do not change the position of the central geometric axis of the gear train and do not cause problems to the couplings.

Like the carrier, the housing measurement of particularly accurate and responsible PGTs is done on a three-coordinate measuring machine.

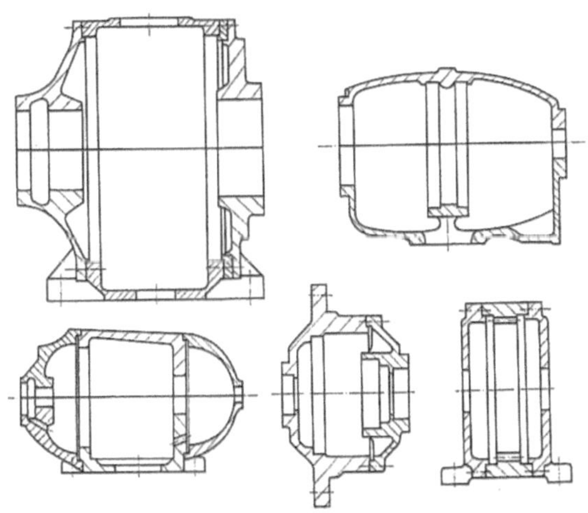

FIGURE 15.11
Housing design solutions.

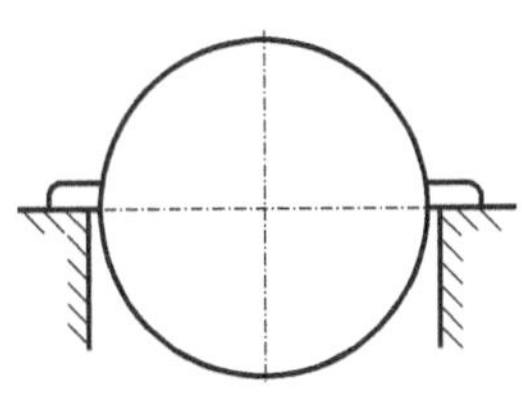

FIGURE 15.12
PGT mounting to the seats where temperature deformations do not affect the position of central geometric axis.

16

Approaches for Design (Project) Calculations of AI-Planetary Gear Train

For design (project) calculation of planetary gear train (PGT), the following initial data are required:

- Nominal power of the engine (input power) P_A,
- Input torque T_A or output toque T_B, i.e., gear train loading, respectively
- Input n_A and output n_B speed, hence
- Speed ratio i and its permissible deviation Δi
- Material and heat treatment of gears, resp. its strength σ_{FG} and σ_{HG} (tooth root stress limit and pitting stress limit)
- Application factor K_A, taking into account external load fluctuation
- Mesh load factor $K_{\gamma max}$, taking into account uneven load distribution (power sharing) between planets

Based on these data, the following basic parameters for the gears (sun gear 1, planets 2, and ring gear 3) must be determined first in the design calculation:

- Number of teeth: z_1, z_2, and z_3
- Profile shift coefficients: x_1, x_2, and x_3
- Module m
- Tooth face width: b_1, b_2, and b_3

With the geometric parameters as defined and some clarifications, the verification calculations of the gear train elements—gears, bearings, shafts, etc., are subsequently performed [115, 175], and, if necessary, the relevant changes are made.

There are different design calculation approaches, depending on the requirements for the gear train:

- Kinematic
- Overall dimension
- Optimization

16.1 Kinematic Approach

In this approach, the gear train requires that the speed ratio i should be preserved with certain accuracy. In this case, the predominant requirement is to determine the number of teeth z_1 and z_3 of central gears so as to achieve the assigned speed ratio i. Two cases are possible:

- Practical "absolute" adherence to the given speed ratio is required.
- Gear ratio may vary within a certain range Δi (usually 3%–5%).

For the realization of an acceptably accurate (± 3%÷5%) speed ratio, very detailed tables (given in the literature [153, 229]) can be used. In addition to the speed ratio with the corresponding teeth number of gears, the known specific conditions for the PGTs (for assembly, coaxiality, and adjacent) are considered, giving also the possible number of planets k. It is also readily possible to compile a short computing program (software) which, for a set number of teeth of the sun gear 1 ($z_{1min} \le z_1 \le z_{1max}$), determines the number of teeth of the ring gear z_3, which produces the smallest deviation from the given speed ratio (with or without complying with the aforementioned conditions). It should be noted that with the AI-PGT in question, it is not always possible to achieve a precisely defined speed ratio (see Chapter 2).

Easier to meet is the coaxiality condition by verifying the following dependences:

$$\begin{aligned}
&i = n_A/n_B;\ \Delta i \le [\Delta i] = 0.03 \div 0.05;\\
&z_2 = \frac{z_3 - z_1}{2} - (0.5 \div 1.5) \quad \text{by}\,(4.8);\\
&\frac{z_1 + z_3}{k} = \text{an integer} \quad \text{by}\,(4.1);\\
&i_0 \le i_{0max} = f(k) \quad \text{by Figure 4.3;}\\
&d_{a2} < 2a_w \cdot \sin\frac{\pi}{k} \quad \text{by}\,(4.13);\\
&\varepsilon_{\alpha 12} \ge \varepsilon_{\alpha min} \approx 1.2 \quad \text{by}\,(11.3);\\
&b_1/d_1 \le 0.8 \div 1.0 \quad \text{by}\,[158, 175].
\end{aligned} \tag{16.1}$$

To maximize load capacity, it is necessary, through appropriate adjustments, to maximize the load capacity of the external meshing between sun gear 1 and planets 2. As the sun gear 1 and the planets 2 are, as a rule, cemented and hardened, their load capacity is determined by their bending strength (σ_{FG1} and σ_{FG2}). Based on the strength of sun gear σ_{FG1}, which has the largest number of load cycles N_{F1}, and the initial data, the required module m is determined.

16.2 Overall Dimension Approach

In this approach, for some reason, the overall dimensions of the gear train must not exceed certain limits that means the diameter d_3 of ring gear 3 must not exceed these limits. For the value of the speed ratio, some freedom ($\Delta i \le [\Delta i]$) is allowed.

Very often in this case, the teeth number of the ring gear is chosen within the limits:

$$z_3 = 100 \div 150.$$

Following the procedure described in Section 16.1, an acceptable result is obtained with this second approach.

16.3 Optimization Approach

This approach relates to the optimal choice of gear train ratio i, when it can be chosen freely within wide limits. This is the case with the compound (multi-carrier) PGTs, where, in order to obtain generally the most compact arrangement, it is required that the last planetary stage should have the largest specific load capacity (the ratio of the torque T to mass m).

Figure 16.1 illustrates how the load capacity changes (bending and pitting) of a specific gear train expressed though the torque $T_H \equiv T_B$ of the carrier H, being an output torque, depending on the basic speed ratio $i_0 = -z_3/z_1$ and torque ratio $t = T_3/T_1$, respectively, depending on the most often used speed ratio of this type of gear train $i_{1H(3)} = 1 - i_0 = 1 + z_3/z_1$ and on the module m.

From this figure and from any research done [147–149, 155], it is known that the specific load capacity of the considered type of $\overline{\text{AI}}$-PGT is the highest; i.e., the gear is most compact when the basic speed ratio i_0 is within the following most favorable limits [155] at $k = 3$ planets:

$$i_0 \approx 5 \div 6.$$

Contact load capacity depends on the quality of material of gear wheels and their hardness. Load capacity of teeth at bending depends not only on thermal treatment and hardness but also heavily on the module, as shown in the figure.

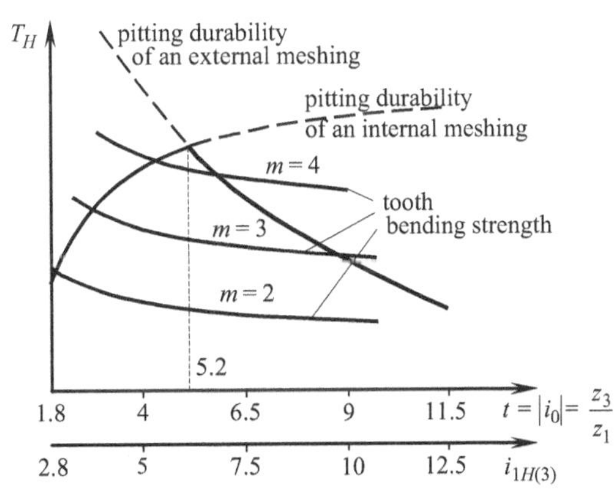

FIGURE 16.1
Load capacity of $\overline{\text{AI}}$-PGT depending on the basic speed ratio i_0, and the torque ratio t, respectively, and on the module m. (Redrawn from Leistner, F., G. Lörsch and O. Wilhelm. *Umlaufrädergetriebe*. 3. Auflage. Berlin: VEB Verlag Technik, 1987.)

From the above-stated, it is clear that the three approaches solve three different problems:

I approach—choosing the number of teeth, i.e., a kinematic problem
II approach—choosing the ring gear diameter, i.e., an overall dimension problem,
III approach—choosing the optimum basic speed ratio i_0, i.e., an optimization problem

In the approach examined here, the optimization is of the speed ratio *i*. The optimization of the load capacity of tooth meshing is considered in Section 18.3.

17

Simplified Verifiable Calculation of $\overline{AI}$-Planetary Gear Train Gears (according to ISO 6336)

17.1 General Concepts and Peculiarities

Nowadays, the authoritative document for load capacity calculation of gears is the international standard ISO 6336 [112]. Due to its too large volume, a simplified calculation is given here. In many cases, this simple calculation is fully acceptable.

The aim of the next few pieces is to illustrate and bring more clarity to the problem.

Figure 17.1 shows schematically the components of a machine unit:

- A driving machine (an engine)—*A*
- A driven (working) machine—*B*
- A gear train—*G*
- Couplings—input C_A and output C_B

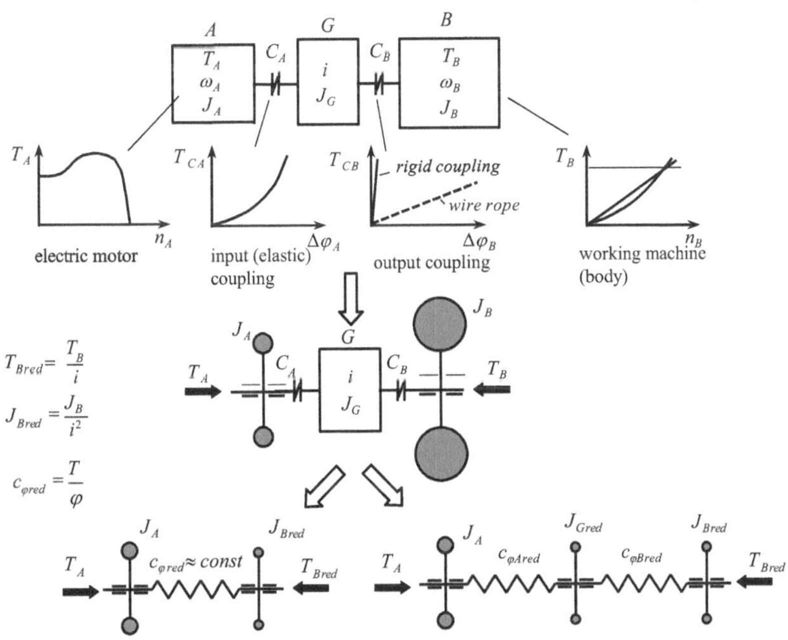

FIGURE 17.1
Machine unit as two- and three-mass model.

where T_A and T_B are the input and output torques of the gear train, J_A and J_B are the mass moments of inertia at the input A and output B of the gear train, $T_{B\,red}$, $J_{G\,red}$, and $J_{B\,red}$ are reduced to the input A output (loading) torque and mass moment of inertia of the gear train G and driven (working) machine B, and $c_{\varphi\,red}$ is the reduced torsional stiffness.

Also, the most commonly used two-mass dynamics model and the less commonly used three-mass model (especially for hoisting machines [74, 100]) are shown. In hoisting mechanisms, the connection between the gear train and the load is markedly elastic (a rope), and this is the reason to work with a three-mass model—engine–train–load.

Figure 17.2 shows different working phases of a machine unit and the corresponding loads [30]:

- Starting load
- Base working load
- Breaking load
- Eventually revers load

As can be seen from the figure, the direction of the braking load T_{Br} depends on where the brake Br is located—before or after the gear train G.

Fluctuation of the load on a gear train, and hence on its gears, depends on the combination of the parameters of driving (A) and driven machine (B), gear train (G), couplings at the input (C_A) and output (C_B) of the gear train (mass-elastic properties of the system), and the operating conditions—load T_B and angular velocity ω. The gear train load can be determined in different ways. In its theoretical determination, a suitable dynamic model is used [74, 191, 193, 246]. The experimental results from a case similar to the calculating one can be used. Some gear trains are calculated with a load prescribed by normative documents (see Section 17.5). The most rough is the calculation using the application factor K_A for lack of more accurate data. Relatively, the most accurate is the calculation described in Section 17.5.

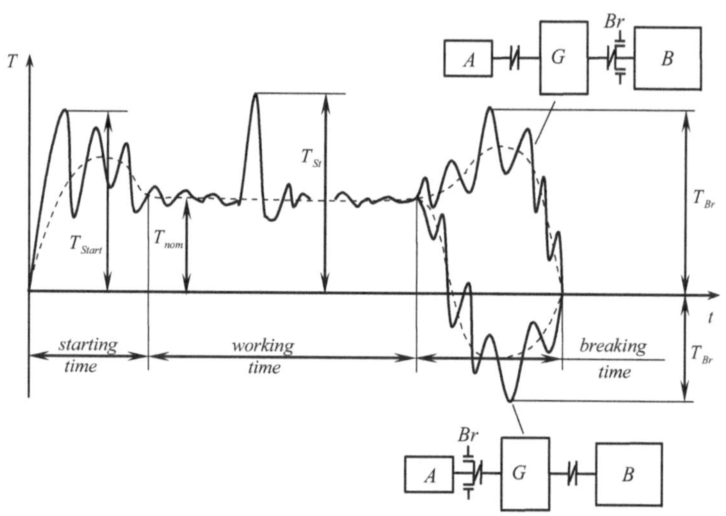

FIGURE 17.2
Exemplary load and its phases on a gear train (G), respectively, on its gears, depending on break (Br) position with respect to driving (A) and driven (B) machine.

It is necessary to emphasize once again that the calculation presented here is very simplified. In responsible cases, ISO 6336 [112] must be used.

According to the normative documents [63, 94, 112, 113, 215], the main work (operational) load is of four types:

- Uniform
- Light shocks
- Moderate shocks
- Heavy shocks

Taking into account the specifics of different driven machines, another classification can be made (Figure 17.3) [30]:

- Uniform (approximately)
- Variable (nonuniform) in bins (duty cycle [113])
- Smoothly or randomly varying
- With single shocks

Gears damages are varied [4, 62, 114, 120, 140], but the ones shown in Figure 17.4 are the most common [30, 178]. Therefore, in the most common cases, load capacity calculations are made only for the two most common damages:

- Tooth root fatigue fracture (calculation of tooth bending strength)
- Tooth flank pitting (calculation of surface durability)

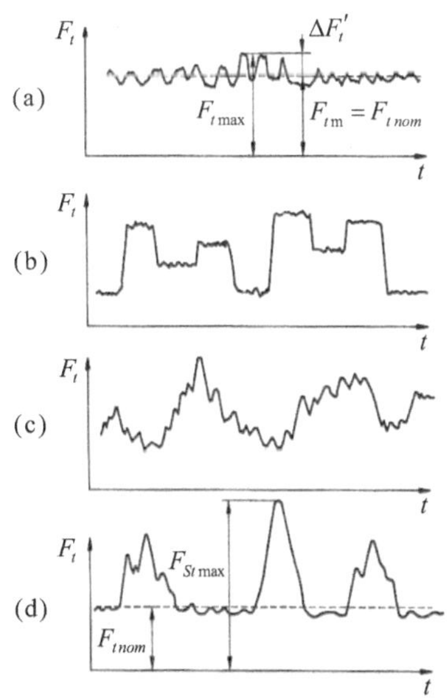

FIGURE 17.3
Typical cases of gears loading: (a) Uniform (approximately); (b) variable (nonuniform) in bins (duty cycle [113]); (c) smoothly or randomly varying; (d) with single shocks (impacts).

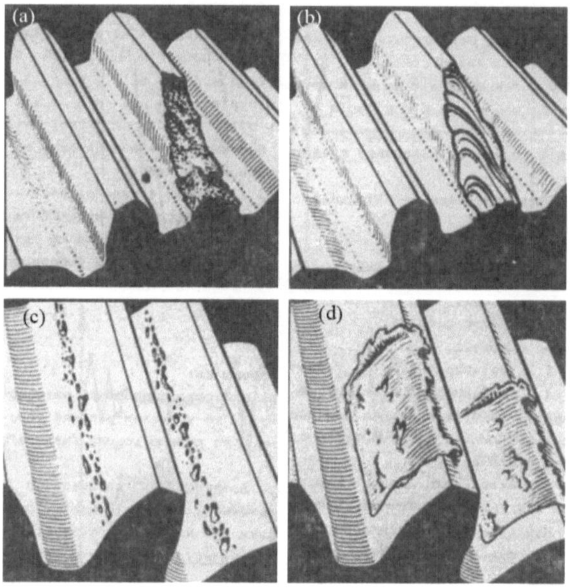

FIGURE 17.4
The most common tooth damages: (a) Tooth fracture due to overload (forced breakage); (b) tooth fracture due to cyclic load (fatigue breakage); (c) tooth flank pitting; (d) tooth flank plastic deformation.

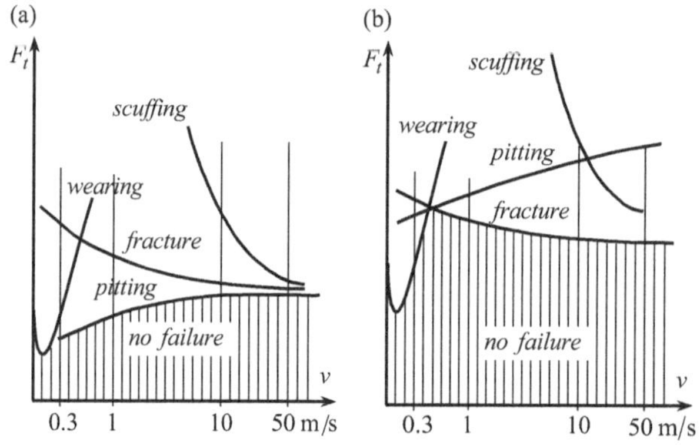

FIGURE 17.5
Failure mode likelihood as a function of load F_t and peripheral velocity v of a gear: (a) Non-hardened; (b) hardened. (Redrawn from Allianz-Handbuch der Schadenverhütung. 3. Auflage. Düsseldorf: VDI-Verlag, 1984.)

Figure 17.5 illustrates the failure mode likelihood as a function of load and peripheral velocity of a given gear [4, 58, 175]. It can be seen that the wear is limited for very low peripheral velocities v when it is not possible to form a load-bearing oil film between the teeth flanks. In turn, scuffing is typical for very high peripheral velocities v combined with heavy load F_t.

Figure 17.6 illustrates the models used to determine contact (σ_{H12} and σ_{H23}) and bending (σ_{F1}, σ_{F2}, and σ_{F3}) stresses in teeth. In the first case, the contact between two straight circular cylinders pressed externally or internally is considered. In the second case, the teeth are considered to be the cantilever beam loaded most unfavorably, when the normal force

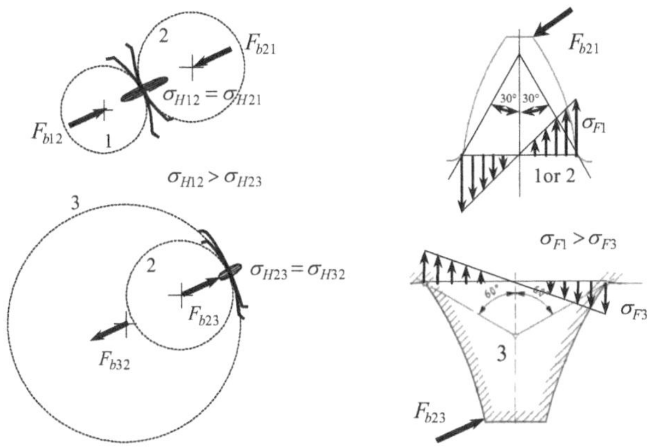

FIGURE 17.6
Models used in load capacity calculations to determine contact stresses σ_{H12} and σ_{H23} in both the external and the internal meshes, and bending stresses of the external teeth σ_{F1} and σ_{F2}, and of the internal teeth σ_{F3}.

F_b acts on the tooth's tip. According to [63, 112, 113, 175], the critical sections of the external and internal teeth are determined by tangents at different angles to their filet curves—30° for external teeth and 60° for internal teeth.

The gears of $\overline{\text{AI}}$-planetary gear train ($\overline{\text{AI}}$-PGT) are nowadays calculated according to the commonly accepted methodology contained in such an important international document as ISO 6336 [112]. It is necessary to bear in mind that the international standard ISO 6336 has been developed on the basis of the German standard DIN 3990 [65], so that these standards are very common and overlap to a great extent. Both standards, which are quite bulky (about 350 pages), have shorter standards derived from the different gearing applications. These are the so-called application standards (Anwendungsnorm), e.g., for industrial gears (ISO 9085 [113] and DIN 3990, Teil 11 [63]) as well as ship gears, high-speed gears of vehicles, etc.

Nowadays, computer programs for geometry and load capacity calculation of gears are increasingly used in practice. Fatal delusion, however, is to assume that with a computer program available, one gear train can be calculated and designed. The computer program is just one "hard worker," really capable of working, but devoid of the ability to think. Therefore, in order for a computer program to be used successfully, the designer must first have mastered the relevant methodology. If he is not skillful at it, he will act blindly, without understanding, with all the resulting negative consequences.

An important place in modern gear calculations have more precise methods of determining the load (especially its dynamic component) [61, 74, 82, 96, 109, 110, 159, 172, 182, 191, 203], taking into account various influential factors—stiffness, backlashes [96, 97, 160, 165, 172] fabrication misalignment [73, 165, 187], tooth modifications [40, 56, 96, 242], and others.

Within this book, a *simplified verifying calculation* of gears, specifically for $\overline{\text{AI}}$-PGT, has been given, paying attention to some of the *features* of this gear train that should be considered.

1. Since $\overline{\text{AI}}$-PGT has generally more than $k = 1$ planets, i.e., it is a torque split gear train, in the calculations, mesh load factor $K_{\gamma max}$ should be used, taking into account the uneven load distribution among the planets, detailed in Chapter 8, where recommended numerical values for it are given. This factor is specific for PGTs and for torque split gear trains in general.

2. Face load factor K_β, considering uneven load distribution over the tooth face width, is in relation to mesh load factor $K_{\gamma\ max}$. Therefore, it is necessary to ensure that the improvement (decreasing) of $K_{\gamma\ max}$ does not lead to a deterioration (increasing) of K_β, so that their product grows and that the situation deteriorates. According to statistics [4], the most common cause for gears damages is this uneven load distribution. Higher values for K_β may be recommended when
   - Planet bearings are two, especially in case of one-wall carrier.
   - Sun gear (or ring gear) if floating with single articulated gear coupling with short sleeve (small ratio L/d, see Figure 8.17).
   - The sun gear shaft, which pliability is relied on to reduce the unevenness of load distribution among the planets ($K_{\gamma\ max}$), is asymmetrically supported at the same shaft diameter on both sides of the gear (see Figure 18.3).

   Recommendations for choosing values of mesh load factor $K_{\gamma\ max}$ are given in Chapter 8. Particular attention should be paid to the influence of the sensitivity of equalizing (distribution) device (see Section 8.6.4).
3. Application factor K_A takes into account the variable external load of a given gear very approximately. It is assumed that for implicated calculations, the same value of this factor is used for all gears of the PGT, although things are far more complex.[1]
4. The teeth of planets, which in some sense are parasitic gears, are always loaded with a completely reversed bending, which forces in calculations to use a lower strength, i.e., with a 30% lower permissible bending stress compared to a pulsating load (see Section 9.1).
5. The teeth of the two central gears (sun gear and ring gear) are loaded with pulsating bending (when PGT rotates in one direction), as in ordinary, non-PGTs. In lifting (hoisting) mechanisms, they are loaded pulsatingly, no matter what direction they are rotating and whether the load is lifted or lowered. The teeth of these gears, of course, are loaded with a completely reversed bending, when, along with the change in the direction of rotation (reversing), the load direction changes, too, e.g., in a crane travel mechanisms.
6. It is also a feature of $\overline{\text{AI}}$-PGT that at the same tangential forces in the two meshings (external and internal) $F_{t12} = F_{t23}$, the contact stress σ_{H23} in the internal meshing between planets 2 and ring gear 3 is considerably lower than the contact stress σ_{H12} in the external one between sun gear 1 and planets 2 (see Section 9.1). This is taken into account in the formula for the contact stress through expressions $(u_{12}+1)/u_{12}$ and $(u_{23}-1)/u_{23}$ and zone factors Z_{H12} and Z_{H23} which consider the shape (curvature) of the mated teeth.
7. Load capacity calculations are assumed to be based on the torque of sun gear T_1, regardless of the working mode of PGT—with $F=1$ or $F=2$ degrees of freedom (i.e., as a reducer or multiplier, as a summation or division differential), and regardless of whether the torque T_1 is an input torque ($T_1 = T_A$), an output (loading) torque ($T_1 = T_B$), or a reactive one ($T_1 = T_C$) on fixed sun gear.
8. The number of load cycles of the three gears is quite different, and the most common order is $N_1 > N_2 > N_3$ (see Section 9.1).

[1] The loading spectra of the individual teeth of all gears are different, and each tooth has its own equivalent load.

9. Here, the case of spur gears ($\beta = 0°$) is considered, but the following methodology may, with the necessary additions, also be used for helical gears, which are increasingly entering PGTs [204].
10. Tooth form factor Y_{FS} here is determined (according to DIN 3990 [65] rather than ISO 6336 [112]) on the assumption that the load (normal force) acts on the tooth's tip, which is more appropriate for a simplified methodology. Factor Y_ε applies the load to the point D of the tooth profile—the outer point of single pear tooth contact (Figure 11.2). It should be explicitly noted that such an important document for calculating internal teeth, such as VDI-Richtlinie [238], also uses the factor Y_{FS}.

New possibilities for significant increase in the load capacity of gears in general, including PGTs, provide the use of gears with asymmetric teeth (see Chapter 38). These teeth increase both the surface durability and the bending strength. The disadvantage is that a special tool is required which hinders the wide implementation of this meshing in general mechanical engineering. More details on this perspective can be found in [71–73] and in the literature cited in Chapter 38.

17.2 Used Parameters

In the presented simplified calculation of gears, the following parameters are used:

Load factors

K_A—application factor, adjusting the nominal load F_t in order to compensate for incremental gear train loads from external sources.

K_v—internal dynamic factor, accounting for the effects of gear tooth accuracy grade as related to speed and load. Considering meshing variable stiffness as parametric excitation too.

K_β—face load factor, taking into account the effects of the nonuniform distribution of load over the gear face on the tooth root stress ($K_{F\beta}$) and on the surface stress ($K_{H\beta}$) due to inaccuracies, deformations, and bearing clearances.

K_α—transverse load factor, $K_{F\alpha}$ and $K_{H\alpha}$, considering the effect of the nonuniform distribution of transverse load between several pairs of simultaneous contacting gear teeth due to inaccuracies in the base pitch p_b (as well as deflection under load, profile modifications, etc.).

$K_{\gamma\ max}$—mesh load factor, considering uneven load distribution between planets due to the gears and carrier inaccuracies.

Factors used in the bending stress calculation

Y_{FS}—form factor, taking into account the influence of tooth form as well as the stress concentration in root fillet (converts the nominal tooth root stress to local tooth root stress).

Y_ε—contact ratio factor, considering transvers contact ratio ε_α $\left(\varepsilon_\beta = 0!\right)$, i.e., double meshing.

Factors used in the contact stress calculation

Z_H—zone factor, accounts for the influence on Hertzian pressure of tooth flank curvature at the pitch point and transforms the tangential load at the reference cylinder to normal load at the pitch cylinder.

Z_E—elasticity factor, considering the influence of material properties—modulus of elasticity E and Poisson's ratio v, $\sqrt{\frac{\text{N}}{\text{mm}^2}}$.

Z_ε—contact ratio factor, considering the influence of sum length of contact line (virtual face width) because double meshing, i.e., the influence of transverse contact ratio $\varepsilon_\alpha \left(\varepsilon_\beta = 0!\right)$.

Safety factors

S_F—safety factor for tooth breakage (calculation of tooth bending strength).

$S_{F\,min}$—minimum required safety factor (tooth breakage).

S_H—safety factor for pitting (surface durability calculation).

$S_{H\,min}$—minimum required safety factor (pitting).

Stresses (all in $\frac{\text{N}}{\text{mm}^2}$)

σ_F—calculated tooth root stress.

σ_{FG}—tooth root stress limit (bending strength).

$\sigma_{F\,lim\,b}$—nominal stress number—bending (tooth root endurance limit).

σ_{FP}—permissible tooth root stress.

σ_H—calculated contact stress at pitch point.

σ_{HG}—pitting stress limit (contact strength).

$\sigma_{H\,lim\,b}$—allowable stress number—contact (pitting endurance limit).

σ_{HP}—permissible contact stress.

Base number of cycles (see Figure 17.9)

$N_{F\,lim\,b}$—base number of cycles—bending (at the knee of Wöhler curve).

$N_{H\,lim\,b}$—base number of cycles—pitting (at the knee of Wöhler curve).

17.3 Calculations Sequence

17.3.1 Required Input Data

1. *Geometrical data*

   m—module, mm

   a_w—operating center distance, mm

z_1, z_2, z_3—number of teeth
x_1, x_2, x_3—profile shift coefficients
b_1, b_2, b_3—tooth face widths, mm
d_1, d_2, d_3—reference circle diameters, mm
$u_{12} = z_2/z_1 > 1$ (very rare $u_{21} = z_1/z_2 > 1$), $u_{23} = z_3/z_2 > 1$—teeth ratio of both meshings
b_{H12}, b_{H23}—operating face widths of both meshings, mm
$\varepsilon_{\alpha12}, \varepsilon_{\alpha23}$—transverse contact ratios of both meshings $(\varepsilon_\beta = 0!)$

2. *Load and operating conditions*
   Given
   P_A—input power (*A*), kW
   T_B—working load (*B*), Nm
   T_1—torque on the sun gear, Nm
   n_1—rotation speed of sun gear, min^{-1}
   L—operating time (life) of PGT, h
   Calculated
   $F_t = 2\frac{T_1}{k \cdot d_1}10^3 = 19.1\frac{P}{k \cdot d_1 \cdot n_1}10^6$—nominal tangential load (force) at reference cylinder for *k* planets, N
   $v = \frac{\pi \cdot d_1(n_1 - n_H)}{60}10^{-3}$—peripheral velocity, m/s
   Number of load cycles of individual gears:
   $N_{F1} = N_{H1} = N_1 = k \cdot 60(n_1 - n_H)L = k \cdot 60 \cdot n_{1\,rel} \cdot L$
   $N_{H2(1)} = 60(n_1 - n_H)\frac{z_1}{z_2}L = 60\, n_{1\,rel}\frac{z_1}{z_2}L \quad N_{F2} = 2N_{H2(1)}$—for external meshing
   $N_{H2(3)} = 60(n_3 - n_H)\frac{z_3}{z_2}L = 60\, n_{3\,rel}\frac{z_3}{z_2}L = N_{H2(1)} \quad N_{F2} = 2N_{H2(1)}$—for internal meshing
   $N_{F3} = N_{H3} = N_3 = k \cdot 60(n_3 - n_H)L = k \cdot 60 \cdot n_{3\,rel} \cdot L$
3. *Materials, heat treatment, hardness* (see Chapter 15)
4. *Grades of accuracy* (see Chapter 15)

17.3.2 Calculation of Bending Stresses and Safety Factors

$$\sigma_{F1} = Y_{FS1} \cdot Y_{\varepsilon12}\frac{F_t}{b_1 \cdot m}K_A \cdot K_v \cdot K_\beta \cdot K_\alpha \cdot K_{\gamma\,max} \le \sigma_{FP1} = \frac{\sigma_{FG1}}{S_{F\,min1}} \tag{17.1}$$

$$S_{F1} = \frac{\sigma_{FG1}}{\sigma_{F1}} \ge S_{F\,min1} = 1.5 \div 1.7(> 2) \tag{17.2}$$

$$\sigma_{F2} = Y_{FS2} \cdot Y_{\varepsilon12}\frac{F_t}{b_2 \cdot m}K_A \cdot K_v \cdot K_\beta \cdot K_\alpha \cdot K_{\gamma\,max} \le \sigma_{FP2} = \frac{\sigma_{FG2}}{S_{F\,min2}} \tag{17.3}$$

$$S_{F2} = \frac{\sigma_{FG2}}{\sigma_{F2}} \ge S_{F\,min2} = 1.5 \div 1.7(> 2) \tag{17.4}$$

$$\begin{cases} \sigma_{F3} = Y_{FS3} \cdot Y_{\varepsilon 23} \dfrac{F_t}{b_3 \cdot m} K_A \cdot K_v \cdot K_\beta \cdot K_\alpha \cdot K_{\gamma\, max} \le \sigma_{FP3} = \dfrac{\sigma_{FG3}}{S_{F\, min3}} & (17.5) \\ S_{F3} = \dfrac{\sigma_{FG3}}{\sigma_{F3}} \ge S_{F\, min3} = 1.5 \div 1.7 (> 2) & (17.6) \end{cases}$$

17.3.3 Calculation of Contact Stresses and Safety Factors

$$\begin{cases} \sigma_{H12} = Z_{H12} \cdot Z_{E12} \cdot Z_{\varepsilon 12} \sqrt{\dfrac{F_t}{b_{H12} \cdot d_1} \dfrac{u_{12}+1}{u_{12}} K_A \cdot K_v \cdot K_\beta \cdot K_\alpha \cdot K_{\gamma\, max}} \le \begin{cases} \sigma_{HP1} = \dfrac{\sigma_{HG1}}{S_{H\, min\, 1}} \\ \sigma_{HP2} = \dfrac{\sigma_{HG2}}{S_{H\, min 2}} \end{cases} & (17.7) \\ S_{H1} = \dfrac{\sigma_{HG1}}{\sigma_{H12}} \ge S_{H\, min\, 1} = 1.1 \div 1.3 (> 1.4) & (17.8) \\ S_{H2} = \dfrac{\sigma_{HG2}}{\sigma_{H12}} \ge S_{H\, min\, 2} = 1.1 \div 1.3 (> 1.4) & (17.9) \end{cases}$$

$$\begin{cases} \sigma_{H23} = Z_{H23} \cdot Z_{E23} \cdot Z_{\varepsilon 23} \sqrt{\dfrac{F_t}{b_{H23} \cdot d_2} \dfrac{u_{23}-1}{u_{23}} K_A \cdot K_v \cdot K_\beta \cdot K_\alpha \cdot K_{\gamma\, max}} \le \begin{cases} \sigma_{HP2} = \dfrac{\sigma_{HG2}}{S_{H\, min2}} \\ \sigma_{HP3} = \dfrac{\sigma_{HG3}}{S_{H\, min 3}} \end{cases} & (17.10) \\ S_{H2} = \dfrac{\sigma_{HG2}}{\sigma_{H23}} \ge S_{H\, min2} = 1.1 \div 1.3 (> 1.4) & (17.11) \\ S_{H3} = \dfrac{\sigma_{HG3}}{\sigma_{H23}} \ge S_{H\, min3} = 1.1 \div 1.3 (> 1.4) & (17.12) \end{cases}$$

17.4 Numerical Data for Factors and Teeth Strength

The values proposed here are taken from the relevant standards [63, 112].

Application factor K_A (Table 17.1)

The values only apply to transmissions which work outside the resonance speed range under relatively steady loading.

Internal dynamic factor $K_v = f(v)$

TABLE 17.1

Typical Values for Application Factor K_A according to the Working Characteristics of the Driving and Driven Machine

Working Characteristics of the Driving Machine	Working Characteristics of the Driven Machine			
	Uniform	**Light Shocks**	**Moderate Shocks**	**Heavy Shocks**
Uniform	1.00	1.25	1.50	1.75
Light shocks (impacts)	1.10	1.35	1.60	1.85
Moderate shocks	1.25	1.50	1.75	≥2.00
Heavy shocks	1.50	1.75	2.00	≥2.25

$$K_v \approx 1+\left(\frac{25}{\frac{F_t}{b}K_A}+0.02\right)\frac{z_1\cdot v}{100}\sqrt{\frac{u_{12}^2}{u_{12}^2+1}} \quad (17.13)$$

Face load factor $K_\beta = f\left(\frac{b_1}{d_1}, HB\right)$

$$\text{For not-hardened gears (hardness} \le 350\,\text{HB)}\ K_\beta = 1+0.4\left(\frac{b_1}{d_1}\right)^2 \quad (17.14)$$

$$\text{For hardened gears}\left(\text{hardness} > 350\,\text{HB}\right)\ K_\beta = 1+0.5\left(\frac{b_1}{d_1}\right)^2 \quad (17.15)$$

Transverse load factor $K_\alpha = f(HB)$

For not-hardened gears (hardness ≤ 350 HB) $K_\alpha = 1.1$

For hardened gears (hardness > 350 HB) $K_\alpha = 1.2$

Mesh load factor $K_{\gamma\,max}$ (Tables 17.2 and 17.3)

Form factor $Y_{FS} = f(z, x)$

TABLE 17.2

Typical Values for Mesh Load Factor $K_{\gamma\,max}$ Depending on How the Load Is Equalized Between k = 3 Planets (see Chapter 8).

Way of Load Equalizing	$K_{\gamma\,max}$
No purposive actions	1.5 ÷ 2.0 (2.5)
Purposive accuracy	1.3 ÷ 1.5
Purposive compliance	1.2 ÷ 1.4
Kinematic flexibility (radial float)	1.1 ÷ 1.2
Complex way	1.05 ÷ 1.1

TABLE 17.3

Typical Values for Mesh Load Factor $K_{\gamma\,max}$ Depending on Number of Planets k

k	3	4	5	6	7	8
$K_{\gamma\,max}$	1.1 ÷ 1.2	1.15 ÷ 1.3	1.2 ÷ 1.4	1.25 ÷ 1.5	1.3 ÷ 1.6	1.35 ÷ 1.7

Values depend on the type and parameters of the tool and can be determined by the following diagrams for external (Figure 17.7) and internal (Figure 17.8) teeth [63, 238].

Contact ratio factor (bending) $Y_\varepsilon = f(\varepsilon_\alpha)$

$$Y_{\varepsilon 12} = 0.25 + \frac{0.75}{\varepsilon_{\alpha 12}} \text{ and } Y_{\varepsilon 23} = 0.25 + \frac{0.75}{\varepsilon_{\alpha 23}} \tag{17.16}$$

Zone factor Z_H

Factor values can be taken from Figure 17.9.

Elasticity factor Z_E in $\sqrt{\frac{\text{N}}{\text{mm}^2}}$

$Z_E = 190$ for steel/steel; $Z_E = 189$ for steel/cast steel

$Z_E = 188$ for cast steel/cast steel; $Z_E = 164$ for steel/iron

More detailed tables can be found in the standards.

Contact ratio factor (pitting) $Z_\varepsilon = f(\varepsilon_\alpha)$

$Z_\varepsilon = \sqrt{\frac{4-\varepsilon_\alpha}{3}}$ for spur gears $(\beta = 0°)$

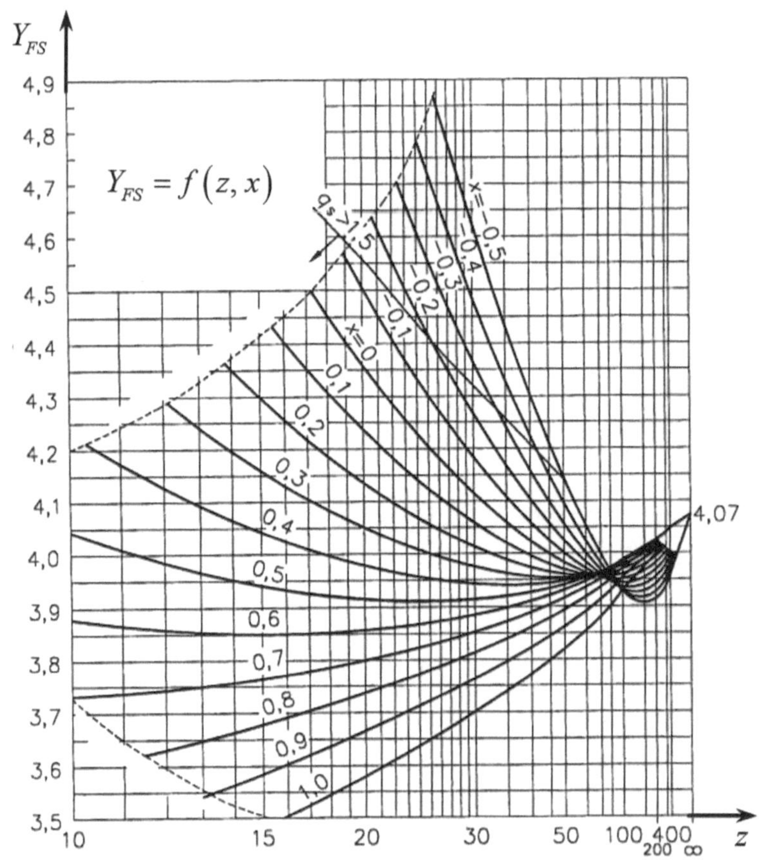

FIGURE 17.7

Form factor Y_{FS}, considering tooth form and stress concentration in root fillet of external teeth machined by rack without protuberance (undercut) with $\alpha_0 = 20°$; $h^*_{a0} = 1.0$; $c^*_0 = 0.25$; $\rho^*_{f0} = 0.38$. (Extracted from DIN 3990, Teil 11 Tragfähigkeitsberechnung von Stirnrädern—Anwendungsnorm für Industriegetriebe. Detail-Methode. 1989.)

Stress limits σ_{FG} and σ_{HG}

In ISO 6336 [112], 40 pages to material selection and stress limits determination are devoted. A more general idea of the stress limits of materials used for gears can be obtained from Figure 17.10 [158]. Stress—number of cycles diagram (Wöhler curve) obtained experimentally. Figure 17.11 shows results of its experimental determination for bending stress [216].

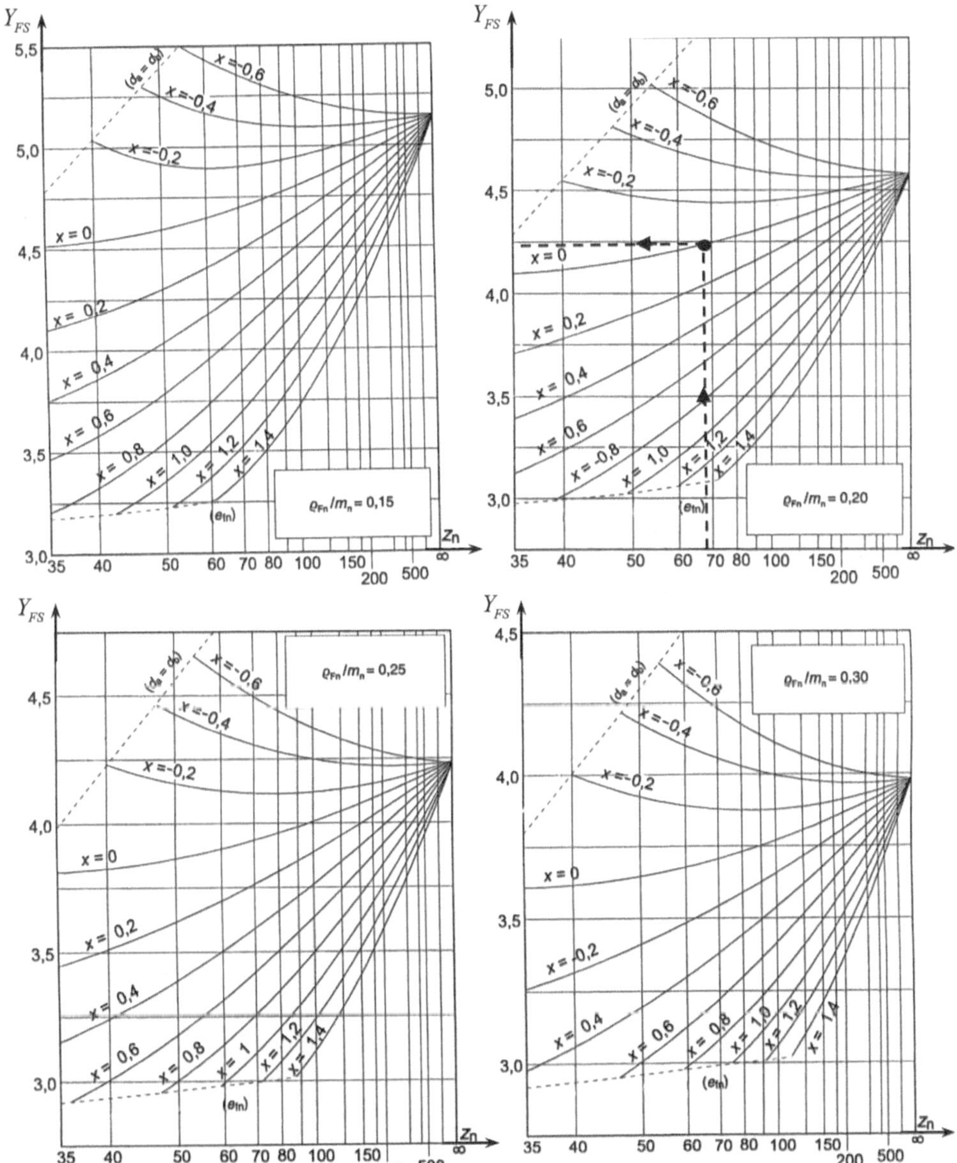

FIGURE 17.8

Form factor Y_{FS}, considering tooth form and stress concentration in root fillet of internal teeth for different values of radius of root fillet curvature ρ_{Fn} in critical section for following parameters of sharp-cornered cutter: $\alpha_0 = 20°$; $h^*_{a0} = 1.0$; $c^*_0 = 0.25$; $\rho^*_{f0} = 0$. (Extracted from VDI-Richtlinie 2737 Berechnung der Zahnfußtragfähigkeit von Innenverzahnungen mit Zahnkranzeinfluß. 2005.)

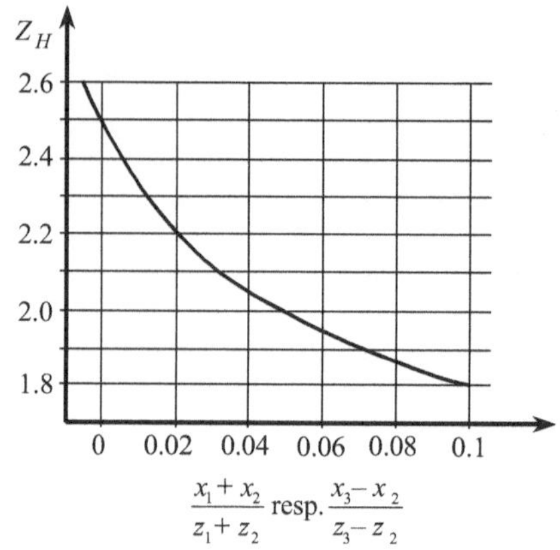

FIGURE 17.9
Values for zone factor Z_H depending on number of teeth and profile shift coefficients of mated gears.

In Section 15.1, the most commonly used steels for $\overline{\text{AI}}$-PGT are given.

Minimum required safety factors $S_{F\,min}$ and $S_{H\,min}$

Recommendations concerning $S_{F\,min}$ and $S_{H\,min}$ are made in ISO 3663, but values are not proposed. For example, ISO 9085 [165] says "If no otherwise between manufacturer and user, the minimum safety factor (tooth breakage) $S_{F\,min} = 1.2$ is applied." Recommended values for these factors one can find in [94, 175]. From manufacturers' and users' experience, especially for PGTs, the following values may be recommended:

$$S_{H\,min} = 1.1 \div 1.3(>1.4) \text{ and } S_{F\,min} = 1.5 \div 1.7(>2).$$

In conclusion, it is necessary to *expressly point out* that the load capacity calculation shown here is highly *simplified*. In case of a more accurate calculation, ISO 6336 [112] and the specialized literature [101, 156–158, 184, 210] should be used. Especially, the load capacity of internal teeth is investigated in detail in [250, 253], as well as in [116, 117].

17.5 Gears Load Capacity under Variable Load

In Section 17.3, it is shown how gear load capacity can be estimated by the application factor K_A. In a number of machines, their load is known from experience, which allows for a load spectrum to be given. For example, the Fédération Européenne de la Manutention (FEM)[2] normative document applies to the electric hoists in Bulgaria (Figure 17.12).

With a specified load spectrum and prescribed operating time (life) L of the gear train in hours, the following can be proceeded:

1. Determine the total number of load cycles $N_{F\Sigma}$ and $N_{H\Sigma}$ for both types of stress as well as the number of cycles N_{Fj} and N_{Hj} of individual load bins for each gear by formulas (9.4) and Section 17.3.

[2] Fédération Européenne de la Manutention (FEM)—European Materials Handling Federation.

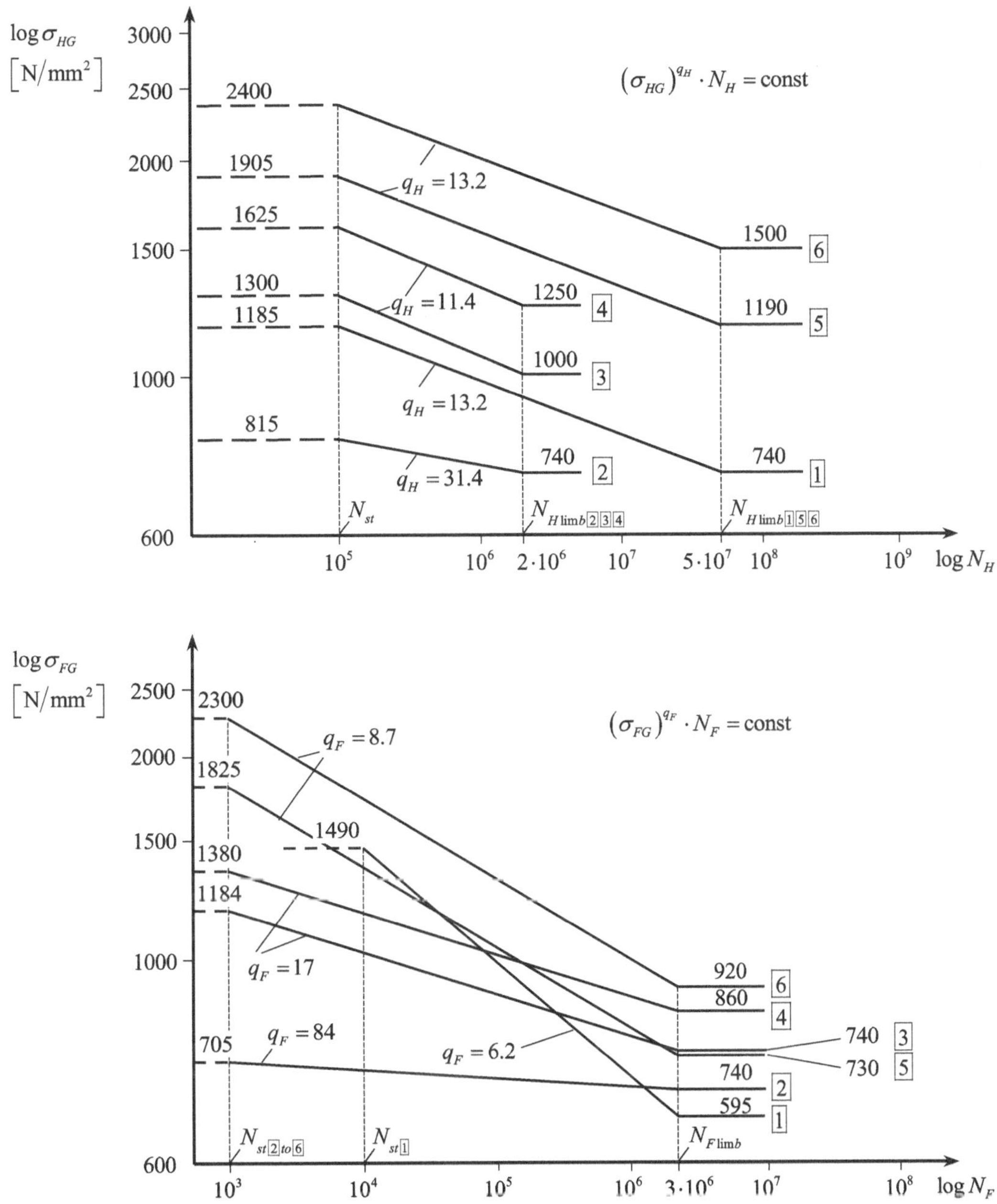

FIGURE 17.10
Tooth root stress limit σ_{FG} and pitting stress limit σ_{HG} for various materials and heat treatment for medium quality (MQ) grade and 1% probability of damage: 1—alloyed through hardening (tempering) wrought steels, through hardened (tempered); 2—through hardening (tempering) and carburizing wrought steels, normalized or through hardened (tempered), nitrocarburized; 3—through hardening (tempering) and carburizing wrought steels, through hardened (tempered), ion-nitrided; 4—nitriding wrought steels (no Al), through hardened (tempered), ion-nitrided; 5—through hardening (tempering) wrought steels, through hardened (tempered), flame or induction hardened; 6—alloyed carburizing wrought steels, carburized and hardened. (Redrawn from Linke, H. und andere. *Stirnradverzahnung—Berechnung, Werkstoffe, Fertigung*. 2. Auflage. München/Wien: Carl Hanser Verlag, 2010.)

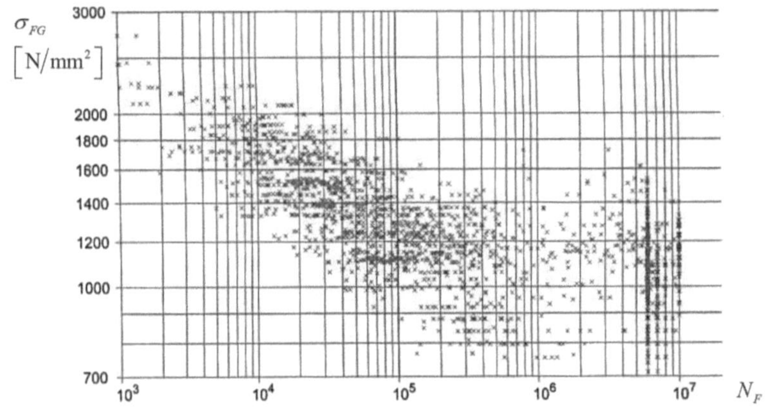

FIGURE 17.11
Experimental determination of Wöhler curve for bending stress in a spur gear. (From Stahl, K. *Vorlesung Maschinenelemente II*. Technische Universität München. 2013. With permission.)

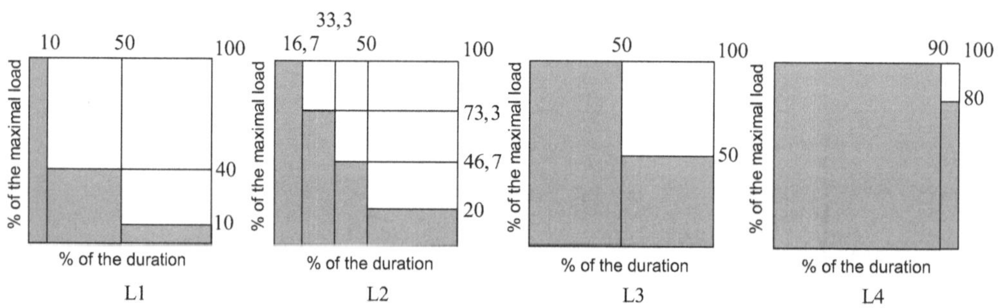

FIGURE 17.12
Load spectrum for hoisting mechanisms according FEM 9.511: light (L1); medium (L2); heavy (L3); very heavy (L4).

2. Determine the stress σ_{Fj} and σ_{Hj} of individual load bins (bin size) for each gear by formulas given in Section 17.3.
3. Determine the equivalent number of load cycles $N_{F\,Eq}$ and $N_{H\,Eq}$ (in which the equivalent stress will be determined) for the two types of stress (Figure 17.13) for each gear, as follows:
   - When the total number of load cycles $N_{F\Sigma}$ or $N_{H\Sigma}$ is smaller than the corresponding base number of cycles [112]
     $N_{F\,lim\,b} = 3 \cdot 10^6 = \text{const}$ and $N_{H\,lim\,b} = 50 \cdot 10^6 = \text{const}$, or $N_{H\,lim\,b} = 2 \cdot 10^6 = \text{const}$ (see Figure 17.10),
     i.e.,
     $N_{F\Sigma} < N_{F\,lim\,b}$ and $N_{H\Sigma} < N_{H\,lim\,b}$,
     the equivalent number of load cycles is equal to the total one (Figure 17.13a)
     $N_{F\,Eq} = N_{F\Sigma}$ and $N_{F\,Eq} = N_{H\Sigma}$.
   - When the total number of load cycles is greater than the corresponding base number of cycles, i.e.,
     $N_{F\Sigma} > N_{F\,lim\,b}$ or $N_{H\Sigma} > N_{H\,lim\,b}$,
     the equivalent number of load cycles is equal to the base one (Figure 17.13b)
     $N_{F\,Eq} = N_{F\,lim\,b}$ and $N_{H\,Eq} = N_{H\,lim\,b}$

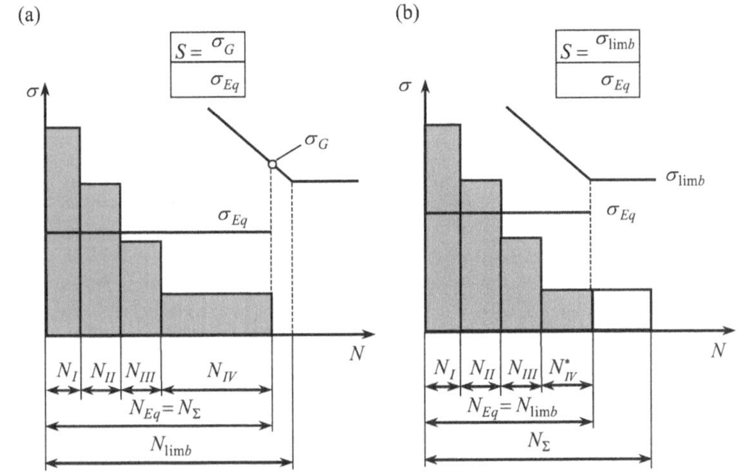

FIGURE 17.13
Equivalent number of load cycles: (a) $N_{Eq} = N_{\Sigma} < N_{lim\,b}$; (b) $N_{Eq} = N_{lim\,b} < N_{\Sigma}$.

As can be seen from Figure 17.13b, in this case, no load is taken into account after $N_{F\,lim\,b}$ and $N_{H\,lim\,b}$.

4. Determine the both types equivalent stress for each gear

$$\sigma_{F\,Eq} = \sqrt[q_F]{\sum_{j=1}^{n} \sigma_{Fj}^{q_F} \cdot \frac{N_{Fj}}{N_{F\,Eq}}}, \tag{17.17}$$

$$\sigma_{H\,Eq} = \sqrt[q_H]{\sum_{j=1}^{n} \sigma_{Fj}^{q_H} \cdot \frac{N_{Hj}}{N_{H\,Eq}}}, \tag{17.18}$$

where q_F and q_H are the slopes of Wöhler curve for both types of stress, taken from Figure 17.10 depending on the material and heat treatment of corresponding gear, and $j = 1 \div n$ is the number of bins of load spectrum diagram (Figures 17.12 and 17.13).

5. Determine the safety factor S_F and S_H for both types of stress for each gear. Equivalent stress is compared with the stress limit σ_{FG} and σ_{HG} for the corresponding equivalent number of cycles at $N_{F\,Eq} = N_{F\Sigma} < N_{F\,lim\,b}$ and $N_{H\,Eq} = N_{H\Sigma} < N_{H\,lim\,b}$ (Figure 17.13a) or with the endurance limit $\sigma_{F\,lim\,b}$ and $\sigma_{H\,lim\,b}$ at $N_{F\,Eq} = N_{F\,lim\,b} < N_{F\Sigma}$ and $N_{H\,Eq} = N_{H\,lim\,b} < N_{H\Sigma}$ (Figure 17.13b)

$$\boxed{S_F = \frac{\sigma_{FG}}{\sigma_{F\,Eq}} = f(N_{F\Sigma})} \quad \text{or} \quad \boxed{S_F = \frac{\sigma_{F\,lim\,b}}{\sigma_{F\,Eq}} = f(N_{F\,lim\,b})}; \tag{17.19}$$

$$\boxed{S_H = \frac{\sigma_{HG}}{\sigma_{H\,Eq}} = f(N_{H\Sigma})} \quad \text{or} \quad \boxed{S_H = \frac{\sigma_{H\,lim\,b}}{\sigma_{H\,Eq}} = f(N_{H\,lim\,b})}. \tag{17.20}$$

In these formulae, the endurance stress limits σ_{FG} and σ_{HG} are taken from the slope part of Wöhler curve for the corresponding number of cycles $N_{F\,Eq} = N_{F\Sigma} < N_{F\,lim\,b}$ and $N_{H\,Eq} = N_{H\Sigma} < N_{H\,lim\,b}$, and $\sigma_{F\,lim\,b}$ and $\sigma_{H\,lim\,b}$ are the so-called long-life fatigue strengths.

Values for minimum required safety factors are given in Section 17.4.

17.6 Durability of Gears

Sometimes, in practice, there is a need to determine, albeit roughly, the lifetime (fatigue life) of a particular gear at a given spectrum of its variable load and, hence, the corresponding safety factor S_N with respect to gear durability [158]. The examination here will be limited only to the more dangerous type of damage to the gears—tooth breakage. The same simplification is used as in the determination of the stress limit σ_{FG} and the safety factor S_F—it is operated with an equivalent load spectrum reaching maximum to the base number of cycles (Figure 17.14)

$$N_{F\,Eq} = N_{F\,lim\,b} = 3\cdot 10^6.$$

Deducing from Palmgren–Miner's simplest linear hypothesis for damage accumulation D in the material of a given gear at a given spectrum of its load and failure probability P (Figure 17.14), the following formula is used [112, 113, 175]:

$$D = \sum_{j=1}^{n} \frac{N_{Fj}}{N_{FGj}} \le 1, \tag{17.21}$$

where D is the relative damage accumulation (Miner sum) in the material. When $D = 1$, a failure could be expected (theoretically with probability P), N_{Fj} is the number of load cycles for bin j, involved in the calculation (e.g., in Figure 17.14 for the fourth bin will be taken N^*_{FIV}), and N_{FGj} is the number of load cycles to failure for bin j for a certain failure probability P; e.g., ISO 6336 [112] recommends $P = 1\%$.

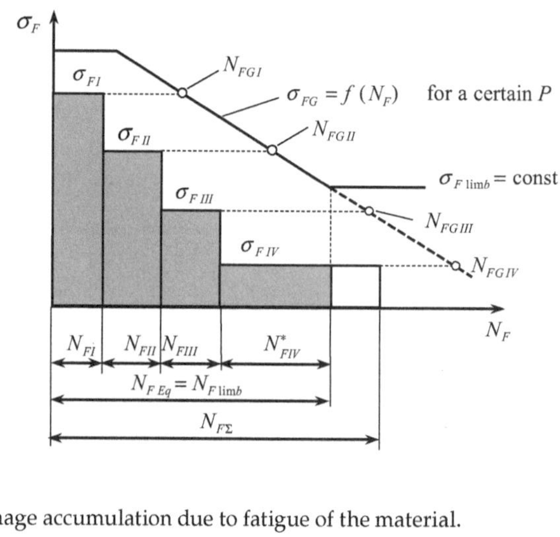

FIGURE 17.14
Determination of the damage accumulation due to fatigue of the material.

If the number of load cycles N_{Fj} for bin j is expressed as a part of the equivalent number of cycles N_{FEq}

$$\varphi_j = \frac{N_{Fj}}{N_{FEq}}, \text{ resp. } N_{Fj} = \varphi_j \cdot N_{FEq}, \tag{17.22}$$

and N_{FEq} equates to the desired number of cycles N_{FL} of a fatigue life of the gear in question $N_{FEq} = N_{FL}$, resp. $N_{Fj} = \varphi_j \cdot N_{FL}$,

then, starting from the Wöhler curve formula for the base number of cycles $N_{F\,lim\,b}$ and endurance limit $\sigma_{F\,lim\,b}$, the number of load cycles to failure N_{FGj} for bin j is determined:

$$N_{FGj} \cdot \sigma_{Fj}^{q_F} = N_{F\,lim\,b} \cdot \sigma_{F\,lim\,b}^{q_F}, \text{resp. } N_{FGj} = N_{F\,lim\,b}\left(\frac{\sigma_{F\,lim\,b}}{\sigma_{Fj}}\right)^{q_F}. \tag{17.23}$$

After substitution for the relative damage accumulation (Miner sum) is obtained:

$$D = \sum_{j=1}^{n} \frac{N_{Fj}}{N_{FGj}} = \sum_{j=1}^{n} \frac{\varphi_j \cdot N_{FL}}{N_{F\,lim\,b}\left(\frac{\sigma_{F\,lim\,b}}{\sigma_{Fj}}\right)^{q_F}} = \frac{N_{FL}}{N_{F\,lim\,b}} \sum_{j=1}^{n} \frac{\varphi_j}{\left(\frac{\sigma_{F\,lim\,b}}{\sigma_{Fj}}\right)^{q_F}}. \tag{17.24}$$

At $D = 1$ (failure could be expected) for the fatigue life (in number of cycles) theoretically is obtained:

$$\boxed{N_{FL} = \frac{N_{F\,lim\,b}}{\sum_{j=1}^{n}\left(\frac{\sigma_{Fj}}{\sigma_{F\,lim\,b}}\right)^{q_F}\varphi_j}}, \tag{17.25}$$

where the slope of Wöhler curve q_F depends on the material and thermal treatment of gears [61, 112, 158] (see Figure 17.10):

$q_F = 8.7$ for carburized and case hardened gears,

$q_F = 6.2$ for through hardened (tempered) gears,

$q_F = 17.0$ for nitrided gears, and

$q_F = 84.0$ for nitrocarburized gears

Similar are data in [120].

The above formula can be converted and N_{FL} expressed by the equivalent stress $\sigma_{F\,Eq}$:

$$\boxed{N_{FL} = \left(\frac{\sigma_{F\,lim\,b}}{\sigma_{F\,Eq}}\right)^{q_F} N_{F\,lim\,b}}. \tag{17.26}$$

If the endurance limit $\sigma_{F\,lim\,b}$ is chosen with 1% failure probability, this probability is also theoretically valid for the fatigue life of the given gear. For some machines, e.g., agricultural, higher failure probability is used ($P > 1\%$).

However, as shown in Figure 17.15, relatively small stress deviations $\Delta\sigma_F$ result in a greater variation ΔN_{FL} in the fatigue life of the given gear. This is due to the fact that, in contrast to the commonly accepted double-logarithmic depicting the Wöhler curve, in fact, in a large number of cycles, there is a very sloping, almost horizontal, line of dependence $\sigma_{FG} = f(N_F)$. By increasing the number of cycles, this deviation of fatigue life increases.

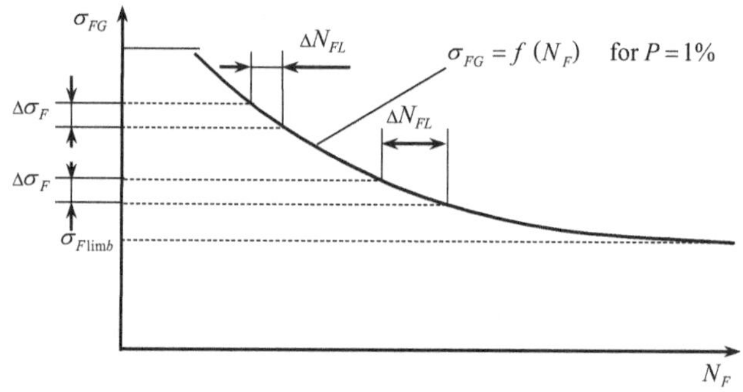

FIGURE 17.15
Fatigue life deviation ΔN_{FL} in different sections of the Wöhler curve for the same deviation in the stress limit $\Delta\sigma_F$.

In general, it should be borne in mind that determining the fatigue life of a gear is quite approximate [158], and this makes the use of the safety factor S_N not very appropriate.

As noted above, tooth fracture is the more dangerous type of damage compared to the occurrence of pitting. Also important is the large difference in the base number of load cycles for both types of damage

$$N_{F\,lim\,b} = 3 \cdot 10^6 < N_{H\,lim\,b} = 50 \cdot 10^6,$$

i.e., the risk of tooth breakage is greater because it occurs faster than the failure due to pitting.

The volume and the objectives of this book do not allow for a more detailed consideration of the issues of durability and reliability of gears and other elements of PGTs.

18

Optimization of $\overline{AI}$-Planetary Gear Train

18.1 Optimization Goal

The optimization in general and in particular of $\overline{AI}$-PGT is an extensive area, the detailed penetration of which cannot be the aim of this relatively short book, intended primarily for industrial engineers and students. This chapter presents only the basic directions in which the optimization of PGTs is performed, and things are given as a prescription, as it is preferred by engineers. For more information, read the specialized literature mentioned in Section 18.5.

Generally, the *purpose of optimization* is to improve the quality of a given product, in this case the $\overline{AI}$-planetary gear train ($\overline{AI}$-PGT), so that it performs its function better and is more competitive, respectively.

Along with the sufficient load capacity (strength), a gear train must be

- Compact, i.e., with a small diameter and width, and with a small volume, respectively
- Inexpensive to manufacture
- With acceptable efficiency and minimal operating costs
- Reliable

This way optimization subjects are

- Diameter D and width B, resp. volume V
- Production costs and price C
- Efficiency η

Optimization can be of

- Arrangement
- Parameters

The *optimization objects* of the PGT in question can be

- Tooth meshings, above all, both external and internal, which are of major importance for the train's load capacity and compactness, as well as for power losses and noise
- Individual gear train elements—gears, shafts, planet pins, bearings, carrier, housing

18.2 Optimization Objectives

There are different performance indices for gear trains in general, the most important of which are the following:

An indicator of the *compactness* of a gear train is the ratio:

$$V/T_B, \tag{18.1}$$

i.e., the constructive volume V that falls on a unit of output torque T_B. The smaller this ratio, the more compact the gear.

Another indicator of compactness is the *specific load carrying capacity*, which is in fact the reciprocal value of the above ratio, given by

$$T_B/V, \tag{18.2}$$

and shows how much output torque T_B is "squeezed" from a unit of volume V. More often, the analogical indicator *power density* P_B/V is used.

Economical spending of materials is characterized by the ratio given by

$$m/T_B, \tag{18.3}$$

where m is the gear train mass.

A very important indicator of a product is undoubtedly its price C. For comparative purposes, it is more convenient to work with relative values, whereby the price is related to the mass of the gear train m given by

$$C/m. \tag{18.4}$$

This indicator is called a *kilogram cost (price)* and shows the price per kilogram of gear train weight. It has been established that the cost of PGTs is proportional to their mass.

As is known, the kilogram cost of PGTs is higher than that of ordinary non-planetary gear trains. Thanks, however, to the higher compactness, resp. a smaller mass, in case of proper design and optimization of the arrangement, their price may be significantly lower than that of comparable non-planetary gear trains. This is, moreover, the goal of optimization [53, 125, 126].

18.3 Meshing Optimization

18.3.1 Optimization Regarding Load Capacity

The external meshing between the sun gear and the planets is most often the weaker one compared to the internal meshing between the planets and the ring gear.

In the case of unhardened gears, the teeth pitting durability has proven to be decisive (see Figure 17.5a). Therefore, optimization should be directed by correcting the tooth geometry to achieve a greater pressure angle α_{w12} of external meshing between sun gear 1 and planet 2. But one needs to be careful not to get an unacceptably low value of the contact ratio $\varepsilon_{\alpha 12}$ which is important for the silent operation of the gear train. For this reason, it should be within the range of 24°÷27° [147–149].

In the case of a hardened sun gear and planets, the bending strength of teeth is decisive (Figure 17.5b). Typically, for these gears, the same carburizing steel and heat treatment with the same bending strength σ_{FG} is used (Figure 18.1). Due to the completely reversed bending load of the planet teeth in nonreversing PGTs (the most often case), their strength should be chosen by 30% lower than that of the sun gear teeth (Figure 18.2).

In reversed PGT, when the load direction changes, the teeth of the sun gear and ring gear are also reversely loaded. Planet teeth in this case are reversely loaded too; i.e., they are *always* reversely loaded. Only in hoisting mechanisms of cranes, hoists, etc. regardless of whether the load is lifted or lowered, the teeth of sun and ring gear are loaded pulsatingly. In this case, their bending strength is not reduced by 30%, as of the planets.

Further, the case of bending load, as the most common and more dangerous case compared to the contact stress, is considered.

Depending on the number of teeth z_1 and z_2 of sun gear 1 and planets 2, as well as the number of planets k, for the same service life (in hours), the number of their load cycles is different (see Sections 9.1 and 17.3), whereby, as a rule, $N_{F1} > N_{F2}$.

Figure 18.2 shows the three possible cases that are obtained for the number of bending load cycles N_{F1} and N_{F2} with respect to base number of cycles $N_{F\,limb}$.

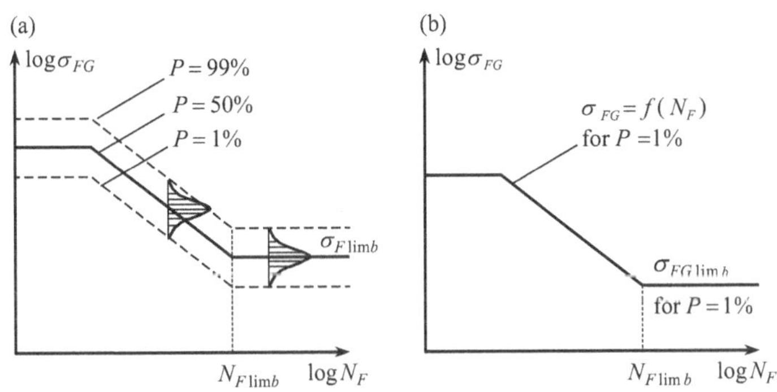

FIGURE 18.1
Wöhler curve for tooth root stress limit (bending strength): (a) Full probability diagram with probability of damage P; (b) resultant diagram with probability of damage P=1%.

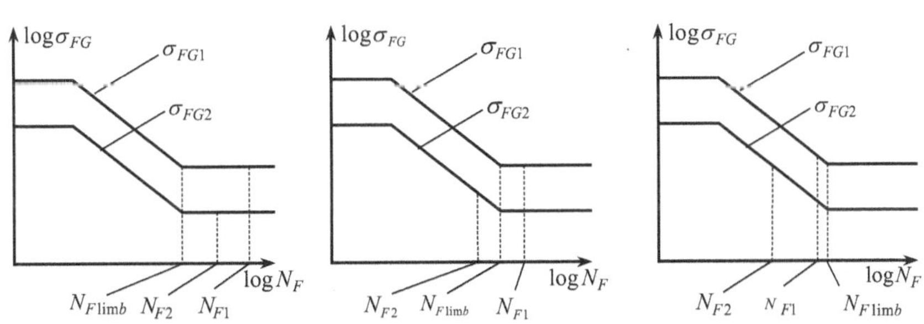

FIGURE 18.2
Possible cases of number of load cycles N_{F1} and N_{F2} of sun gear 1 and planet 2 under pulsating (sun gear 1) and completely reversed (planet 2) bending load.

I case—This is the most common case of long-time operating where the number of load cycles of the two gears is greater than the base number of cycles $N_{F\,lim\,b}$, i.e.,

$$N_{F1} > N_{F\,lim\,b} \text{ and } N_{F2} > N_{F\,lim\,b}.$$

II case—This is an intermediate case where only the number of cycles of the sun gear exceeds the base number of cycles, i.e.,

$$N_{F1} > N_{F\,lim\,b}, \text{ but } N_{F2} < N_{F\,lim\,b}.$$

III case—This is the relatively rare case of short-time operation of the gear train where the calculations are made for limited life (finite life fatigue), because the number of cycles of both gears is less than the base number of cycles, i.e.,

$$N_{F1} < N_{F\,lim\,b} \text{ and } N_{F2} < N_{F\,lim\,b}.$$

If it is required that the two gears have (theoretically!) the same durability $L_1 = L_2$ in hours, although $N_{F1} > N_{F2}$, for the three load cases, the following formulae are obtained, giving the correlation between the required tooth form factors Y_{FS1} and Y_{FS2} of the two wheels (see Chapter 17):

I case

$$\frac{Y_{FS2}}{Y_{FS1}} = 0.7\frac{b_2}{b_1}; \tag{18.5}$$

II case

$$\frac{Y_{FS2}}{Y_{FS1}} = 0.7\frac{b_2}{b_1}\sqrt[q_F]{\frac{N_{F\,lim\,b}}{N_{F2}}}; \tag{18.6}$$

III case

$$\frac{Y_{FS2}}{Y_{FS1}} = 0.7\frac{b_2}{b_1}\sqrt[q_F]{\frac{N_{F1}}{N_{F2}}}; \tag{18.7}$$

where Y_{FS1} and Y_{FS2} are the form factors, taking into account the influence of tooth form as well as the stress concentration in root fillet (Figure 17.7), b_1 and b_2 are the tooth face widths, and q_F is the slope of Wöhler curve (Section 17.4 and Figure 17.9).

These formulae apply when PGTs are nonreversible or reversible but without changing the direction of tooth loading (hoisting mechanisms). In a reverse PGT with a change in load direction, where both the sun and the ring gear teeth are reversed bending loaded, the same formulas are used, but without the multiplier 0.7.

Since the tooth form factors Y_{FS1} and Y_{FS2} depend, in addition to the number of teeth, on both wheels, which are unchanged, even on their shift coefficients x_1 and x_2, observance of the above dependencies can be made by varying the latter until the desired result is obtained. Observance of the above dependencies is not always possible, but the direction in which the solution is to be sought is clear from the formulae. When looking for appropriate shift coefficients x_1 and x_2, it should not be forgotten that the coefficient x_3 of ring gear 3 depends on it, and it must be within certain rational limits [175], i.e., $x_3 = 0 \div +0.65$ (see Sections 5.1 and 5.7).

Gears load capacity depends, of course, not only on shifting but on two other very important factors (see Sections 17.1 and 17.4):

- Uneven load distribution among the planets, considered in calculations by mesh load factor $K_{\gamma\,max}$
- Uneven load distribution over the tooth face width, considered in calculations by face load factor K_β

The issue of equalization of load distribution among the planets is discussed in detail in Chapter 8.

Ensuring a more uniform load distribution over the tooth face, and hence K_β reducing, can be done in several ways.

For gears with a kinematic equalizing device (whether a single or double-articulated gear coupling is used), the sleeve should be as long as possible (see Section 8.6.4 and Figures 8.15 and 8.17).

When the gear train is without a kinematic equalizing device, but relies on the pliability of one of the elements (most often the sun gear shaft), an increase in the inter-bearing distance (bearing straddle) and a possible thinner shaft is used. More special case is when the sun gear is closer to one of the bearings. In order to ensure its parallel displacement at bending of the shaft as a result of the uneven planet load $K_{\gamma\,max} > 1$, the shaft diameters on both sides of the sun gear are different (Figure 18.3) and can be determined by the following formula [32] from the requirement for identical deformations $f_A = f_c$ at both ends of the shaft

$$\frac{d_A}{d_c} = \sqrt[4]{\frac{\left(c+\frac{b}{2}\right)a^3}{\left(a+\frac{b}{2}\right)c^3}} > 1. \tag{18.8}$$

There is another possibility to provide a more uniform load distribution over the tooth face $\left(K_\beta \approx 1\right)$—to use self-aligning planet bearings as follows:

- In case of small load and narrow planet, a single-row radial ball bearing can be used, which allows for certain inclination of the planet.
- In case of heavy load, it is preferable to use only one self-aligning bearing, e.g., a double-row ball (spherical) or double-row roller with barrel rollers.

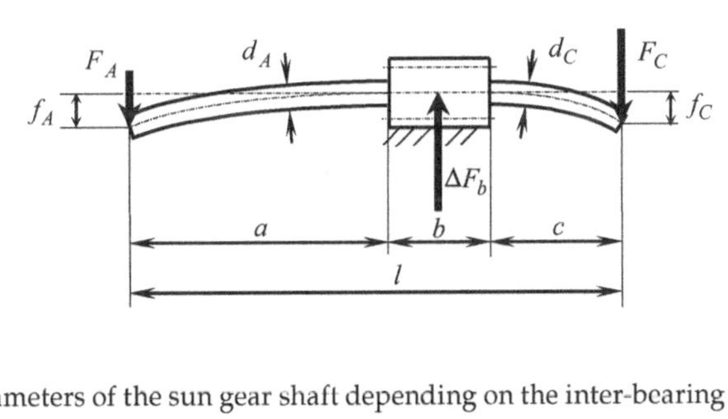

FIGURE 18.3
Determination of the diameters of the sun gear shaft depending on the inter-bearing distance.

The practice of using two bearings in a planet really exists, but it requires great caution. It is necessary for each planet to select bearings with equal radial clearance. With unequal radial clearance, the planets operate in a tilled position, which means a conspicuous unevenness of load distribution over the tooth face, respectively, K_β increasing. Also, one of the two bearings is loaded more and may become out of order. This is particularly true for a single-wall carrier (Figure 9.11a), which deforms more than the two-wall one (Figure 9.11b).

In a small-diameter planet, when there is no place for bearings in it, it is practiced to bear the planet pin in the carrier (Figure 9.6b). The advantage of the case is that the bearings' inner races are loaded rotating rather than stationary, as in the bearings located inside the planet. On the other hand, however, the two bearings do not allow self-alignment of the planet as well as increase the axial dimensions of the carrier.

More details on tooth geometry optimizing to reduce dynamic loads can be found in [56] and in [125, 171] to increase load capacity.

18.3.2 Optimization Regarding Teeth Wearing

Besides the two predominant gear damages—fracture and pitting, there is another damage characteristic of low-speed heavy-loaded gears. This is the wear of the teeth due to the inability to form a load-bearing oil film due to the low speed and high load. The optimization in this case is expressed in the alignment of the specific sliding of sun gear 1 and planet 2 teeth [46], i.e., $\theta_1 \approx \theta_2$,

$$\theta_1 = -\frac{(u_{12}+1)(\tan\alpha_{a2} - \tan\alpha_{w12})}{\tan\alpha_{w12} - u_{12}(\tan\alpha_{a2} - \tan\alpha_{w12})}, \tag{18.9}$$

$$\theta_2 = -\frac{(u_{12}+1)(\tan\alpha_{a1} - \tan\alpha_{w12})}{u_{12}\cdot\tan\alpha_{w12} - (\tan\alpha_{a1} - \tan\alpha_{w12})}, \tag{18.10}$$

where

$u_{12} = z_2/z_1$ is the teeth ratio of mated gear pair—sun gear 1 and planet 2, α_{a1} and α_{a2} are the profile angles of tip circle of the respective gear, and α_{w12} is the operating pressure angle of external meshing.

In internal meshing between the planet and the ring gear, the issue of specific sliding does not cause any particular difficulty.

Equalizing of specific sliding $\theta_1 \approx \theta_2$ (as far as possible) is achieved, varying with the shift coefficients x_1 and x_2 for a given number of teeth z_1 and z_2, unless other circumstances impede this process.

18.3.3 Optimization Regarding Efficiency and Heating

$\overline{\text{AI}}$-PGT has a very high efficiency. The record is η=99.4%. As a result of the high compactness of the gear, a circumstance favorable in itself, its cooling surface may in some cases be insufficient. This leads to complication of the arrangement—adding gills to the housing surface, cooling, or even using a forced oil circulation. Therefore, efficiency increasing, resp. reducing losses in both meshings, and above all in the external, is also subject of optimization. Generally, in the usual low-speed PGTs with rolling bearings, meshing loss is dominant. It is quite different in the case of high-loaded high-speed turbine PGTs with sliding (plane) bearings. The losses in the sliding bearings are dominant, but these gears are not considered here.

The smallest losses in external meshing are obtained when the transverse contact ratios of in-front-of-pitch-point $\varepsilon_{\alpha 1}$ and of beyond-of-pitch-point $\varepsilon_{\alpha 2(1)}$ are equal (see Chapter 11 and Figure 11.2), i.e.,

$$\varepsilon_{\alpha 1} = \varepsilon_{\alpha 2(1)}.$$

Since these ratios influence not only the number of teeth on both gears z_1 and z_2 that are unchanged but also their shift coefficients x_1 and x_2, compliance with the above dependencies can be achieved by varying x_1 and x_2 until a desired result is obtained, of course, if other circumstances do not impede this.

More details on tooth geometry optimizing to reduce meshing loss can be found in [59, 60, 108].

When it comes to optimizing the efficiency and heating of PGT, it should be borne in mind that the oil type, quantity, and level in the housing in case of conventional splash (oil bath) lubrication affect the hydraulic losses, hence the efficiency and heating of the gear train. Adversely, optimization of hydraulic losses can only be done experimentally for each individual case.

18.3.4 Optimization Regarding Noise Reduction

When a gear train is required to be as quiet as possible, this is achieved above all by higher accuracy of the machining. Along with this, silent work is also affected by the transverse contact ratio ε_α, which depends on the number of teeth (the bigger, the better!), as well as on the profile shift of the gears. In principle, the contact ratio of the external meshing is smaller than that of the internal one:

$$\varepsilon_{\alpha 12} < \varepsilon_{\alpha 23}$$

and above all, it is subject to optimization.

Another way to reduce noise is tooth tip corner chamfering [120]. However, it is most efficient for a given load [175].

Also, beneficial effect on the noise level has the height modification of the tooth—increase of the addendum coefficient:

$$h_a^* > 1.$$

For the reduction of noise in high-speed PGTs, it is recommended that the number of teeth z_1 and z_3 of the center gears should not be divisible by the number of planets, but only their sum [236], i.e.,

$$\frac{z_1}{k} \neq \text{an integer}, \frac{z_3}{k} \neq \text{an integer, but } \frac{z_1 + z_3}{k} = \text{an integer.}$$

An important factor in the fight for noise reduction is the determination of the modal frequency of gear train as a whole, taking into account its attachment to the seating (floor or metal structure of the respective equipment). Modern software allows for this [52, 125, 131, 164, 171, 223, 231, 256]. It is preferable if the results are tested experimentally [170].

18.4 PGT Elements Optimization

This concerns the optimization of sun gear and its shaft, planets and their bearing, ring gear and its attachment, as well as the housing.

Since this design optimization of individual elements is discussed in Chapter 15, here it will not be repeated.

18.5 Multi-Objective Optimization

In the multi-objective optimization of PGT, it is not possible to obtain a mathematical expression of the goal function, so it is not possible to determine the private derivatives due to the complex dependencies as well as the strict requirements (standards) that must be complied with.

In order not to go beyond the engineering direction of this book, the various methods for multi-objective optimization are not considered. More details on this extensive question can be found in the specialized literature [52, 125, 131, 164, 171, 220, 231].

Multi-objective optimization is especially suited to serial products where significant cost savings can be achieved. This, of course, is also important in single pieces, but not to such a large extent.

For single gear train optimization, efforts must be directed primarily into two directions:

- Optimization of the meshings, which are of major importance for the compactness of the gear train and hence for its price
- Optimization of the efficiency in order to avoid difficulties with cooling

In optimization, there is at least one very important situation that should not be forgotten. As shown in Figure 18.4, from an engineering (rather than mathematical!) point of view, the optimization of a parameter only makes sense when the function's optimum is clearly expressed (Figure 18.4a) and not when a significant change of x negligibly improves Δy of function y (Figure 18.4b).

Additional information on optimization of gear trains, including planetary, can be found in [53, 72, 132, 138, 202, 205, 218, 219, 227, 230, 244].

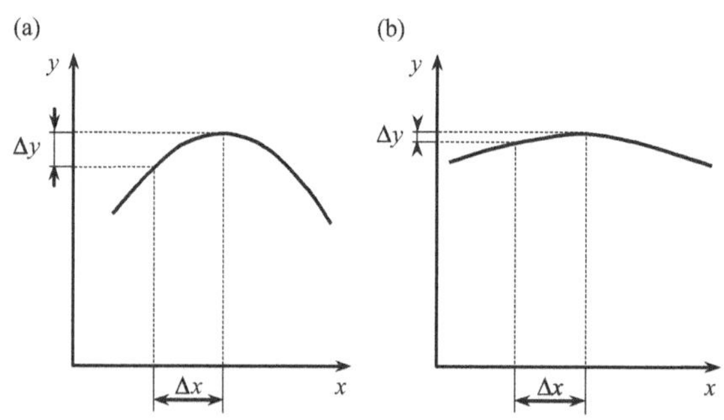

FIGURE 18.4
Examples of a pronounced (a) and poorly expressed (b) function optimum.

19

Advantages, Disadvantages, and Application of $\overline{AI}$-Planetary Gear Train

The studied $\overline{AI}$-planetary gear train ($\overline{AI}$-PGT) has a number of very important advantages and, of course, also some disadvantages.

19.1 Advantages

1. High compactness, expressed in
   - Smaller overall dimensions and a smaller volume, respectively.
     - Easier toothing, hence higher accuracy.
     - Lower peripheral velocities, hence lower noise level.
   - A smaller mass, two or three (up to five) times smaller than that of non-planetary gear trains.
   - A smaller and cheaper foundation for stationary PGTs, which is particularly important from an economic point of view in industrial buildings with many gear trains leading to a smaller built-up area due to the favorable symmetrical shape of the gear train and the shafts coaxiality.
   - The same applies to hoisting machines and to vehicles (gearboxes), which leads to lightening and lowering the price of the construction.

   This compactness is due to the following reasons:
   - Using the principle of torque splitting (between several planets).
   - Placing all elements inside the ring gear.
   - The higher load capacity of the internal meshing as well as the other gears, which due to their small size allows for more precise machining and heat treatment, leading to a further increase in their load capacity.
2. Very high efficiency—the highest of all transmissions, not only the gear trains. In the case of a turbine PGT with chevron gears, efficiency of 99.4% has been achieved! The explanation is that the losses in PGTs are at all determined by the relative power P_{rel} in the gear train, and in the case of an $\overline{AI}$-PGT, it is always less than the input (and transmitted) power P_A, i.e., always $P_{rel} < P_A$ (see Chapter 10 and Figure 10.1).
3. Relatively good kinematic capabilities—speed ratio = $3 \div 8$ (12), satisfying the most common needs. Also, the ability to sum up and split the movement, as well as a step-by-step and stepless change of speed ratio in combination with another gear train.

4. $\overline{\text{AI}}$-PGT is the simplest and most common type of power PGT. The reason is its simpler manufacturing than other PGTs. This is especially true for planets that have one, not two rims.
5. Relative noiselessness of the gear train, due to several reasons: lower peripheral velocity, smaller dynamic forces due to smaller dimensions, and more accurate gear manufacture.
6. The coaxiality of the input and output shafts is in some cases too desirable. It makes the $\overline{\text{AI}}$-PGT suitable for motor-reducer (geared motor) drives, for installation in drums (in hoists, belt conveyors, etc.), in wheels as well as in automatic transmissions of vehicles and steering systems of continuous track vehicles as tractors, tanks, excavators, etc.

19.2 Disadvantages

1. Geometric gear calculations are more complex, especially profile shift, since the parameters of the external and internal meshing cannot be independently selected.
2. The uneven load distribution between planets is a complex and specific problem whose solution is not easy.
3. Teeth of the planets are bending loaded less favorably (completely reversed loading) than those of the other gears.
4. The arrangement is more complicated, with lots of details and with increased risk in the event of an accident, resulting in a total breakdown of the gear train when a single tooth is broken, which unlike the non-planetary gears cannot be discarded into a safe place.
5. Requirements for the PGT manufacturing are very high, and they are not attainable for every plant. High production technology is required.
6. Kilogram price ($/kg) of PGTs is higher than that of non-planetary ones. Only the maximum lightening can be achieved at a lower price than a comparable non-planetary gear train.
7. The compactness of PGTs, which in principle is their important advantage, means, on the other hand, a small cooling surface and, in some cases (poor quality and low efficiency, or even not so low efficiency), may lead to gear train heating.

19.3 Application

Planetary gears have *many different capabilities* to solve different tasks.

1. Realizing a permanent speed ratio $i = \text{const}$ at work like
   - A reducer—$|i| > 1$
   - A multiplier—$|i| < 1$

2. Realizing a step change in speed ratio for single- and twin-motor drive
   - Three speed ratios (gears) at single-motor drive (Figure 19.1)
   - Two, three, and four speed ratios (gears) at twin-motor drive (Figures 19.2 and 19.3)
3. Realizing a non-step (smooth) change in speed ratio (i = var) (Figure 19.4)
4. Realizing
   - Movement (power) summing (Figure 19.4)
   - Movement (power) division (Figures 19.5 and 19.6)

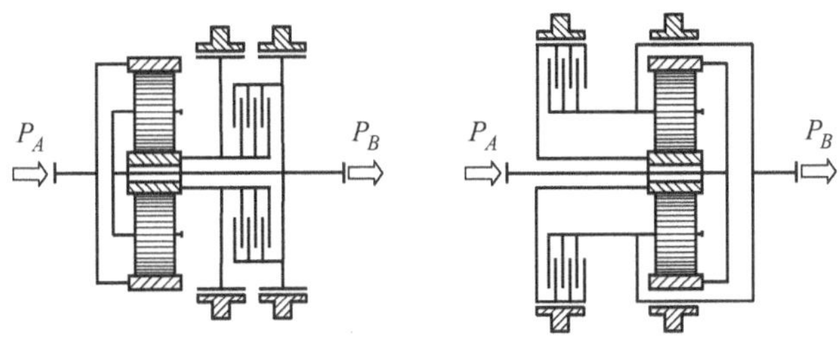

FIGURE 19.1
Tank steering mechanisms with three speed ratios (three gears): (1) At fixed sun gear $i > 1$; (2) at blocked gear train $i = 1$; (3) at fixed output shaft $i = \infty$.

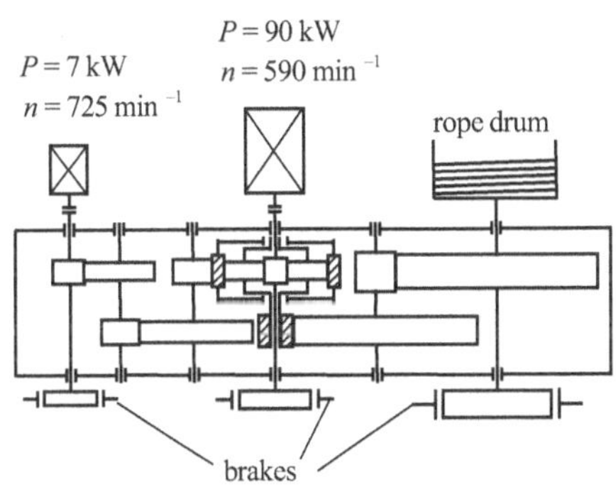

FIGURE 19.2
Kinematic scheme of the reducer of a two-velocity-hoisting mechanism (container crane in nuclear power plant "Kozloduy", Bulgaria, designed by "Bulmachinery Enterprises", Radomir, Bulgaria, realizing two speed ratios and two hoisting velocities [32]): $i = 141$ at main velocity (only the main motor works); $i = 2228$ at micro velocity (only the small motor works).

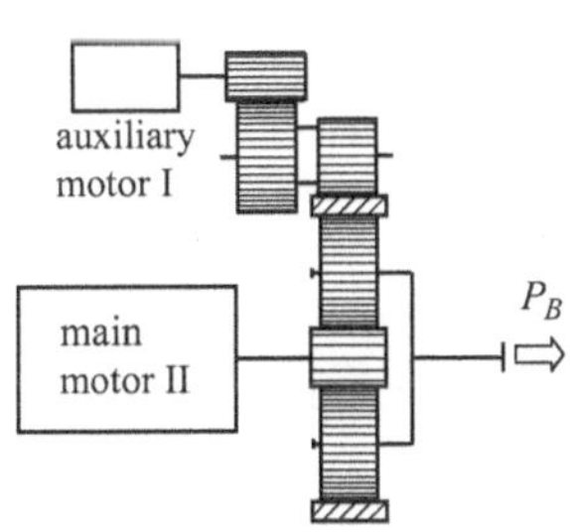

FIGURE 19.3
Ropeway reducer, realizing three speeds.

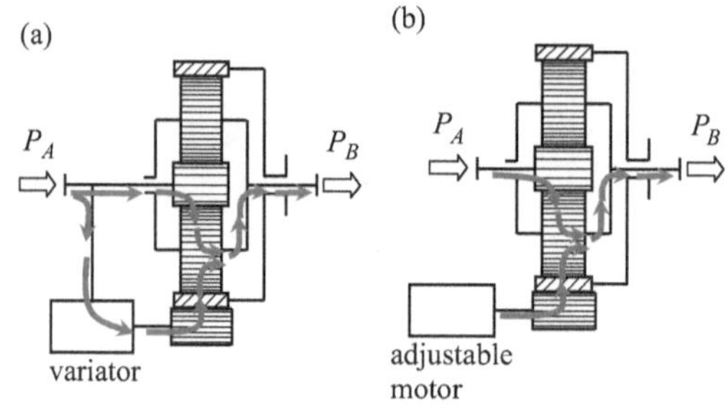

FIGURE 19.4
Non-step (smooth) change in speed ratio, resp. movement (power) summing: (a) Mechanically or hydraulically; (b) electrically.

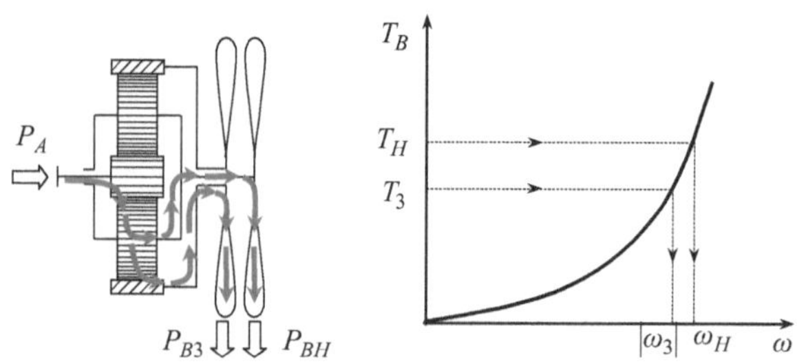

FIGURE 19.5
Movement and power division in twin-propeller airplane engine (Figure 7.1).

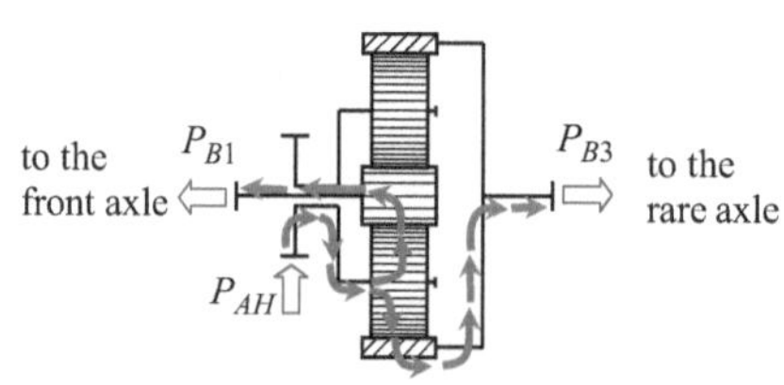

FIGURE 19.6
Movement and power division in inter-axle differential of vehicle.

5. Sharing a very large output torque between two meshings on one gear of a working machine (Figure 19.7)
6. Realizing very high angular velocities (Figure 19.8)

Application examples

Below are some examples of application of the PGT in question:

1. Realizing a permanent speed ratio i = const
   a. As a reducer (the most common case)
      - The most often $\omega_3 = 0$
      - Less often $\omega_1 = 0$—in the case of propeller airplanes or continuous track vehicles steering systems

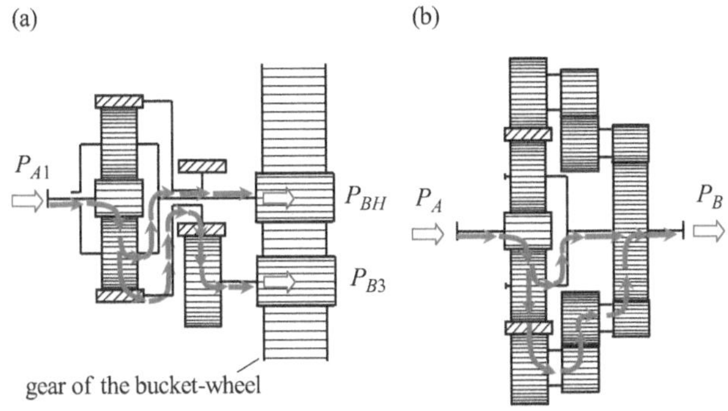

FIGURE 19.7
Torque splitting in drive of (a) Bucket-wheel of an excavator and (b) rope drum of a hoisting mechanism.

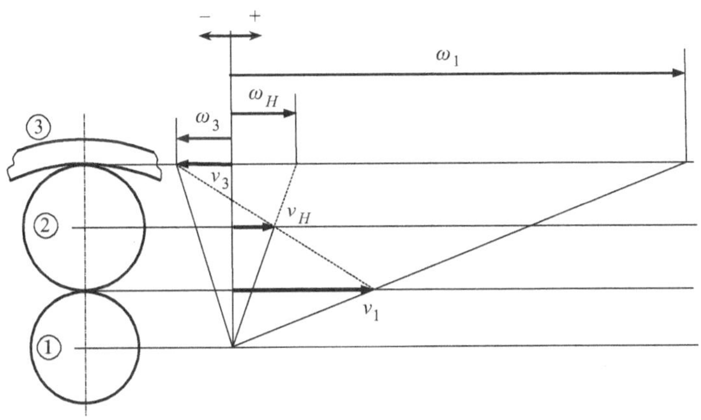

FIGURE 19.8
Realizing a very high angular velocity ω_1 of sun gear 1 in a centrifuge drive (ring gear 3 and carrier H are input).

  - The most rare $\omega_H = 0$—at very high angular velocities ω_A in turbine PGTs and the very high centrifugal force on the planets
 b. As a multiplier (less often)
  - The most often $\omega_3 = 0$—in centrifuges, water turbines
  - Less often $\omega_1 = 0$—as high gear in vehicles' change-gears
  - The most rare $\omega_H = 0$—in high-speed turbines and centrifuges
2. Step change in speed ratio for single- (Figure 19.1) and twin-motor (Figures 19.2 and 19.3) drive
 a. Two-speed (two gears) PGT (Figure 19.2).
 b. Three-speed (gears) PGT. Figure 19.3 shows a ropeway drive reducer.
  I gear—the auxiliary electric motor works at 30% of the total power of the drive ($P_I = 0.3P_\Sigma$).
  II gear—the main electric motor works ($P_{II} = 0.7P_\Sigma$).
  III gear—both motors work.

c. Four-speed drive—applied in wagon-tippers, metallurgy, etc. Scheme is the same (Figure 19.3), the four gears (speed ratios) being as follows:

I gear—the auxiliary electric motor works.

II gear—both motors work in different directions.

III gear—the main motor works.

IV gear—both motors work unidirectionally.

3. Non-step (smooth) change in speed ratio (Figure 19.4). This can be done in different ways:
   a. Mechanically—via a variator
   b. Hydraulically—via a hydromotor
   c. Electrically—via an adjustable electric motor
4. Movement (power) summing

   The above examples of two-speed (Figure 19.2), three-speed, and four-speed (Figure 19.3) twin-motor drives are a practical example of movement and power summing when the two motors work. Similar is the situation in the non-step change of speed ratio (Figure 19.4).
5. Movement (power) division

   Movement and power division occurs in twin-propeller airplane engines (Figure 19.5), in automatic transmissions with hydrotransformer (hydromechanical transmissions) and in inter-axle differentials of vehicles (Figure 19.6). Each of the three central elements (sun gear, ring gear, or carrier) can be driving (input).
6. Transmitting very large output torques

   In fact, there is a division of motion and power here as in the previous examples. The economic benefit here is rather great, as the large and expensive gear can be made smaller, because it is loaded in half. Figure 19.7a shows the drive of the bucket wheel of an excavator, whereby the large gear (of the bucket-wheel) is driven by two small gears (splitting the power flow). This results in a smaller tooth module and a smaller diameter of the large, complicated, and costly toothed wheel.
7. Realizing very high angular velocities

   Such angular velocities are required in ultracentrifuges, and these are achieved by summing up two movements—of the carrier and the ring gear, wherein the sun gear is the output. The velocity diagram (Figure 19.8) shows how the high angular velocity of the sun gear ω_1 is obtained at low angular velocities of the ring gear ω_3 and carrier ω_H.

20

Examples for Application of $\overline{AI}$-Planetary Gear Train

In this section, examples are given of the various arrangements of $\overline{AI}$-planetary gear trains ($\overline{AI}$-PGT) which have application in different technical areas. The examples are accompanied by comments aiming at facilitating the designer and suggesting successful and time-tested technical solutions.

20.1 PGTs with General Application

Figure 20.1 shows an arrangement which mostly relies on the very precise execution in order to achieve approximate uniform (even) load distribution of gear wheels (mesh load factor $K_{\gamma\, max}$!). Central elements (carrier and sun gear) bearings along with their clearance can help to some extent the equalizing of load distribution between planets, since they create radial mobility necessary for the equalizing movements (Figure 8.3). With sun gear, the situation is a bit different. Its equalizing movements are connected with a slight inclination

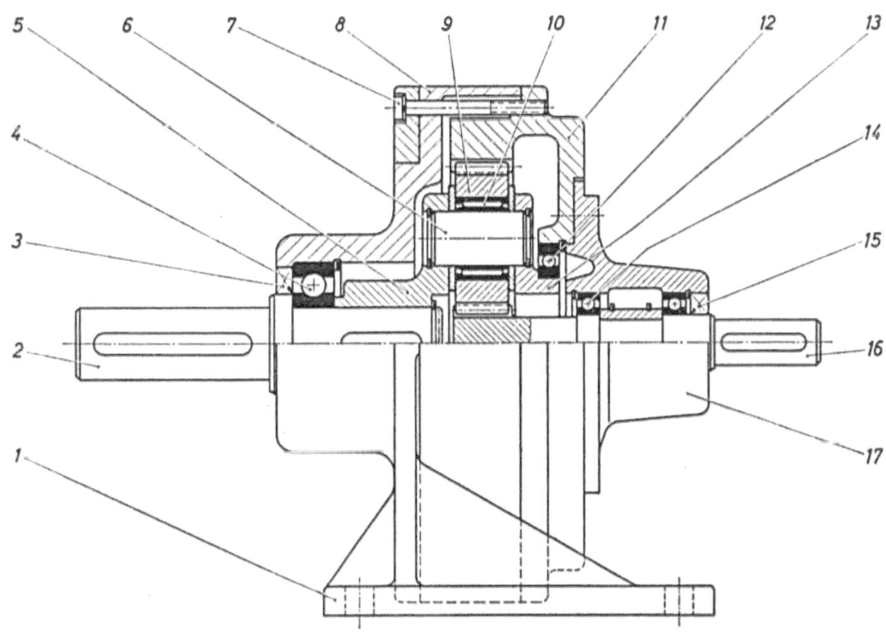

FIGURE 20.1
$\overline{AI}$-PGT with detachable split carrier and a sun gear cantilevered in the top and in the housing, respectively: 1—seat; 2—output shaft; 3—seal; 4—bearing; 5—carrier with hub; 6—planet pin; 7—screw; 8—housing; 9—planet; 10—bearing (needle); 11—ring gear; 12—bearing; 13—carrier; 14—bearing; 15—seal; 16—input shaft; 17—top.

due to the cantilever bearing in the top which leads to uneven load distribution on the tooth face width (face load factor K_β!).

In Figure 20.2, a gear train with a sun gear seated in the carrier is shown. In this arrangement, the equalizing of load distribution between planets is also achieved with purposive accuracy and possibly at the expense of the clearances of tapered roller bearings which can be adjusted. The upper half of the image is for speed ratio $i = 4$, and the lower one—for $i = 13$.

In Figure 20.3, a PGT is shown having a monolithic carrier and a kinematic equalizing (distribution) device—a floating ring gear 3. Due to the very short gear coupling, i.e., the ratio L/d is very small (see Section 8.6.5), sensitivity ξ of this equalizing device is very low. One cannot expect that at the operating gear loading, the equalizing device will work effectively (Figure 8.17). Ring gear 3 will make a single movement only in the beginning, when applying the increasing load, and then, due to the low sensitivity ξ of this equalizing device, will remain unchanged. In order to get uniform load distribution over the tooth face, the two planet bearings must be selected to have the same radial clearance. The same applies for the arrangement shown on Figure 20.4. Here, the sun gear is seated in a monolithic carrier, to which an oil spreading disc is attached.

In Figure 20.5, a gear train with a floating sun gear 1 (the element with the lowest mass) is shown, where the sensitivity of the equalizing device ξ is slightly higher than that of

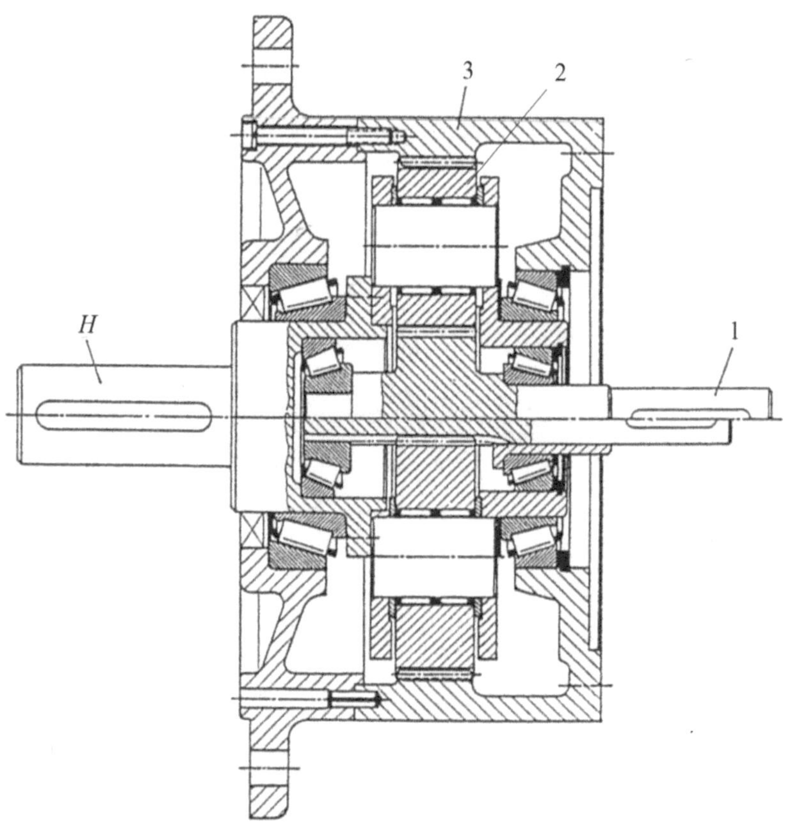

FIGURE 20.2
AI-PGT with a detachable split carrier and a sun gear seated in the carrier.

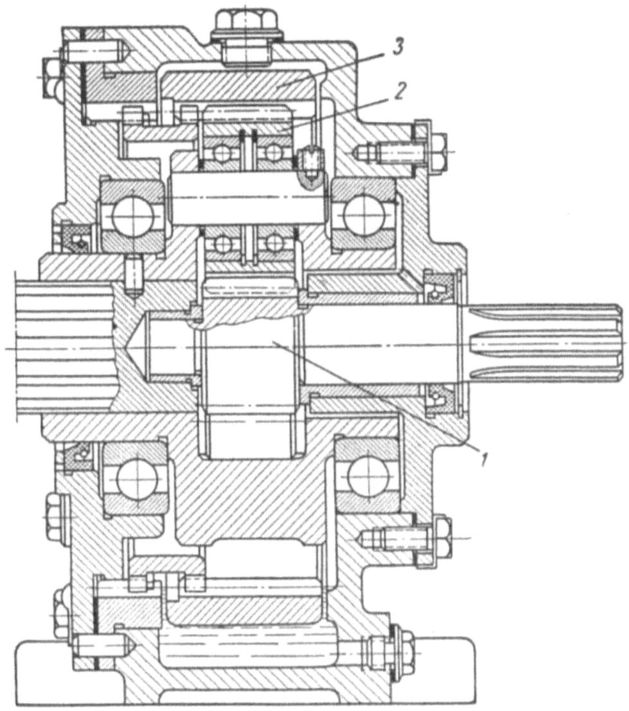

FIGURE 20.3
$\overline{AI}$-PGT with a monolithic carrier, a floating ring gear, and a sun gear seated both in the carrier and in the housing.

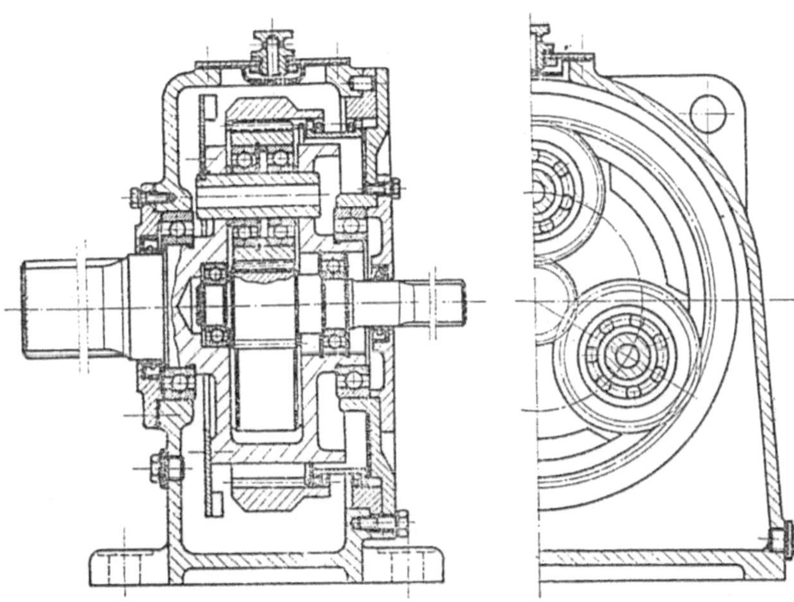

FIGURE 20.4
$\overline{AI}$-PGT with a monolithic carrier, a floating ring gear, and a sun gear seated in the carrier.

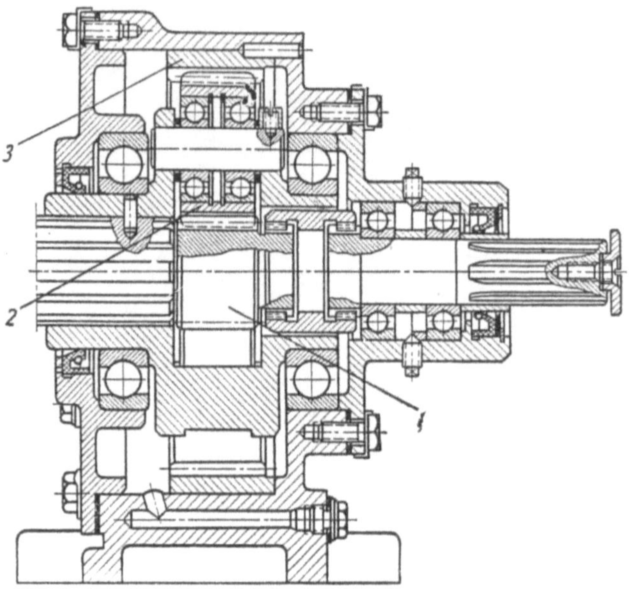

FIGURE 20.5
AI-PGT with a monolithic carrier and a floating sun gear.

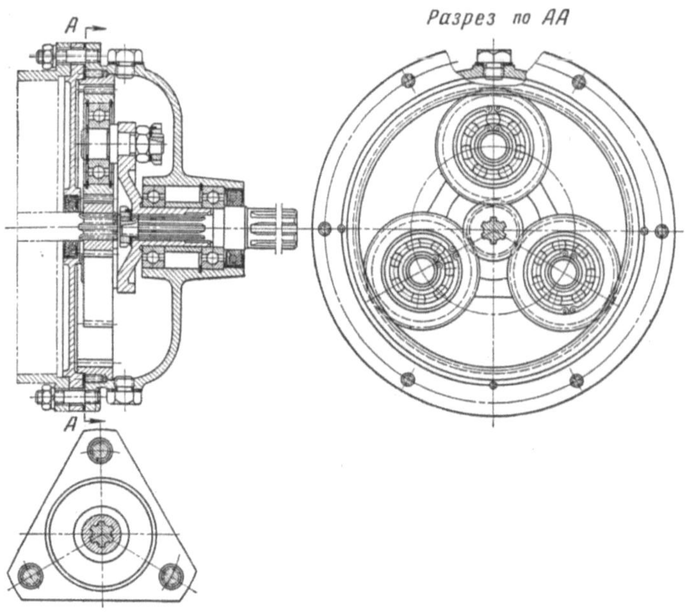

FIGURE 20.6
AI-PGT with a single-wall carrier and a cantilever-seated sun gear.

the previous arrangements due to the bigger ratio L/d. Here, again the two planet bearings must be selected with the same radial clearance.

Figure 20.6 shows a not particularly recommended arrangement due to the many additional details (nuts, washers) and the need to cut threads.

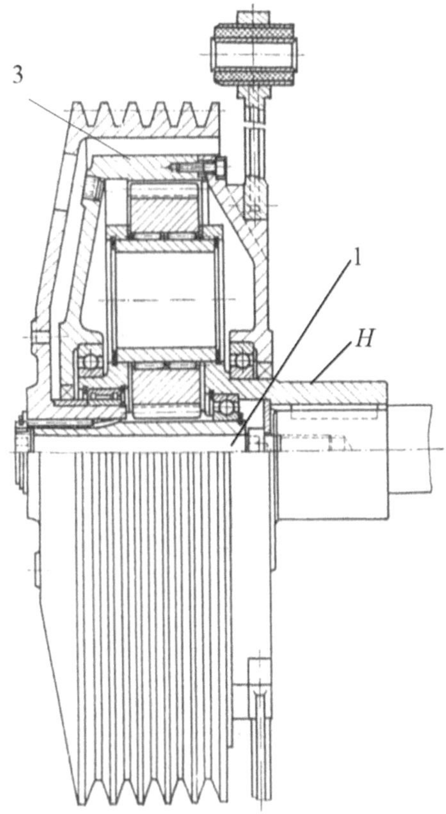

FIGURE 20.7
$\overline{AI}$-PGT embedded into a belt pulley.

What is advantageous here is the smaller unevenness of the load over the tooth faces due to the narrow planets and ability of the single-row ball bearing of the planets to self-align within small limits.

This is one of the cases when ensuring the necessary oil quantity requires deeper planets immersion in oil bath (see Chapter 13).

On the arrangement shown in Figure 20.7, the PGT is embedded into a belt pulley. In conformity with the common practice, the fixed element is ring gear 3, the driving element is sun gear 1 (connected with the belt pulley), and the output element is carrier H.

In Figure 20.8, a compound gear train is shown consisting of a worm and a planet stage. This gear train performs very large speed ratios. The PGT has a pliable ring gear. The sun gear is seated in the housing and carrier, and the carrier—in the housing and on the sun gear shaft.

20.2 PGTs with a Specific Function

There is a great variety of PGTs that are specially designed for a specific application in different areas of industry. Figure 20.9 shows a wire rope drum of a hoisting mechanism

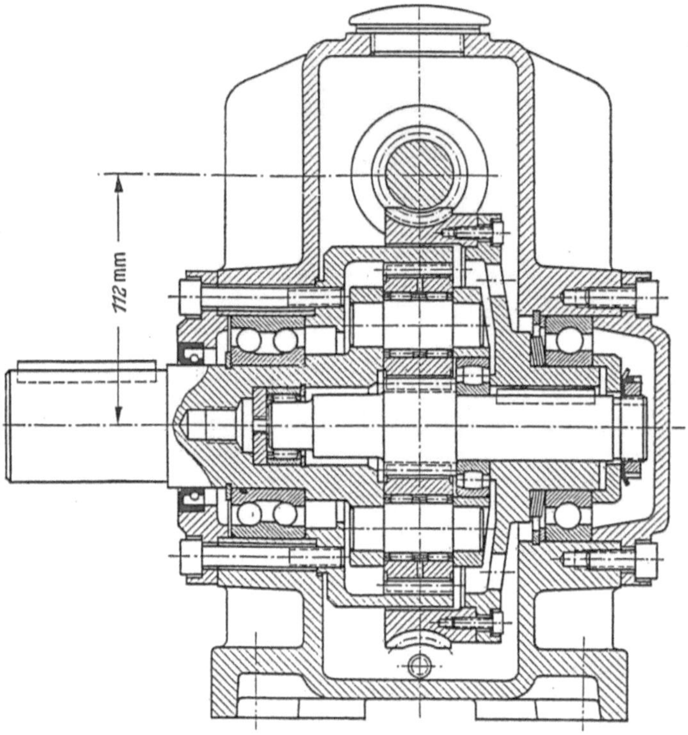

FIGURE 20.8
A compound two-stage gear train, consisting of $\overline{\text{AI}}$-PGT with a pliable ring gear and a worm gear train.

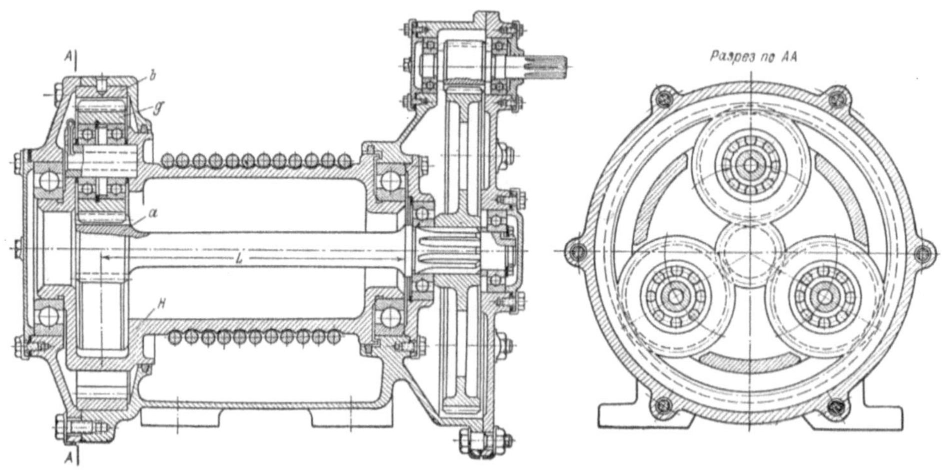

FIGURE 20.9
$\overline{\text{AI}}$-PGT embedded in a hoisting mechanism (winch).

(winch) with embedded PGT. The drum outside functions as a carrier. Equalizing of planet load is done by the long cantilever shaft of sun gear which bends to a certain extent (purposive compliance). Generally, in this case full load equalization is impossible (see Section 8.6).

In Figure 20.10, a compound gear train for a cement mill is shown. The first stage is bevel, and the second one—planetary with a vertical shaft. The vertical central (main) axis of the PGT allows for the use of a floating sun gear (pos. 6) and a carrier (pos. 5). Both elements are connected with double articulated gear couplings (pos. 3 and 7), which contributes to the higher sensitivity ξ of equalizing devices. The sun gear coupling (pos. 7), however, has small ratio L/d, so its sensitivity ξ cannot be high. Sensitivity is better with the carrier (pos. 3).

This example shows that sometimes two of gear train elements are made floating.

In Figure 20.11, again the kinematic scheme is shown of the hoisting mechanism reducer of a 200-tonne container crane for Kozloduy nuclear power plant (NPP), manufactured by "Bulmachinery Enterprises"—Radomir (Figure 19.2). The reducer can perform two speed ratios, $i = 141$ and $i = 2{,}228$, for fast and slow speed (main and microspeed). The two speed ratios are performed by the work of only one of the motors. At fast speed (Figure 20.11a), the big motor works driving the sun gear of the planetary stage at fixed ring gear. At slow speed (Figure 20.11b), the small motor works and the ring gear is driving, whereas the sun gear is fixed. In both cases, the output element is the carrier. Figure 20.12 shows a planetary stage arrangement, and Figure 20.13—a reducer with removed top. From Figure 20.12, it can be seen that only one self-aligning (spherical) double-row roller bearing of planets is used.

Figure 20.14 shows the kinematic scheme of the reducer of the main hoisting mechanism of 190-tonne overhead (bridge) crane with embedded PGT. Twin-motor drive is desirable with cranes with big load capacity (and in casting cranes, and generally, in those working with dangerous loads—it must be obligatory). The twin-motor drive is so made that if one of the motors gets out of order, the other one must be able to complete the required

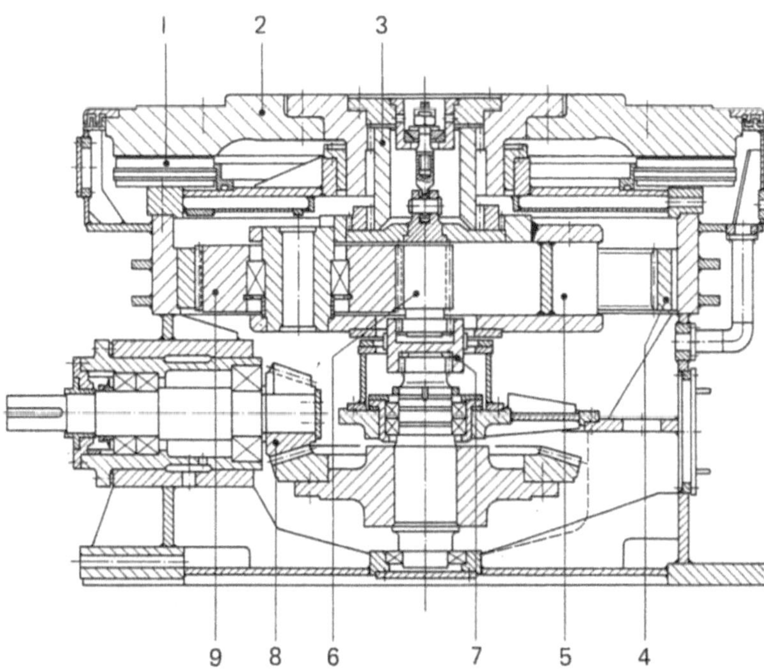

FIGURE 20.10
$\overline{AI}$-PGT embedded into a mechanism for cement mill rotation.

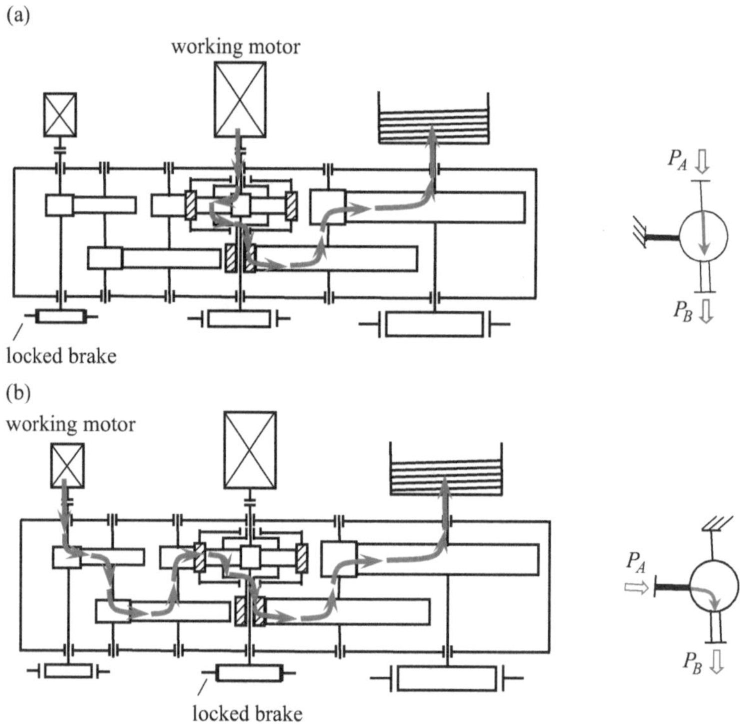

FIGURE 20.11
Operation of $\overline{\text{AI}}$-PGT as a two-speed gear train stage of a hoisting mechanism for Kozloduy NPP [32]: (a) At fast speed, speed ratio $i = 141$; (b) at slow speed, speed ratio $i = 2228$.

operation alone. Usually, when no differential is used, the two motors drive the input shaft, and hence motors that have equal power and angular velocity are selected, each with power equal to 65%–85% of the total necessary power, so that they can work independently in emergency mode, though overloaded, i.e.,

$$P_{AI} = P_{AII} = (0.65 \div 0.85) P_B \text{ and } \omega_{AI} = \omega_{AII}.$$

With this approach, in normal operation, the two motors work at less than nominal power, which leads to lower electrical efficiency.

In the arrangement shown in Figure 20.14, *said disadvantage is avoided* thanks to the use of $\overline{\text{AI}}$-PGT. The two motors are selected with the same power $P_{AI} = P_{AII}$ (equal to 50% of the required for the hoisting mechanism) and angular velocity $\omega_{AI} = \omega_{AII}$, and in normal operation (Figure 20.14), they work at their nominal power:

$$P_B = P_{AI} + P_{AII}. \tag{20.1}$$

When the two motors work, the angular velocity ω_H of carrier H of PGT is determined according to the known formula (7.9) for summing up movements:

$$\omega_H = \frac{\omega_1 - i_0 \cdot \omega_3}{1 - i_0} = \frac{\omega_1 + t \cdot \omega_3}{1 + t}, \tag{20.2}$$

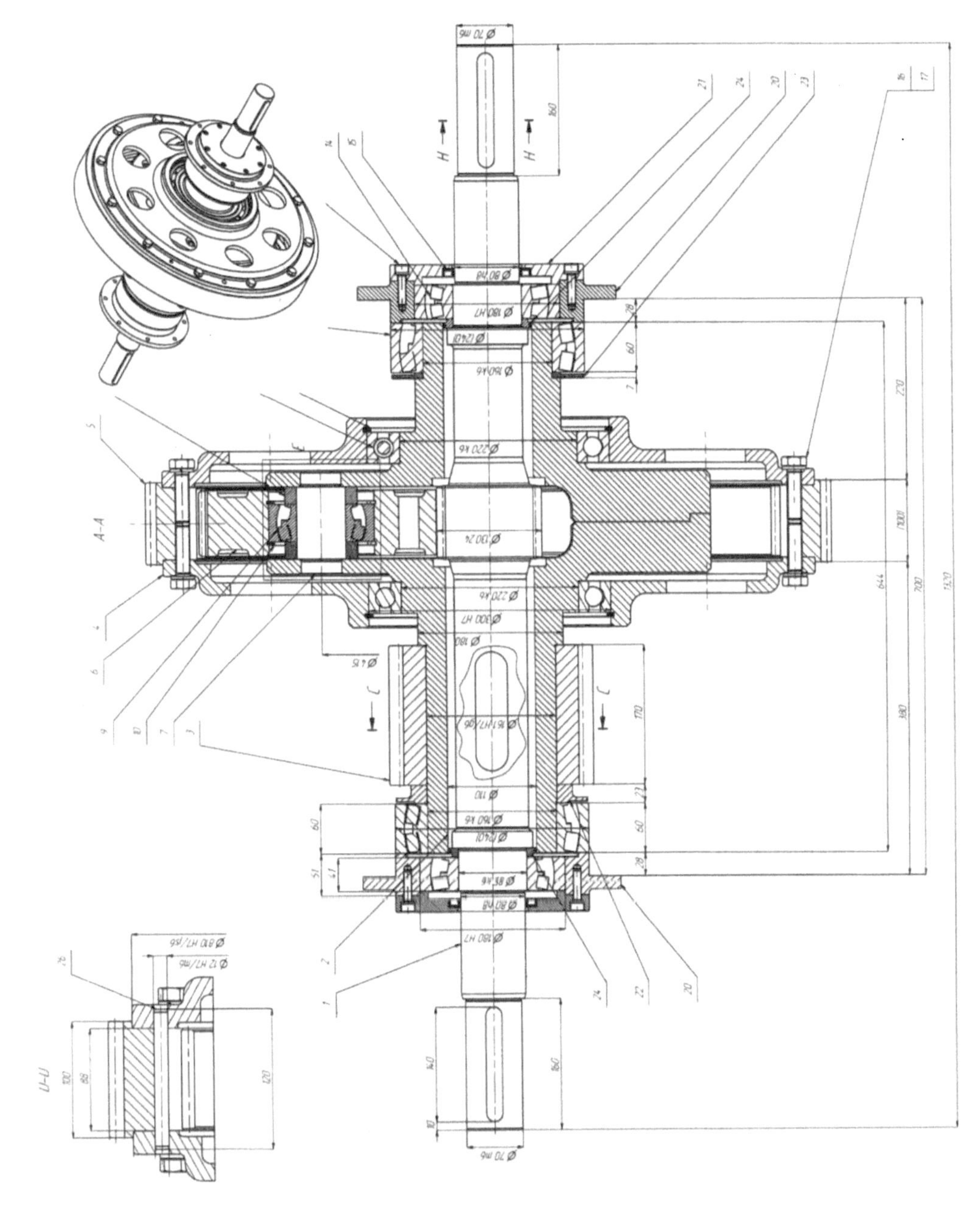

FIGURE 20.12
Assembly drawing of $\overline{AI}$-PGT from Figure 20.11. (Courtesy of "Bulmachinery Enterprises", Radomir, Bulgaria.)

where

$$\omega_1 = \omega_{AI}$$

$$\omega_3 = \frac{\omega_{AII}}{i_{45}} = \frac{\omega_{AII}}{-\frac{z_5}{z_4}}.$$

FIGURE 20.13
The reducer of the hoisting mechanism of 200-tonne crane of Figure 20.15. (Courtesy of "Bulmachinery Enterprises", Radomir, Bulgaria.)

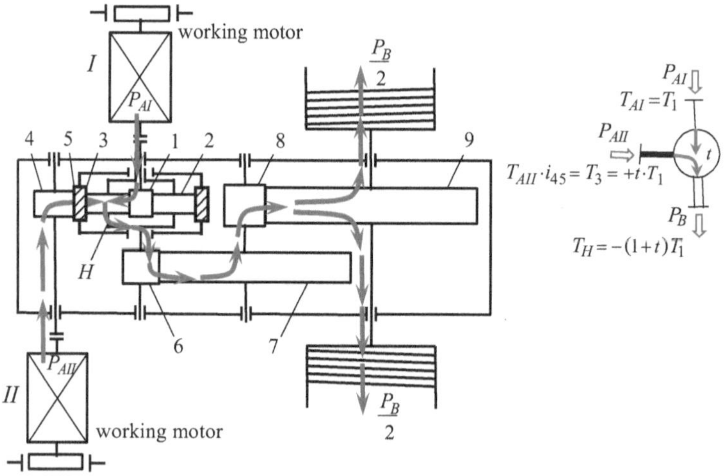

FIGURE 20.14
Reducer of hoisting mechanism of overhead (bridge) crane with twin-motor drive with embedded $\overline{\text{AI}}$-PGT operating as a summation differential.

Obviously, the directions of rotation of both sun gear and ring gear must be the same, which means that the motor shafts must rotate in opposite directions, i.e., $\omega_{AI} = -\omega_{AII}$.

The angular velocity ω_B of output shaft B of the reducer, and the wire rope drum, respectively, is obtained from the following formula, taking into account the speed ratios of the last two stages:

$$\omega_B = \frac{\omega_H}{i_{67}\cdot i_{89}} = \frac{\omega_H}{\dfrac{z_7}{z_6}\cdot\dfrac{z_9}{z_8}}. \tag{20.3}$$

In case of a failure of one of the motors, the other one continues to work at its nominal power with no overloading (Figures 20.15 and 20.16), at unchanged output torque T_B of the reducer, i.e., of the drum shaft.

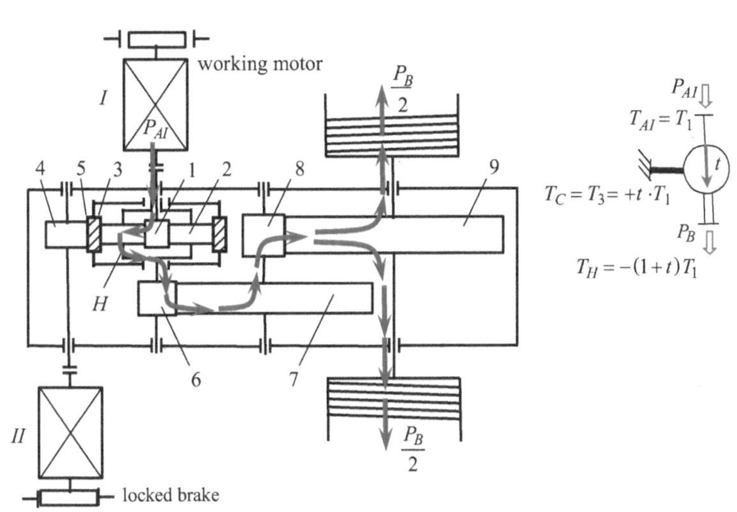

FIGURE 20.15
The reducer of Figure 20.14 with motor II off, where the PGT works with fixed ring gear.

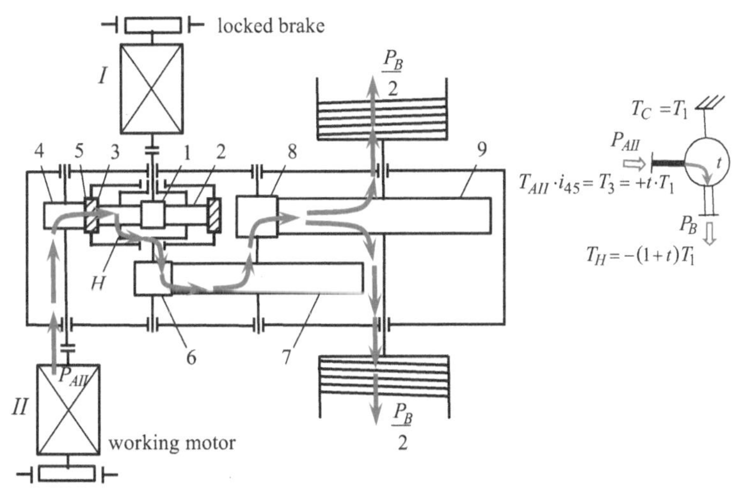

FIGURE 20.16
The reducer of Figure 20.14 with motor I off, where the PGT works with fixed sun gear.

$$P_B = P_{AI} \text{ or } P_B = P_{AII}. \tag{20.4}$$

In Figure 20.15, the case where only motor I works is shown, which motor drives sun gear 1 of the PGT, i.e., $\omega_1 \equiv \omega_{AI} > 0$, $\omega_3 = 0$. The angular velocity ω_H of carrier H is

$$\omega_H = \frac{\omega_{AI}}{i_{1H(3)}} = \frac{\omega_1}{i_{1H(3)}} = \frac{\omega_1}{1 + \frac{z_3}{z_1}}, \tag{20.5}$$

i.e., it is lower than velocity when two motors work, and ω_B of output shaft B is obtained as above, but at different ω_H:

$$\omega_B = \frac{\omega_H}{i_{67} \cdot i_{89}} = \frac{\omega_H}{\frac{z_7}{z_6} \cdot \frac{z_9}{z_8}}. \tag{20.6}$$

In Figure 20.16, the case when only motor II works is shown, which motor drives gear wheel 4 and hence ring gear 3 of the PGT, i.e., $\omega_4 \equiv \omega_{AII}$ at $\omega_1 = 0$. The angular velocity ω_H is obtained as follows:

$$\omega_H = \frac{\omega_{AII}}{i_{45} \cdot i_{3H(1)}} = \frac{\omega_4}{-\frac{z_5}{z_4}\left(1+\frac{z_1}{z_3}\right)}, \tag{20.7}$$

and ω_B is determined from the above formula, but at different ω_H.

In order that the angular velocity ω_B of the output shaft (of the drum and carrier H, respectively) should be the same in both emergency cases, the following condition of equality of speed ratios of the two lines of power flow from the motor to the drum must be met:

$$\omega_B = \frac{\omega_{AI}}{\left(1+\frac{z_3}{z_1}\right)} = \frac{\omega_{AII}}{-\frac{z_5}{z_4}\left(1+\frac{z_1}{z_3}\right)}. \tag{20.8}$$

Since the angular velocities of the two motors are the same, but with different directions $(\omega_{AI} = -\omega_{AII})$, the following relation between the individual speed ratios is obtained, which must be observed:

$$i_{1H(3)} = -i_{45} \cdot i_{3H(1)}, \tag{20.9}$$

i.e., the relation between the number of teeth is

$$\left(1+\frac{z_3}{z_1}\right) = \frac{z_5}{z_4}\left(1+\frac{z_1}{z_3}\right), \tag{20.10}$$

or ultimately, the condition is

$$\frac{z_3}{z_1} = \frac{z_5}{z_4}. \tag{20.11}$$

This condition allows for the selection of the corresponding number of teeth of gear wheels, and the speed ratios, respectively.

Since power and angular velocities of both motors are the same $(P_{AI} = P_{AII} = 0.5P_B$ and $\omega_{AI} = \omega_{AI})$, when only one of the motors works, the angular velocity ω_B of the output shaft of the reducer (the drum, resp.) will be two times lower than the one of both operating motors.

Figure 20.17 shows the kinematic scheme of driving of the two drums of a two-line hoisting mechanism of two-rope clamshell bucket of a bulk-handling crane. Operation of this clamshell (Figure 20.18) requires that the holding rope 1 (caught to the upper sheaved block of the clamshell) and the closing rope 2 (caught to the lower sheaved block) should move both at the same speed and independently of each other. Moreover, it is desirable that at hoisting of the loaded clamshell, the load should be distributed between the two ropes.

At hoisting and lowering of the clamshell, loaded or unloaded (Figure 20.18b and d), the two ropes must move at equal velocity (to keep the clamshell from opening and spilling

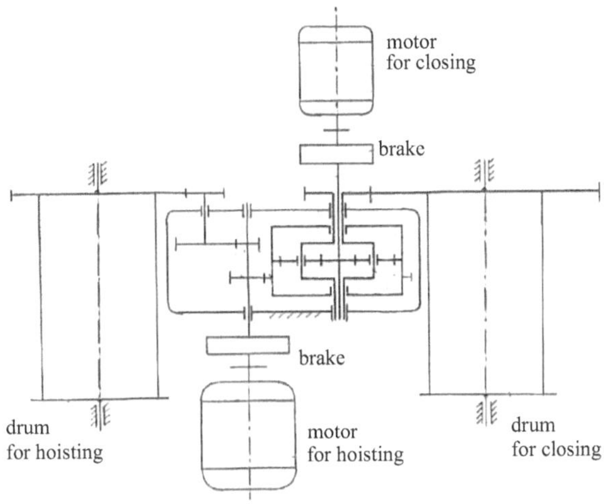

FIGURE 20.17
Kinematic scheme of hoisting mechanism of two-rope clamshell with embedded $\overline{AI}$-PGT in the reducer.

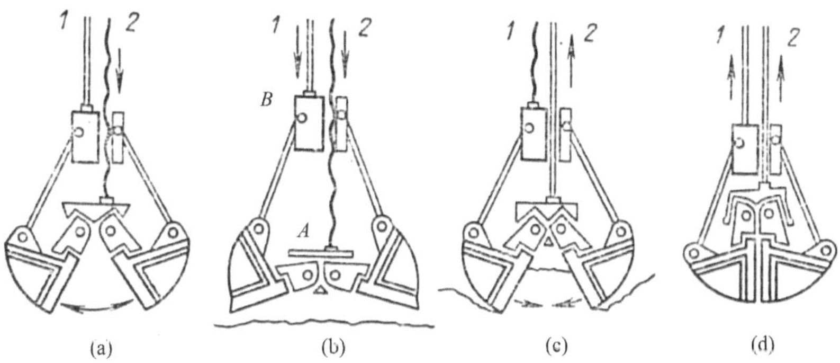

FIGURE 20.18
Different steps of operating cycle of two-rope clamshell (1—holding rope, 2—closing rope, A—lower sheaved block, B—upper sheaved block): (a) Opening; (b) lowering of open clamshell; (c) closing (digging, filling); (d) hoisting (lowering, resp.) of closed clamshell.

material). In this case (Figure 20.19), only the motor for hoisting works, driving the two drums. The PGT operates with fixed sun gear.

At closing of the clamshell (Figure 20.18c), the closing rope 2 moves (upward) with fixed holding rope 1. In this case (Figure 20.20), only the motor for closing works, and the PGT operates with fixed ring gear.

At opening (unloading) of the fixed clamshell, the closing rope 2 moves (downward) at fixed holding rope 1 (Figure 20.18a). In this case (Figure 20.20), only the motor for closing works, and the PGT operates with fixed ring gear.

The arrangement allows for opening of the clamshell with moving holding rope (hoisting or lowering), which reduces the time of operating cycle and increases efficiency of the machine. In this case (Figure 20.21), both motors work, and the PGT operates as a summation differential. Closing rope 2 must move downward with respect to the holding rope.

Speed ratios (the number of teeth, resp.) of the gear train individual stages depend on the diameters of drums and pulley system ratios of both ropes. Provided that both drums have the same diameters and both pulley systems have the same ratio (taken as an example

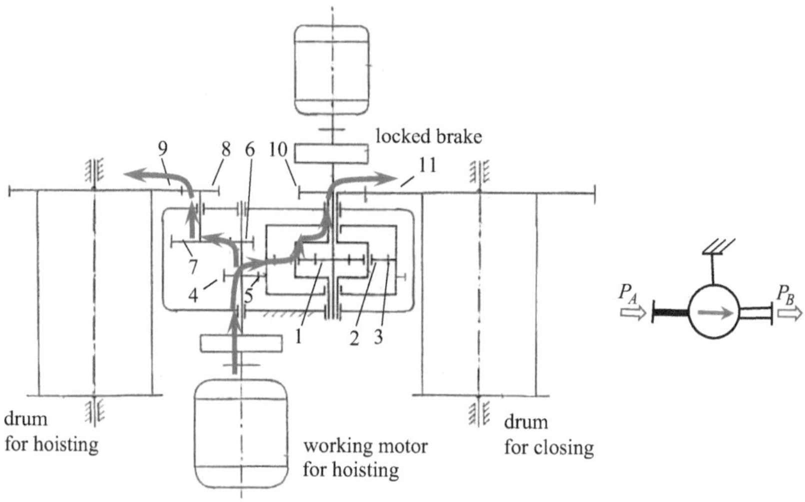

FIGURE 20.19
Operation of the hoisting mechanism of Figure 20.17 at hoisting and lowering of clamshell (holding and closing ropes move at equal velocity).

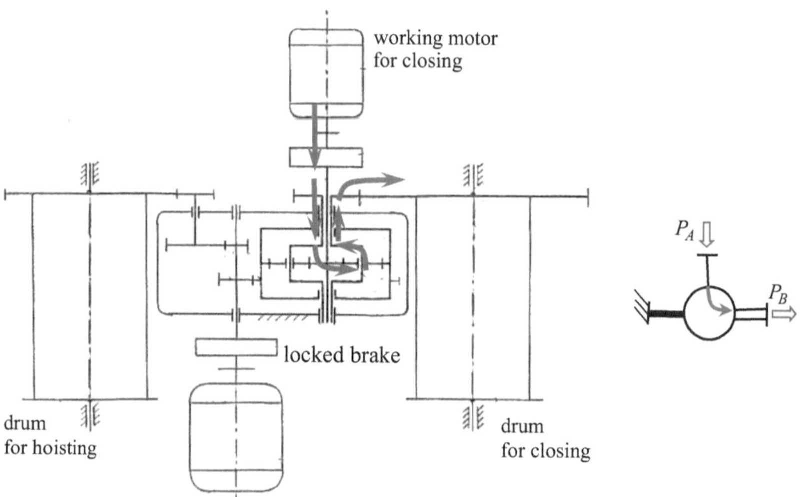

FIGURE 20.20
Operation of the hoisting mechanism of Figure 20.17 at opening and closing of fixed clamshell (only closing rope moves).

only, without being obligatory), to move both ropes at the same velocity (Figure 20.19), drum angular velocities must be equal. This requires that the following kinematic condition should be met:

$$i_{67} \cdot i_{89} = i_{45} \cdot i_{3H(1)} \cdot i_{1011} = \frac{z_7}{z_6} \cdot \frac{z_9}{z_8} = \frac{z_5}{z_4}\left(1 + \frac{z_1}{z_3}\right)\frac{z_{11}}{z_{10}}. \tag{20.12}$$

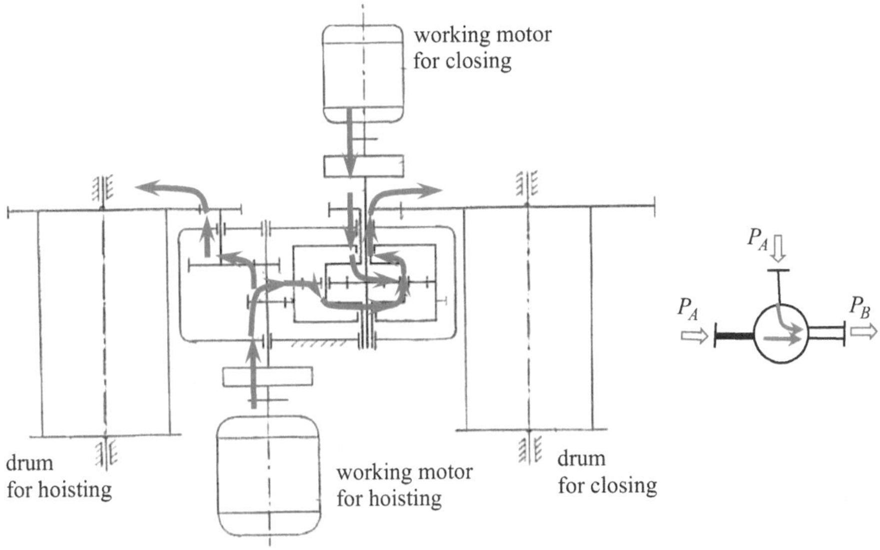

FIGURE 20.21
Operation of the hoisting mechanism of Figure 20.17 at opening of moving clamshell (holding and closing rope move at different velocity).

As it is easy to see, this condition means equality of speed ratios of the two power flows to the two drums after the power division started at gear wheel 4.

20.3 PGTs in Vehicles

This section is not concerned with the automatic transmission (gear) boxes of vehicles, which are compound multicarrier PGTs. Only applications are shown of $\overline{AI}$-PGT as a simple gear train.

Figure 20.22 shows a planetary gear reducer embedded in the steering wheel hub of a tractor. The arrangement has input sun gear, floating carrier as output element and fixed ring gear cuts into the reducer housing. Since the sleeve of gear coupling is longer than in previous examples (Figures 20.3 and 20.4), i.e., sensitivity ξ is higher, a better load equalization among the planets can be expected. In this arrangement, the sleeve of gear coupling is cut to make room for the planets. Due to the small size of planets, their bearings (straight roller bearings) have no races.

When embedding the PGT in driving and chain wheels, in drums, and in other rotating parts, it is more convenient if the carrier is fixed, and the ring gear is output element. An example is the gear train shown in Figure 20.23, where the sun gear shaft is cantilever supported, and the planets are seated on needle roller bearings without races.

Figure 20.24 also shows a PGT, embedded in the hub of the driving wheel of a vehicle, with sun gear supported by a cantilever and planets seated on needle roller bearings without races, but with fixed ring gear.

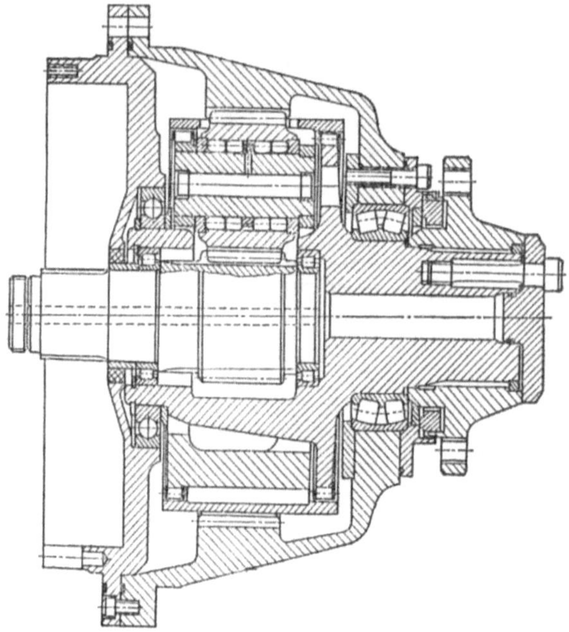

FIGURE 20.22
PGT with fixed ring gear and floating carrier embedded in the steering wheel hub of a tractor.

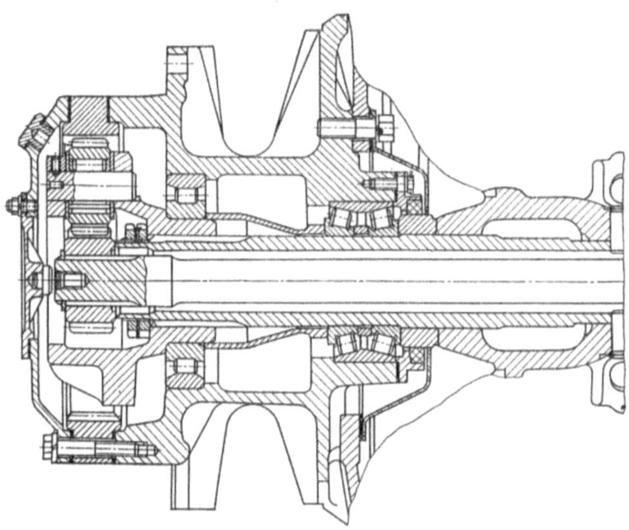

FIGURE 20.23
PGT with fixed carrier and cantilevered sun gear with pliable long shaft embedded in the hub of a tractor chain wheel.

Figure 20.25 shows a PGT of a vehicle with a fixed sun gear. The input shaft supported by a cantilever is connected with the ring gear which is with a thin rim; i.e., it is pliable. Here, as well as with the gear train of Figure 20.24, the output element is the carrier.

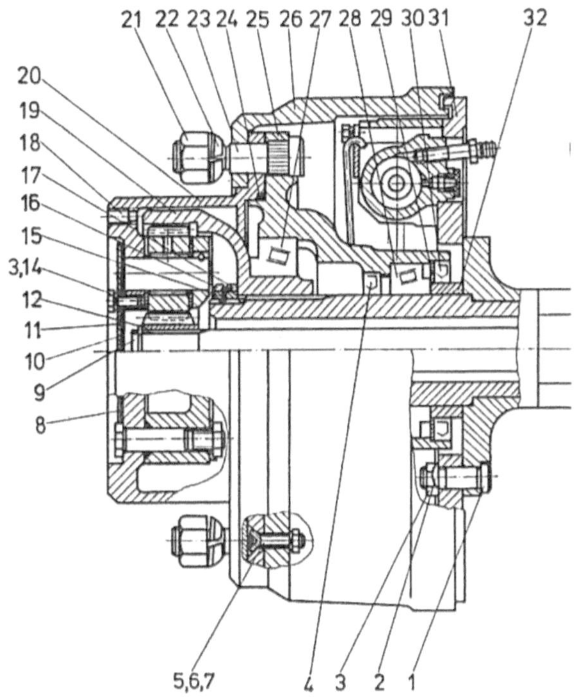

FIGURE 20.24
PGT with fixed ring gear and cantilevered sun gear, embedded in the hub of the driving wheel of a vehicle.

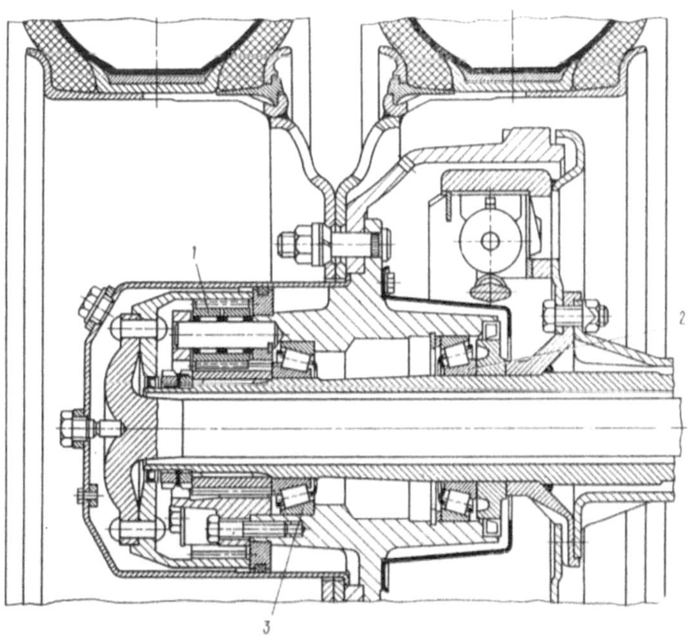

FIGURE 20.25
PGT with fixed sun gear and pliable ring gear, embedded in the hub of the driving wheel of a truck.

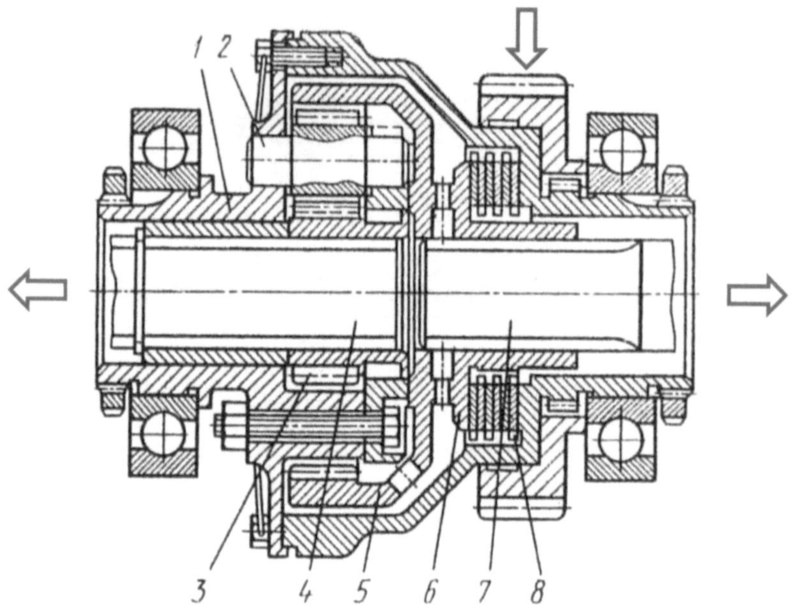

FIGURE 20.26
Asymmetric inter-axle differential with increased friction.

Figure 20.26 shows asymmetric inter-axle differential of a vehicle with two driving axles. The carrier shaft is the input shaft. The sun gear shaft (with smaller torque) transmits the movement to the front axle, and the ring gear shaft (with bigger torque)—to the rare axle. Due to their small size, planet teeth are cut on the pins, which are seated in the carrier of plain bearings.

Part III

Other Types of Simple Cylindrical Planetary Gear Trains

This part reviews some of the other types of simple, i.e., single-carrier cylindrical planetary gear trains that are used, but much less than the $\overline{AI}$-PGT considered in Part II. Therefore, these gear trains are considered here in a much shorter way, referring to the conditions set out in Part II.

21

AI-Planetary Gear Train

The AI-planetary gear train (AI-PGT) shown in Figure 21.1 has much in common with the $\overline{\text{AI}}$-PGT presented in detail in Part II. It can be assumed that the $\overline{\text{AI}}$-PGT is derived from the AI-PGT by simplifying the planets—instead of two rims (as in AI-PGT), just one rim is used.

From the two variants of AI-PGT depicted in Figure 21.1, much more commonly used is the first variant. Of course, this PGT, like $\overline{\text{AI}}$-PGT, can work with both $F=1$ and $F=2$ degrees of freedom, i.e., as a reducer or multiplier in the first case, and as a summation or a division differential in the second case, respectively.

AI-PGT as well as $\overline{\text{AI}}$-PGT is negative-ratio gear train because its basic speed ratio i_0 is negative. For example, for the more commonly used variant (Figure 21.1a),

$$i_0 = \frac{\omega_{1rel}}{\omega_{3rel}} = \frac{\omega_1 - \omega_H}{\omega_3 - \omega_H} = -\frac{z_3}{z_2'} \cdot \frac{z_2}{z_1} < 0. \tag{21.1}$$

Assembly condition of the same variant is given by the following formula [161]:

$$\boxed{\frac{z_1 \cdot z_2' + z_2 \cdot z_3}{k \cdot \delta} = \text{an integer}}, \tag{21.2}$$

where k is the number of planets, and δ is the largest total divisor of the planet teeth number z_2 and z_2'.

Adjacent condition is important only for the larger of the two rims (rim 2) of the planet. It is the same as for $\overline{\text{AI}}$-PGT, and its compliance is not a problem (see Section 4.3).

The tooth geometry of AI-PGT is simpler than that of the $\overline{\text{AI}}$-PGT. This is due to the fact that the geometries of the two meshings (external and internal) are not connected, and therefore, it is easier to define their parameters independently. For this purpose, the formulae given in Sections 5.1–5.3 may be used. In general, for the tooth geometry of AI-PGT, what is stated in Chapter 5 applies.

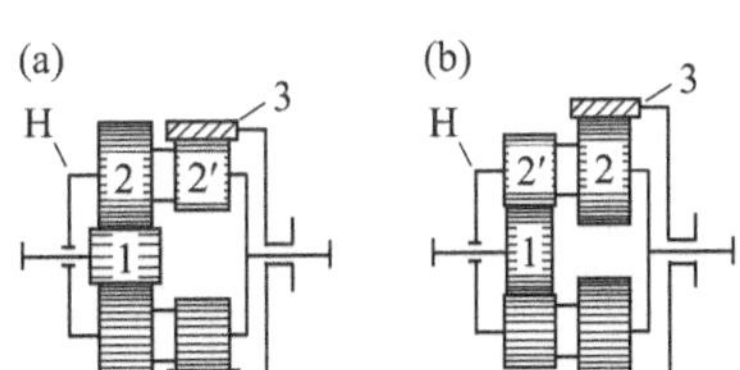

FIGURE 21.1
Two possible variants of AI-PGT.

Torque ratio t, which in this gear is also used to determine the ideal external torque (without taking into account the losses), just like in $\overline{\text{AI}}$-PGT, is the ratio of unidirectional ideal external torques T_1 and T_3 (Figure 21.2) given by

$$\boxed{t = \frac{T_3}{T_1} = +\frac{z_3}{z_2'} \cdot \frac{z_2}{z_1} = -i_0 > +1}. \tag{21.3}$$

Through this torque ratio, the three ideal external torques T_1, T_3, and T_H can be expressed in the same way as in $\overline{\text{AI}}$-PGT (7.25):

$$\boxed{T_3 = +t \cdot T_1} \text{ and } \boxed{T_H = -(T_1 + T_3) = -(1+t)T_1}, \tag{21.4}$$

where they are also in constant ratio (7.28)

$$\boxed{T_1 : T_3 : T_H = T_1 : +t \cdot T_1 : -(1+t)T_1 = +1 : +t : -(1+t)}, \tag{21.5}$$

regardless of the mode of operation of the gear train with $F = 1$ or $F = 2$ degrees of freedom, such as reducer or multiplier; also as a summation or a division differential (see Section 7.4.1).

By the ideal external torques, the *nominal tangential forces* acting on both meshings (external and internal) are calculated as follows (6.1):

$$F_{t12} = F_{t21} = 2{,}000\frac{T_1}{k \cdot d_1} \text{ and } F_{t23} = F_{t32} = 2{,}000\frac{T_3}{k \cdot d_3}. \tag{21.6}$$

These nominal forces are determined by the same simplifying assumptions as in Section 6.1 (T_1 and T_3 are in Nm, d_1 and d_3 in mm).

For AI-PGT kinematic investigation, of course, the different methods of Willis, Kutzbach, Swamp, as well as the torque method, can be used (see Chapter 7). The modified symbol of Wolf (see Section 7.4.1 and Figure 7.4) can be used with the torque method. At this PGT, the dependences from Section 7.4.1 about speed ratio $i_k \equiv i$ (7.30), torque transmit ratio i_T (7.31), and efficiency η (7.32) are in force:

$$\boxed{i_k \equiv i = -\frac{T_B}{T_A}}; \boxed{i_T = \frac{T_B'}{T_A'}}; \boxed{\eta = -\frac{i_T}{i_k}}, \tag{21.7}$$

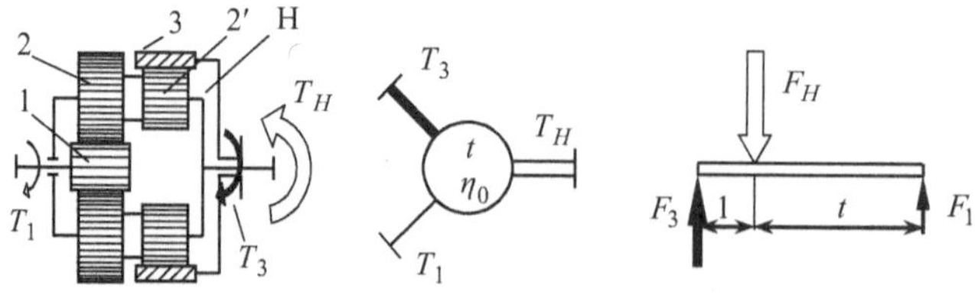

FIGURE 21.2
AI-PGT—kinematic scheme, ideal external torques, modified symbol of Wolf, and lever analogy.

where T_A and T_B are the input and output ideal torques, and

T_A' and T_B' are the input and output real torques.

For the most commonly used case of gearing as a reducer with fixed ring gear 3 ($\omega_3 = 0$), the speed ratio $i_k \equiv i_{1H(3)}$ is determined by the following formula (for the variant of Figure 21.1a):

$$\boxed{i_{1H(3)} \equiv i_k = 1 + t = 1 - i_0 = 1 + \frac{z_3}{z_2'} \cdot \frac{z_2}{z_1} > 1}. \tag{21.8}$$

Figure 21.3 shows a velocity diagram of AI-PGT. It is similar to that of $\overline{\text{AI}}$-PGT (Figure 7.2) and very clear. Here $\omega_{1(H)} = \omega_{1rel}$, $\omega_{2(H)} = \omega_{2rel}$ and $\omega_{3(H)} = \omega_{3rel}$ are relative angular velocities of sun gear 1, planets 2, and ring gear 3 with respect to the carrier H.

For the other possible working modes as a reducer or multiplier, what is stated in Section 7.4.2 is in force, where the analogy is complete.

This gear train can also operate with $F = 2$ degrees of freedom as a summation or a division gear train, as set out in Section 7.4.3.

The question of load sharing in AI-PGT is considerably more complex than that of $\overline{\text{AI}}$-PGT due to the presence of more negative influence factors leading to uneven load distribution between planets. For example, the toothing of the two planet rims must be angularly spaced very close to each other (Figure 21.4) on all the planets of a given PGT, which is not technically easy to execute, and therefore, deviations are unavoidable. In addition, the assembly of planets should be done in a certain position using marked teeth [147]. Mesh load factor $K_{\gamma\,max}$, considering uneven load distribution between the planets, must be chosen with higher value comparing to $\overline{\text{AI}}$-PGT. The issue of load distribution between the planets in this PGT is generally the same as set out in Chapter 8.

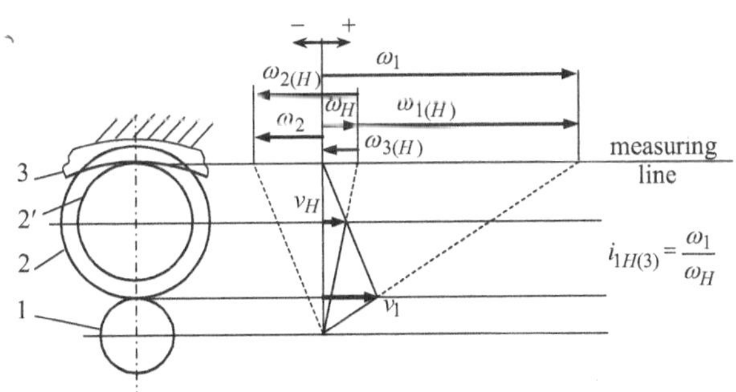

FIGURE 21.3
Velocity diagram of AI-PGT.

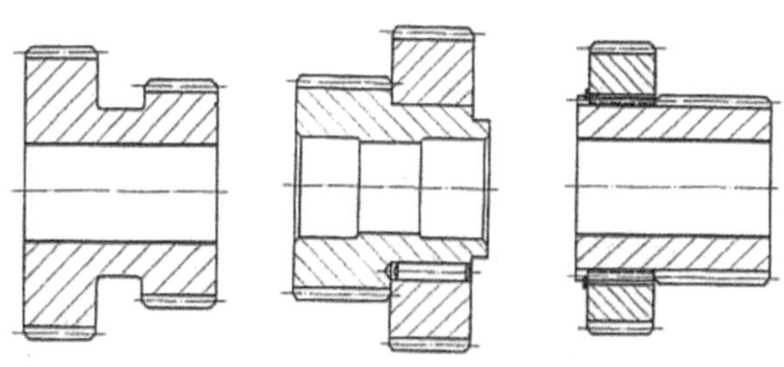

FIGURE 21.4
Possible design of planets for AI-PGT.

Loading of the elements of AI-PGT is generally the same as set out in Chapter 9. The differences are limited to the following:

- The bending teeth load is pulsating on all gears, including planets (not completely reversed as on the planets of $\overline{\text{AI}}$-PGT). It is completely reversed (zero-mean cyclic) when changing the direction of rotation also changes the load direction (e.g., in travel mechanisms, but not in hoisting mechanisms).
- The forces in the two meshings (both external and internal) act in two different parallel planes and create a moment, leading to deformation of the carrier, to tilting of the planets, and to non-uniform distribution of load over the tooth face (face load factor K_β!).

Regarding types of power, types of losses, efficiency, lubrication, heating and cooling, the design and fabrication of the gear elements as well as the load capacity calculations of gears, the broadly stated in the relevant chapters about $\overline{\text{AI}}$-PGT generally applies. Basic loss factor ψ_0 and basic efficiency η_0 of this gear train are defined in a similar way to the $\overline{\text{AI}}$-PGT. If the simplified formula (11.7) is used, the only difference will be that the number of teeth of the internal meshing z_2' is used instead z_2, i.e.,

$$\boxed{\psi_0 = 0.15\left(\frac{1}{z_1}+\frac{1}{z_2}\right)+0.2\left(\frac{1}{z_2'}-\frac{1}{z_3}\right)}. \tag{21.9}$$

Basic efficiency in this case will be determined by (11.9), i.e.,

$$\boxed{\eta_0 = 1-\psi_0 = \eta_{13(H)}}. \tag{21.10}$$

By using the torque ratio t from (21.3), the same formulae for efficiency from Section 12.3 can be applied for the three modes of work with $F = 1$ degree of freedom at $\omega_H = 0$, $\omega_3 = 0$ and $\omega_1 = 0$. For more precise determination of efficiency, see Chapter 11.

Figure 21.5 shows two AI-PGT arrangements—built into the pulley block of a hand-driven chain winch (a) and an airplane gear train (b).

An interesting Bulgarian solution for the use of the AI-PGT is depicted in Figure 21.6 [186]. It is designed for the driving axle of electric trucks (forklifts), replacing the usual differential. An electric motor with rotor and inductor rotating in opposite directions is used. There are AI-PGTs in the wheel-pair hubs, one of which (in this case the left one) is with fixed carrier (i.e., it works as a "pseudo-PGT"), and the other one (the right one) is with fixed ring gear.

In this solution, in straight traveling as well as in turning, on both wheel pairs (left and right) acts the same torque. This is ensured by the same speed ratios of the two PGTs, i.e.,

$$\underbrace{i_{13(H)} = \frac{z_3}{z_2'}\cdot\frac{z_2}{z_1}}_{\text{left PGT}} = \underbrace{i_{1H(3)} = 1+\frac{z_3}{z_2'}\cdot\frac{z_2}{z_1}}_{\text{right PGT}}. \tag{21.11}$$

Of course, depending on the necessary speed ratios, other types of PGTs can be used—single-carrier or compound.

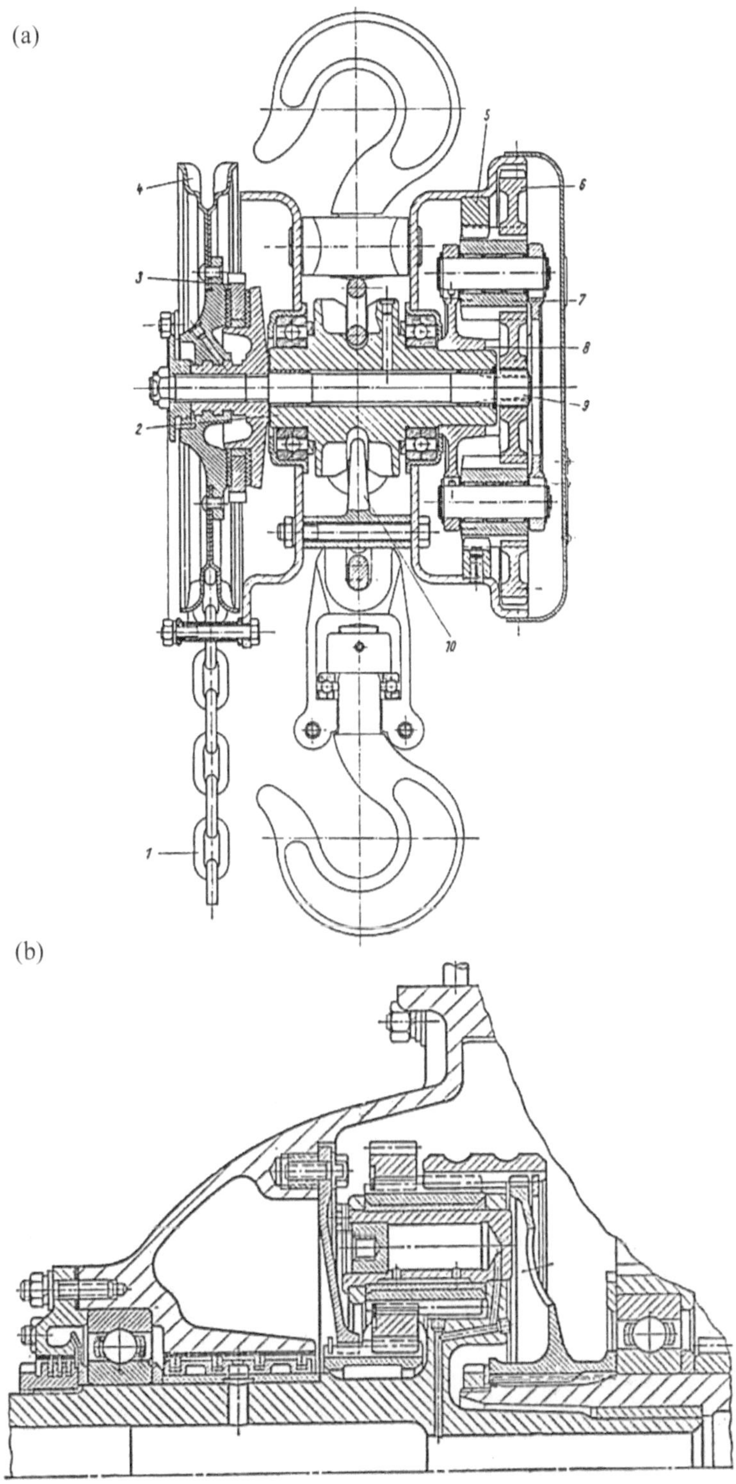

FIGURE 21.5
Arrangements of AI-PGT: (a) Built into the pulley block of a hand-driven chain winch; (b) airplane gear train.

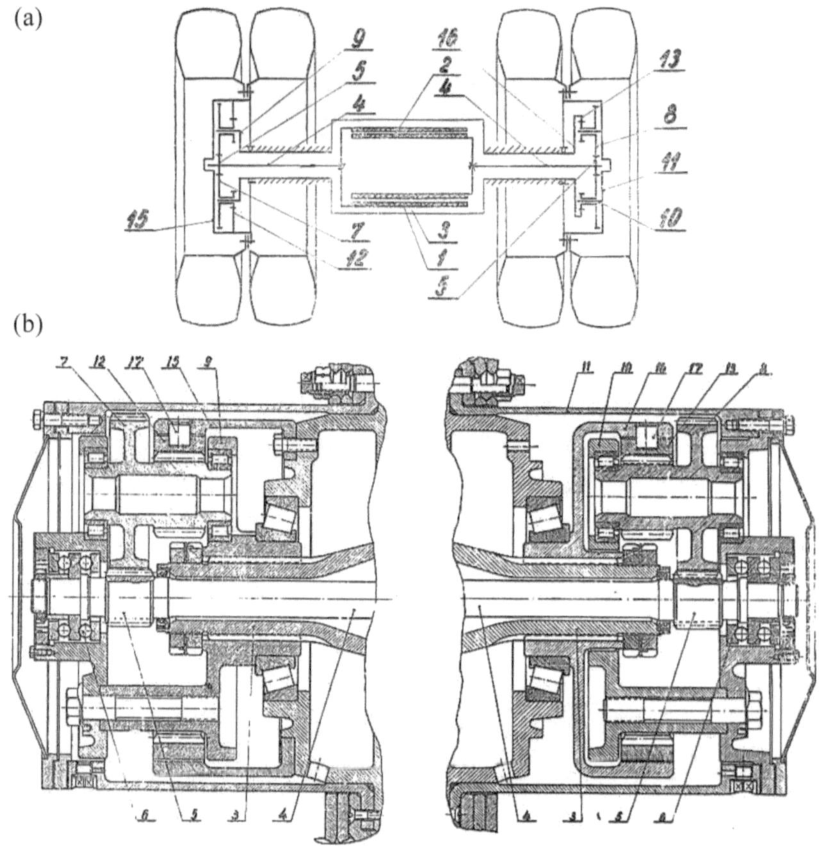

FIGURE 21.6
Driving axle of an electrical truck with an electrical motor with rotating rotor and inductor and with AI-PGT in wheel-pair hubs [186]: (a) Kinematic scheme; (b) arrangement.

In comparison with the $\overline{\text{AI}}$-PGT, the AI-PGT has the following advantages, disadvantages, and application:

Advantages

1. AI-PGT has larger kinematic capabilities; i.e., it can perform larger speed ratios and more intermediate values, such as
   $i_{1H(3)} = 8 \div 15$ are real limits.
   $i_{1H(3)} \approx 35$ can take place.
2. The teeth of the planets are not completely reversed loaded but pulsatingly when rotating in one direction. Also, the same side of the teeth is always loaded in the reversing of the hoisting mechanisms.

Disadvantages

1. Planet manufacturing is considerably more complicated due to both rims and especially because of the need for their exact angular positioning.

2. The problem of planet load equalization is considerably more complex because of the greater number of negative influence factors that cause the greater risk of failure.
3. Since the forces act on a planet in two parallel planes and create a moment, they cause a greater deformation of the carrier, which in turn leads to a more nonuniform distribution of load over the tooth face (face load factor K_β!).
4. The quality manufacturing of this gear train is possible only in a well-equipped factory with qualified staff.

Application

The application is limited. It is used much less frequently than $\overline{\text{AI}}$-PGT and mainly as a reducer. Not particularly recommended.

22

AA- and $\overline{\text{AA}}$-Planetary Gear Trains

Figure 22.1 shows the two possible variants of this type of positive-ratio planetary gear train (PGT)—with two- or one-rim planets. Undoubtedly, the second variant is simpler in design and technological terms. While AA-PGT has greater kinematic options for the desired speed ratio, the capabilities of $\overline{\text{AA}}$-PGT are more limited due to the close number of teeth z_1 and z_3 of central gears 1 and 3.

These two PGTs are used to obtain very large speed ratios of the order of 1,000–1,500 and only as reducers. Carrier H is always input, one of the central gears is fixed, and the other one is output.

With these gear trains, depending on whether the small central gear 1 or the large central gear 3 is fixed (with driving carrier H), coinciding or different directions of rotation of the input and output shafts are obtained, i.e., positive or negative speed ratio

$i_{H3(1)} > 0$ or $i_{H1(3)} < 0$.

The two types of trains in question are positive ratio since their basic speed ratio i_0 is positive:

$$i_0 = \frac{\omega_{1rel}}{\omega_{3rel}} = \frac{\omega_1 - \omega_H}{\omega_3 - \omega_H} = \left\{ \begin{array}{c} \dfrac{z_3}{z_2'} \cdot \dfrac{z_2}{z_1} \\ \dfrac{z_3}{z_1} \end{array} \right\} > 0. \tag{22.1}$$

Assembly conditions for both types are as follows [161]:

for AA-PGT

$$\boxed{\frac{z_2 \cdot z_3 - z_1 \cdot z_2'}{k \cdot \delta} = \text{an integer}}, \tag{22.2}$$

for $\overline{\text{AA}}$-PGT

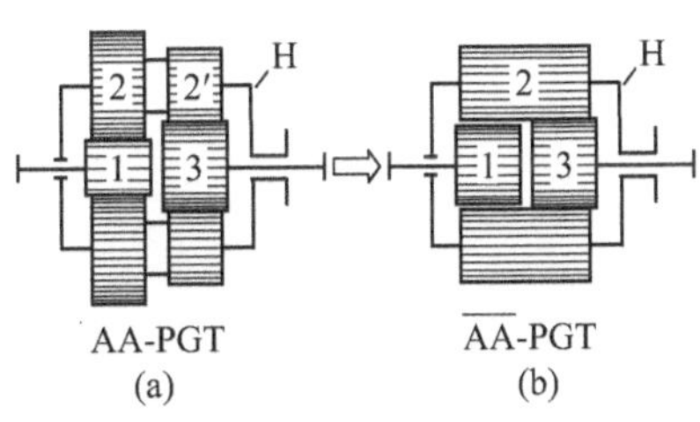

FIGURE 22.1
Two variants of positive-ratio PGT with two external meshings.

$$\boxed{\frac{z_3 - z_1}{k} = \text{an integer}}, \tag{22.3}$$

where k is the number of planets, and δ is the largest total divisor of the planet teeth number z_2 and z_2'.

Of course, the adjacent condition is in force, and its observance is not a problem (see Section 4.3).

With regard to tooth geometry, the outlined for external teeth and meshing has been described in Section 5.1. Obviously, it is a relatively simple case of external rather than internal meshing.

The ideal external torques T_1 and T_3, on both central gears 1 and 3, are determined by the basic speed ratio i_0. The particular in this case is that these torques are opposite (as opposed to AI-and $\overline{\text{AI}}$-PGT):

$$T_3 = -i_0 \cdot T_1 = \left\{ \begin{array}{c} -\frac{z_3}{z_2'} \cdot \frac{z_2}{z_1} T_1 \\ -\frac{z_3}{z_1} T_1 \end{array} \right\} < 0. \tag{22.4}$$

This in turn leads to the next feature of the gear train in question. Whether the basic speed ratio i_0 is large or small ($i_0 > 2$ or $i_0 < 2$), the greatest ideal external torque T_3 always acts on the large central gear 3. As for the other external torques T_1 and T_H, it is found that they are unidirectional and which of them is larger depends on the basic speed ratio i_0 as follows [136]:

at $i_0 < 2 \rightarrow T_H < T_1$,

at $i_0 > 2 \rightarrow T_H > T_1$.

These two features are easily established using the sum of ideal external torques (see Section 2.4 and Figure 2.5), depending on i_0

$$\sum T_i = T_1 + T_3 + T_H = T_1 - i_0 \cdot T_1 + T_H = 0, \tag{22.5}$$

from which it follows that

$$T_H = (i_0 - 1)T_1. \tag{22.6}$$

As a result, the modified symbol of Wolf changes as shown in Figure 22.2, as well as the formula for the torque ratio t, which in these PGTs also is the ratio of the unidirectional external torques. For the simpler and more common case of planets with one rim (Figure 22.1b), when the basic speed ratio $i_0 = z_3/z_1 < 2$ (usually close to 1), for *torque ratio t*, is obtained:

$$\boxed{t = \frac{T_1}{T_H} = \frac{1}{i_0 - 1} = \frac{1}{\frac{z_3}{z_1} - 1} = \frac{z_1}{z_3 - z_1} = \frac{z_1}{\Delta z_{31}}}. \tag{22.7}$$

From formula (22.7), it can be seen that in the case of choosing a difference in the number of teeth $\Delta z_{31} = 1$, large speed ratios can be obtained, but not as large as at AA-PGT with two-rim planets.

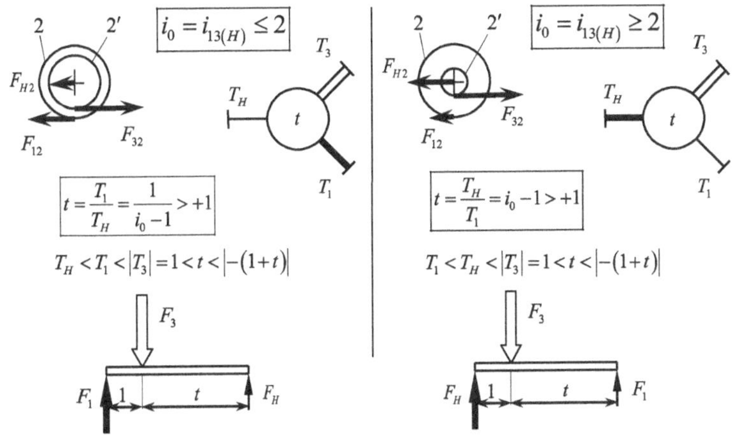

FIGURE 22.2
Change of the ideal external torques ratio depending on the size of the basic speed ratio i_0 at AA-PGT.

Like the other PGTs (AI and $\overline{AI}$) and here, the ideal external torques are in a constant ratio, regardless of how the PGT is operating (Figure 22.2).

Knowing the ideal external torques T_1 and T_3, on both central gears 1 and 3, it is easy to determine the *nominal tangential forces* on the two meshings as follows:

$$F_{t12} = F_{t21} = 2{,}000\frac{T_1}{k \cdot d_1} \text{ and } F_{t23} = F_{t32} = 2{,}000\frac{T_3}{k \cdot d_3} \tag{22.8}$$

with the simplifying prerequisites set out in Section 6.1 (T_1 and T_3 in Nm, d_1 and d_3 in mm).

Since AA- and $\overline{AA}$-PGTs are practically only used as reducers, two cases are possible—with fixed small central gear 1 and large central gear 3. In both cases, the carrier H is always input. When $i_0 < 2$, which is the most common case, the corresponding formulae for the speed ratio of AA- and $\overline{AA}$-PGT are as follows [161]:

At $\omega_1 = 0$

$$i_{H3(1)} = \frac{\omega_H}{\omega_3} = -\frac{T_3}{T_H} = -\frac{-(1+t)T_H}{T_H} = 1 + t = 1 + \frac{1}{i_0 - 1} = \frac{1}{1 - \frac{1}{i_0}} = \left\{ \begin{array}{c} \dfrac{1}{1 - \dfrac{z_2'}{z_3} \cdot \dfrac{z_1}{z_2}} \\[2ex] \dfrac{1}{1 - \dfrac{z_1}{z_3}} = \dfrac{z_3}{\Delta z_{31}} \end{array} \right\} > 0. \tag{22.9}$$

At $\omega_3 = 0$

$$i_{H1(3)} = \frac{\omega_H}{\omega_1} = -\frac{T_1}{T_H} = -t = -\frac{1}{i_0 - 1} = \frac{1}{1 - i_0} = \left\{ \begin{array}{c} \dfrac{1}{1 - \dfrac{z_3}{z_2'} \cdot \dfrac{z_2}{z_1}} \\[2ex] \dfrac{1}{1 - \dfrac{z_3}{z_1}} = -\dfrac{z_1}{|\Delta z_{13}|} \end{array} \right\} < 0. \tag{22.10}$$

In the first case ($\omega_1 = 0$), the directions of rotation of the input and output shafts coincide, and the speed ratio $i_{H3(1)}$ is positive. In the second case ($\omega_3 = 0$), these directions are opposite, and the speed ratio $i_{H1(3)}$ is negative. This change in the direction of rotation of the output shaft is very obvious in Figure 22.3. The same results for both speed ratios can also be obtained by other methods, such as the analytical method of Willis—an appropriate task for the reader.

For these types of gears, very high-speed ratios ($i = 1{,}000 \div 1{,}500$) can be easily obtained by using a small trick—the basic speed ratio is selected close to 1, i.e., $i_0 \approx 1$. Because the difference $1 - 1/i_0$ or $1 - i_0$ (resp. $\Delta z_{31} = z_3 - z_1 > 0$) is a denominator, the abovementioned high values for the speed ratio are obtained. The largest speed ratio can be obtained with AA-PGT with two-rim planets.

And with the gear trains under consideration, especially in AA-PGT (with two-rim planets), the internal loads (load sharing among planets) are nonuniform. However, this is of little importance since both PGTs are used as kinematic not as power trains.

For the simpler and more commonly used case of one-rim planet ($\overline{AA}$-PGT), when the basic speed ratio is $i_0 = z_3/z_1 < 2$ (most commonly close to 1), for the efficiency η determination for both cases ($\omega_1 = 0$ or $\omega_3 = 0$), the following formulae are used [161]:

At $\omega_1 = 0$

$$\boxed{\eta_{H3(1)} = \frac{i_0 - 1}{i_0 - \eta_0} = f(\mu_z)}. \tag{22.11}$$

At $\omega_3 = 0$

$$\boxed{\eta_{H1(3)} = \frac{i_0 - 1}{\dfrac{i_0}{\eta_0} - 1} = f(\mu_z)}. \tag{22.12}$$

In these formulae, η_0 is the basic efficiency of PGT which can be determined either by the following simplified formula based on formula (11.7):

$$\eta_0 \approx 1 - 0.15\left(\frac{1}{z_1} + \frac{2}{z_2} + \frac{1}{z_3}\right), \tag{22.13}$$

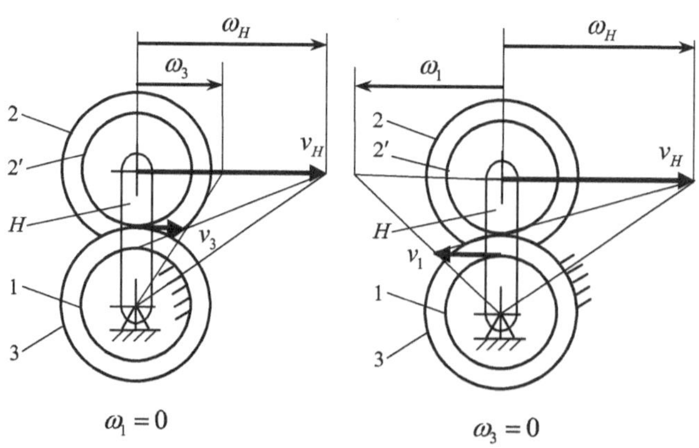

FIGURE 22.3
Change of the direction of rotation of the output shaft at various fixed central gears 1 and 3.

or with greater precision, as set out in Chapter 11, depending on the coefficient of friction in teeth meshing μ_z, defining the scattering of PGT's efficiency η.

These large kinematic capabilities of the PGTs under consideration are, however, combined with a low efficiency which decreases as the speed ratio $i_{H3(1)}$ increases (Figure 22.4). As shown in the figure, there is a difference depending on whether the PGT works as a reducer ($i_{H3(1)} > 1$) or as a multiplier ($i_{3H(1)} < 1$). When working as a multiplier, there is a self-locking [55] which starts from a certain speed ratio. This self-locking may in some cases be desirable. The speed ratio, after which starts the self-locking ($\eta_{3H(1)} < 0$), depends to a great extent on the friction conditions in the two meshings (on lubrication, on the teeth roughness) and can vary widely, which determines the scattering shown in Figure 22.4. This is one of the traps of this type of PGT.

To avoid self-locking, the basic speed ratio i_0 must be outside the following limits [147, 161]:

$$\boxed{\eta_0 < i_0 < \frac{1}{\eta_0}}. \tag{22.14}$$

This dependence is available if PGT works with one degree of freedom with output carrier.

Conversely, the basic transmission ratio should be within the specified limits if self-locking is desirable.

Figure 22.5 shows an AA-PGT embedded in a belt pulley that drives the carrier.

Finally, it is appropriate to define clearly the characteristics of the two types of PGT in question:

1. The correlation of ideal external torques T_1 and T_H changes according to the basic speed ratio i_0 as follows:
   at $i_0 < 2$ obtains $T_H < T_1$.
   at $i_0 > 2$ obtains $T_H > T_1$.
2. In both cases, the torque T_3 remains the biggest one (Figure 22.2).
3. In $\overline{AA}$-PGT, always $i_0 < 2$, and therefore, T_H is the smallest torque.
4. Changing the torque correlation results in a change in the modified symbol of Wolf as shown in Figure 22.2.
5. In $\overline{AA}$-PGT, unlike the $\overline{II}$-PGT (Chapter 23), when from two- to one-rim planet is proceeded, the direction of rotation of the output shaft does not change, and the sign of gear ratio i does not change, respectively.

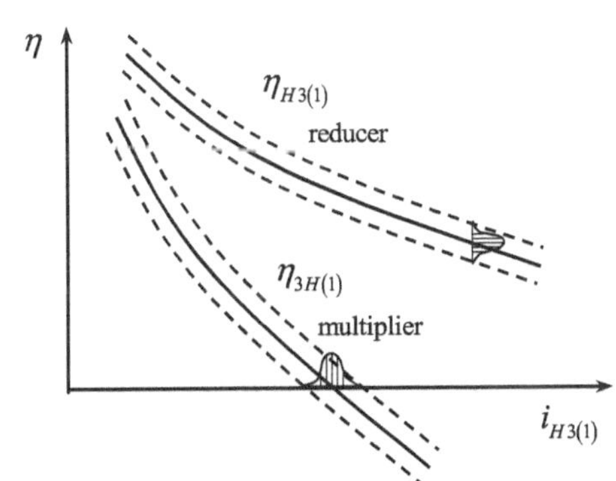

FIGURE 22.4
Efficiency of AA- and $\overline{AA}$-PGT at work as a reducer and as a multiplier.

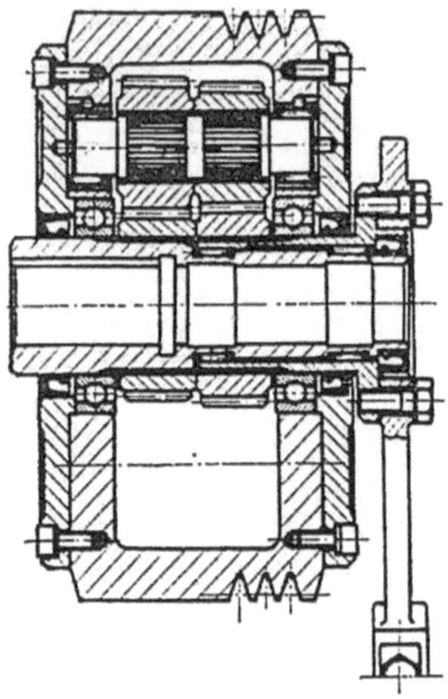

FIGURE 22.5
AA-PGT embedded in a belt pulley.

Advantages, disadvantages, and application of AA - and $\overline{AA}$-PGTs are as follows:

Advantages

1. Very large kinematic capabilities—speed ratios of the order of $i = 1{,}000 \div 1{,}500$. Because of the low efficiency, however, usually $i = 30 \div 100$ is selected, and high values are very rarely used.
2. Only external meshing is used, which is simpler for calculation and manufacturing as compared to the internal one (compared to II- and $\overline{II}$-PGTs, see Chapter 23).
3. Manufacturing of $\overline{AA}$-PGT with one-rim planet is substantially simplified.
4. Self-locking is possible, which in some cases is desirable.

Disadvantages

1. Significantly larger overall diameter compared to II- and $\overline{II}$-PGTs (Chapter 23). Also, a larger mass inertia moment, which is important for transmissions requiring rapid acceleration.
2. Low efficiency, which decreases as the speed ratio increases and results in heating, which makes these PGTs unsuitable for continuous operation.
3. The instantaneous speed ratio of all gear trains due to unavoidable inaccuracies is not constant ($i \neq$const) but changes. In very large speed ratios ($\overline{AA}$-PGT), however, due to unavoidable geometrical variations, the output shaft rotates unevenly and in some cases even backward [195] (Figure 22.6). In $\overline{AI}$-PGT, though with $i \neq$ const, such backward rotation is not possible.
4. Harder manufacturing of two-rim planets of AI-PGT (see Chapter 21).

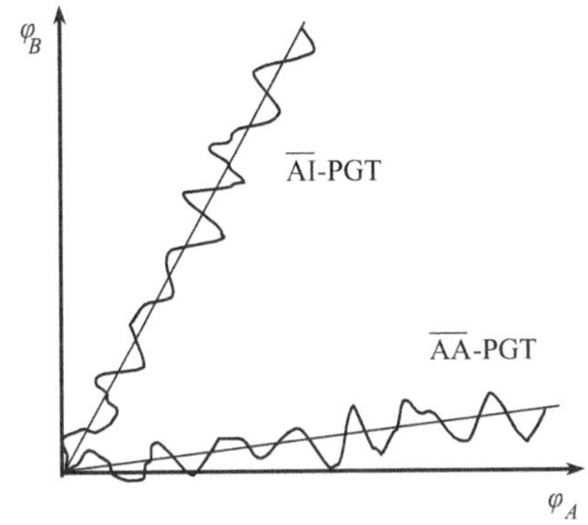

FIGURE 22.6
Backward rotation of output shaft B of $\overline{\mathrm{AA}}$-PGT at very large speed ratio ($i \approx 1,500$).

5. Self-locking in some cases is undesirable.
6. Increased noise due to high speed of the planets.

Application

1. Very limited, only as low-power, kinematic gear trains with significant and very large speed ratios, when low efficiency is tolerable or unimportant, and even self-locking may be desirable.
2. Unavailability for continuous work due to heating as a result from the low efficiency.
3. Mostly in metal-cutting machines and in instrumentation, where efficiency is irrelevant, and work is not continuous.

23

II- and $\overline{II}$-Planetary Gear Trains

Figure 23.1 shows possible variants of this type of positive-ratio planetary gear train (PGT), not only depending on the number of planet rims—one or two, but also depending on the number of planets—$k > 1$ or $k = 1$. Thus, a total of four variants are obtained. $\overline{II}$-PGT may be thought to have been derived from the II-PGT by simplifying the planets—instead of two rims, only one is used. With II-PGT (similar to the AA-PGT), the possibilities for obtaining a desired speed ratio are greater than those with $\overline{II}$-PGT. However, $\overline{II}$-PGT is simpler from design and technological point of view.

II- and $\overline{II}$-PGTs are very similar to AA- and $\overline{AA}$-PGTs. They are also used only as reducers to obtain very large speed ratios—in the order of 1,000 ÷ 1,500. The input of these gear trains is always the carrier H, as one of the central gears 1 or 3 is fixed and the other one is the output.

With these PGTs, not only depending on which central gear (1 or 3) is fixed, but also depending on whether gear II or $\overline{II}$ is used, opposite or coincident directions of rotation of the input and output shafts are obtained, i.e., a negative or positive speed ratio $i_{H3(1)} < 0$ or $i_{H3(1)} > 0$, and $i_{H1(3)} < 0$ or $i_{H1(3)} > 0$, respectively.

The situation with basic speed ratio i_0 of both PGTs, II and $\overline{II}$, is more specific [161]. In addition to the positive value, it must be greater than 1, which is why for each type of both gear trains it defines differently:

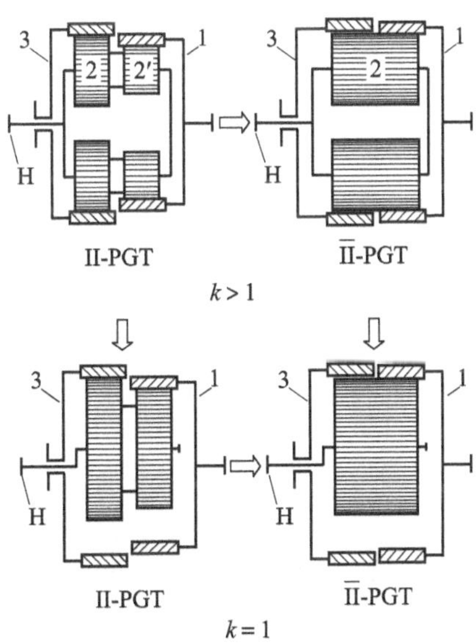

FIGURE 23.1
Four variants of positive-ratio PGT with two internal meshings.

$$i_0 = \left\{\begin{array}{l} = i_{31(H)} = \dfrac{\omega_{3rel}}{\omega_{1rel}} = \dfrac{\omega_3 - \omega_H}{\omega_1 - \omega_H} = \dfrac{z_2}{z_3} \cdot \dfrac{z_1}{z_2'} \\ = i_{13(H)} = \dfrac{\omega_{1rel}}{\omega_{3rel}} = \dfrac{\omega_1 - \omega_H}{\omega_3 - \omega_H} = \dfrac{z_3}{z_1} \end{array}\right\} > +1. \tag{23.1}$$

This peculiarity that at II-PGT $i_0 = i_{31(H)} > 1$, though $z_3 > z_1$, can be easily verified by numerical examples, and it is important to know it.

Assembly condition for both types of PGT in question, when the number of planets is more than = 1, is given by the following formulae [161]:

for II-PGT

$$\frac{z_1 \cdot z_2 - z_2' \cdot z_3}{k \cdot \delta} = \text{an integer}, \tag{23.2}$$

for $\overline{\text{II}}$-PGT

$$\frac{z_1 - z_3}{k} = \text{an integer}, \tag{23.3}$$

where k is the number of planets, and δ is the largest total divisor of the planet teeth number z_2 and z_2' at II-PGT (Figure 23.1).

Observance of the adjacent condition is not a problem for these gears, too (see Chapter 4.3).

About tooth geometry, the above-stated of internal teeth and meshing in Section 5.3 applies.

However, there is a need for an addition in relation to the possible interference outside the meshing zone in case of a small difference in the number of teeth of ring gears 1 and 3 and the one-rim planet 2 in gear trains with only one planet (Figure 23.1) [63, 67, 68, 212]. This difference is

$$\Delta z_{12'} = z_1 - z_2' < 10, \text{ resp. } \Delta z_{12} = z_1 - z_2 < 10 \tag{23.4}$$

$$\Delta z_{32} = z_3 - z_2 < 10. \tag{23.5}$$

Figure 23.2 shows the interference of the tooth tips outside the meshing zone. In order to obtain large speed ratios, the difference in the number of teeth Δz should be reduced, possibly reaching $\Delta z = 1$. This case requires special calculations using the capabilities of the profile shifting [67, 68, 212].

Figure 23.3 shows the impossibility of radial mounting of the planet. However, this is not a significant obstacle as the planet can be mounted axially.

A very practical diagram [65, 158, 175] is provided in Figure 23.4, whereby, avoiding detailed calculations, it can be ascertained whether or not there is interference with the teeth tips and the possibility of radial mounting. There is no interference when the intersection of the tip circles of the two gears (on the diagram labeled 1 and 2!) with radii r_{a1} and r_{a2} lies in zone 1. Zone 4 should be strictly avoided, whereas zones 2 and 3 may be used but after calculations.

The ideal external torques T_1 and T_3 on both ring gears 1 and 3 are determined by the basic speed ratio i_0, and they are with opposite directions, like in AA- and $\overline{\text{AA}}$-PGT (unlike the AI- and $\overline{\text{AI}}$-PGT)

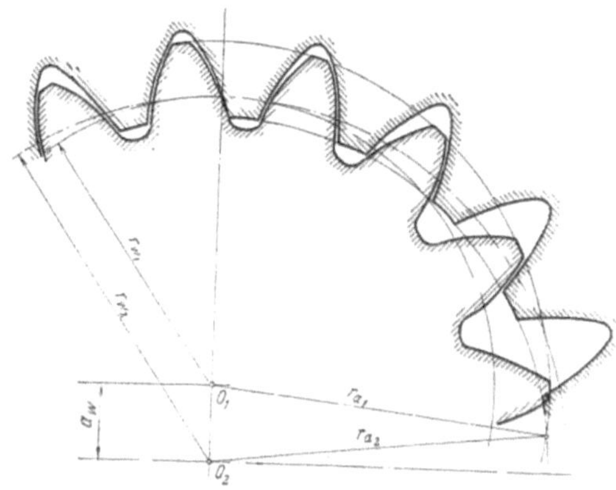

FIGURE 23.2
Interference of the planet and ring gear teeth tips outside the meshing zone in case of a small difference in the number of planet and ring gear teeth $\Delta z < 10$.

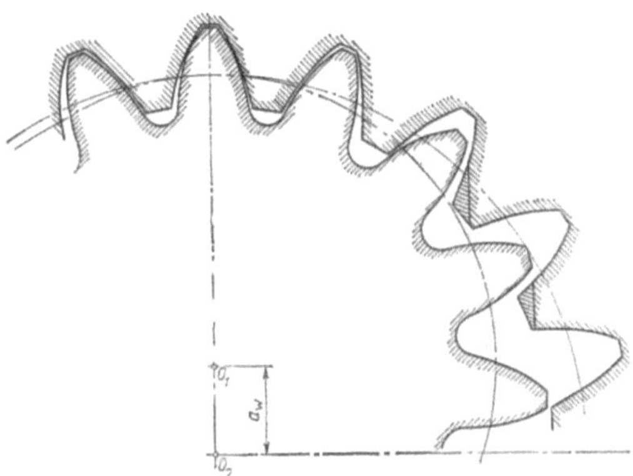

FIGURE 23.3
Impossibility of radial mounting with a small difference in the number of planet and ring gear teeth $\Delta z < 16$.

$$\left.\begin{aligned} T_1 &= -i_0 \cdot T_3 = -i_{31(H)} \cdot T_3 = -\frac{z_2}{z_3} \cdot \frac{z_1}{z_2'} T_3 \\ T_3 &= -i_0 \cdot T_1 = -i_{13(H)} \cdot T_1 = -\frac{z_3}{z_1} T_1 \end{aligned}\right\} < 0. \tag{23.6}$$

Both types of gears (II and $\overline{II}$) have some peculiarities (Figure 23.5), too. Depending on the value of the basic speed ratio i_0 (resp. torque ratio t), the ideal external torque T_1 and T_3 changes its correlation, which is reflected in determining torque ratio t, and the structural symbol of Wolf. For II-PGT, $T_3 < |T_1|$, is obtained and for $\overline{II}$-PGT—$|T_3 > T_1|$. For both gears, T_H is the lowest torque, equal to $T_H = (i_0 - 1)T_3$, resp. $T_H = (i_0 - 1)T_1$.

As shown in Figure 23.5, the same dependences for torque ratio t are obtained for both types of gears. This follows from the equilibrium of ideal external torques:

$$\sum T_i = T_1 + T_3 + T_H = 0,$$

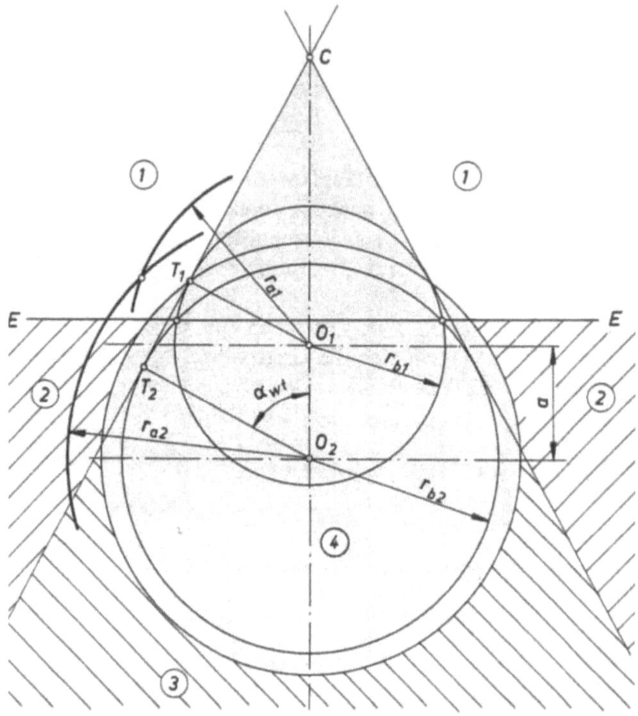

FIGURE 23.4
Diagram for determining the presence or absence of interference of the planet and ring gear teeth outside the respective meshing zone: when the intersection of top circles of the two gears with radii r_{a1} and r_{a2} lie in: zone 1—no risk of interference and radial mounting is possible; zone 2—no risk of interference, radial mounting is only possible after additional calculations; zone 3—there is a risk of interference, and radial mounting is possible, but additional calculations are required for both; zone 4—unacceptable zone—there is interference, and radial mounting is impossible. (Extracted from DIN 3993, Teil 1 bis 4 Geometrische Auslegung von zylindrischen Innenradpaaren mit Evolventenverzahnung.)

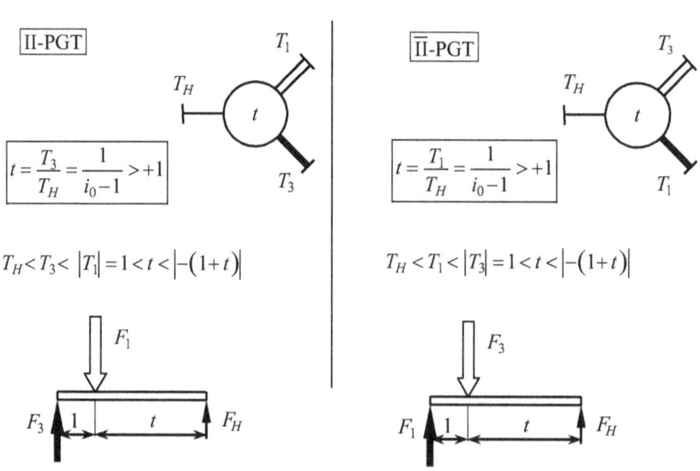

FIGURE 23.5
Both possible variants of modified symbol of Wolf and torque ratio t for both PGT types —II and $\overline{\text{II}}$.

and that in one case $T_1 = -i_0 \cdot T_3$ and $T_H = (i_0 - 1)T_3$, and in the other case $T_3 = -i_0 \cdot T_1$ and $T_H = (i_0 - 1)T_1$. Especially for the simpler and more commonly used case of $\overline{\text{II}}$-PGT with one-rim planet, the formula for torque ratio is

$$t = \frac{T_1}{T_H} = \frac{1}{i_0 - 1} = \frac{1}{z_3/z_1 - 1} = \frac{z_1}{z_3 - z_1} = \frac{z_1}{\Delta z_{31}}. \tag{23.7}$$

This is the same expression as for $\overline{\text{AA}}$-PGT (22.5). From it, it can be seen that in $\overline{\text{II}}$-PGT if choosing a difference in the number of teeth $\Delta z_{31} = 1$, large speed ratios may be obtained, but not as large as in II-PGT with two-rim planet.

Similarly to the previous gears ($\overline{\text{AI}}$, AI, AA, and $\overline{\text{AA}}$), the ideal external torques are in constant ratio, regardless of how the gear train works (Figure 23.5).

Knowing the ideal external torques T_1 and T_3 on both ring gears 1 and 3, it is easy to determine the *nominal tangential forces* in the two meshings as follows:

$$F_{t12} = F_{t21} = 2{,}000\frac{T_1}{k \cdot d_1} \text{ and } F_{t23} = F_{t32} = 2{,}000\frac{T_3}{k \cdot d_3}. \tag{23.8}$$

with the simplifying prerequisites set out in Section 6.1 (T_1 and T_3 in Nm, d_1 and d_3 in mm).

Since II- and $\overline{\text{II}}$-PGTs are practically used as a reducers only, two cases are possible—one of the ring gears 1 or 3 should be fixed, always with carrier H as input. The corresponding speed ratios for the two PGTs are determined by the following formulas (Figure 23.5):

At $\omega_1 = 0$

for II-PGT

$$i_{H3(1)} = \frac{\omega_H}{\omega_3} = -\frac{T_3}{T_H} = -\frac{t \cdot T_H}{T_H} = -t = \frac{1}{i_0 - 1} = -\frac{1}{\frac{z_2}{z_3} \cdot \frac{z_1}{z_2'} - 1} < 0; \tag{23.9}$$

for $\overline{\text{II}}$-PGT

$$i_{H3(1)} = \frac{\omega_H}{\omega_3} = -\frac{T_3}{T_H} = -\frac{-(1+t)T_H}{T_H} = 1 + t = 1 + \frac{1}{i_0 - 1} = \frac{1}{1 - \frac{1}{i_0}} = \frac{1}{1 - \frac{z_1}{z_3}} = \frac{z_3}{z_3 - z_1} = \frac{z_3}{\Delta z_{31}} > 0. \tag{23.10}$$

At $\omega_3 = 0$

for II-PGT

$$i_{H1(3)} = \frac{\omega_H}{\omega_1} = -\frac{T_1}{T_H} = -\frac{-(1+t)T_H}{T_H} = 1 + t = 1 + \frac{1}{i_0 - 1} = \frac{1}{1 - \frac{1}{i_0}} = \frac{1}{1 - \frac{z_3}{z_2} \cdot \frac{z_2'}{z_1}} > 0; \tag{23.11}$$

for $\overline{\text{II}}$-PGT

$$i_{H1(3)} = \frac{\omega_H}{\omega_1} = -\frac{T_1}{T_H} = -\frac{t \cdot T_H}{T_H} = -t = -\frac{1}{i_0 - 1} = -\frac{1}{z_3/z_1 - 1} = -\frac{z_1}{z_3 - z_1} = -\frac{z_1}{\Delta z_{31}} < 0. \tag{23.12}$$

It should be borne in mind that in these formulae, the basic speed ratio is different (23.1): $i_0 = i_{31(H)} > 1$ for II-PGT and $i_0 = i_{13(H)} > 1$ for $\overline{\text{II}}$-PGT.

It can be seen that when proceeding from two- to one-rim planets, the direction of rotation of the output shaft changes, and speed ratio i changes, respectively, from positive to negative and vice versa. This is another feature of the two types of PGTs considered here, which distinguishes them from the previous AA- and $\overline{\text{AA}}$-PGTs. With them, a change in speed ratio, and in the direction of rotation of the output shaft, respectively, occurs when the fixed center wheel is changed, but not when proceeding from two- to one-rim planet ($z_2 \equiv z_2'$). However, as seen above, in II- and $\overline{\text{II}}$-PGTs, the change of the fixed ring gear also leads to a change in the algebraic sign of their speed ratio.

There is another peculiarity related to the use of a one-rim planet ($\overline{\text{II}}$-PGT). As can be seen from Figure 23.6, although the number of teeth z_3 of ring gear 3 is greater than the number of teeth z_1 of ring gear 1, as in the case of II-PGT, i.e.,

$$z_3 > z_1,$$

the radius of operating circle r_{w3} of ring gear 3 is smaller than the radius of operating circle r_{w1} of ring gear 1, i.e.,

$$r_{w3} < r_{w1}.$$

This is due to the difference in the two pressure angles (Figure 23.6), whereby

$$\alpha_{w23} < \alpha_{w12}.$$

To avoid overlapping of the figure, it is not shown, but it should be borne in mind that the two operating circles with radii $r_{w2(1)}$ and $r_{w2(3)}$ of planet mesh with both ring gears 1 and 3 pass through pitch points C_{12} and C_{23}. In that, $r_{w2(1)} > r_{w2(3)}$.

The change in the direction of rotation of the output shaft of II- and $\overline{\text{II}}$-PGT with fixed ring gear 3 is illustrated in Figure 23.7.

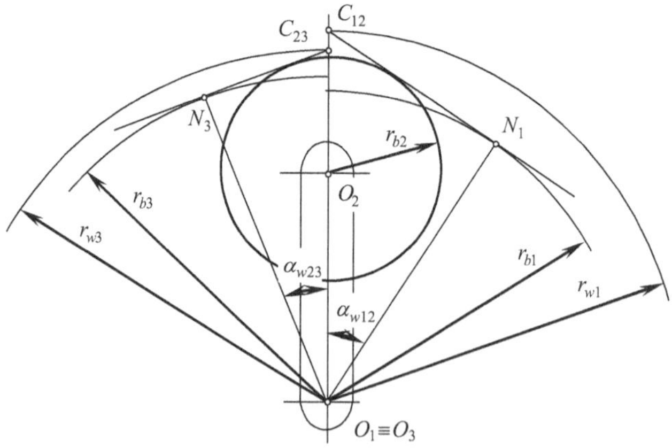

FIGURE 23.6
Two internal meshings of a one-rim planet 2 with both ring gears 1 and 3 with teeth number $z_3 > z_1$, but with operating circle radii $r_{w3} < r_{w1}$.

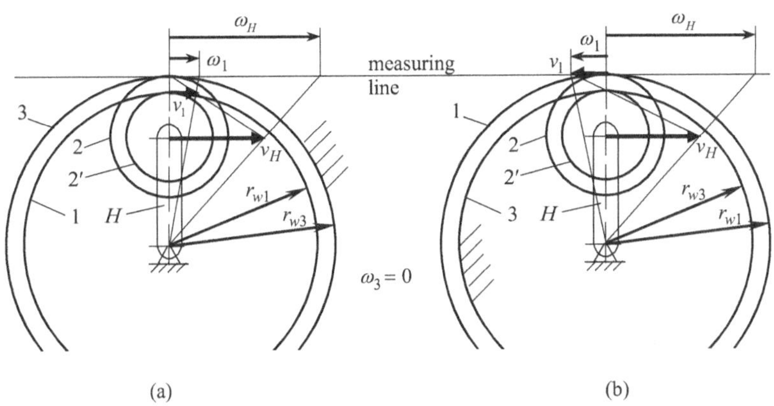

FIGURE 23.7
Change of the direction of rotation of the output shaft of both gear trains when $\omega_3 = 0$: (a) II-PGT, $i_{H1(3)} > 0$; (b) $\overline{\text{II}}$-PGT, $i_{H1(3)} < 0$.

In these types of PGTs, like the AA- and $\overline{\text{AA}}$-PGT, it is easy to obtain very large speed ratios, in the order of 1,000 ÷ 1,500, choosing the basic ratio $i_0 \approx 1$ by very small difference Δz_{31} in the number of teeth z_1 and z_3 of ring gears 1 and 3, even up to $\Delta z_{31} = z_3 - z_1 = 1$ [67, 68, 212].

The internal loads (load sharing among planets) in these PGTs are nonuniform, especially in the case of more than one two-rim planets. But since both gears are used not as power ones but as kinematic ones, the problem of uneven load distribution between planets is not relevant.

For the only case of working as a reducer, efficiency can be determined by the following formula depending on the fixed element—ring gear 1 or 3 [161]:

- For II-PGT at $\omega_1 = 0$, i.e., at $i_{H3(1)}$, as well as for $\overline{\text{II}}$-PGT at $\omega_3 = 0$, i.e., at $i_{H1(3)}$

$$\left.\begin{matrix}\eta_{H3(1)} \\ \eta_{H1(3)}\end{matrix}\right\} = \frac{i_0 - 1}{\dfrac{i_0}{\eta_0} - 1} = f(\mu_z). \tag{23.13}$$

- For II-PGT at $\omega_3 = 0$, i.e., at $i_{H1(3)}$, as well as for $\overline{\text{II}}$-PGT at $\omega_1 = 0$, i.e., at $i_{H3(1)}$

$$\left.\begin{matrix}\eta_{H1(3)} \\ \eta_{H3(1)}\end{matrix}\right\} = \frac{i_0 - 1}{i_0 - \eta_0} = f(\mu_z). \tag{23.14}$$

Particular attention should be paid to the fact that in these formulae, the basic speed ratios are taken as follows (23.1):

$i_0 \equiv i_{31(H)} > 1$ for II-PGT ($z_3 > z_1$);

$i_0 \equiv i_{13(H)} > 1$ for $\overline{\text{II}}$-PGT ($z_3 > z_1$ too!).

In these formulae, η_0 is the basic efficiency of PGT which can be determined either by the following simplified formula based on (11.7)

$$\eta_0 \approx 1 - 0.2\left(\frac{2}{z_2} - \frac{1}{z_1} - \frac{1}{z_3}\right), \tag{23.15}$$

or with greater precision as set out in Chapter 11, depending on the coefficient of friction in teeth meshing μ_z, defining the scattering of PGT's efficiency η (Figure 11.1).

The principal change in efficiency of AA- and $\overline{\text{AA}}$-PGTs working as a reducer or as a multiplier, shown in Figure 22.4, is also valid for II- and $\overline{\text{II}}$-PGTs (their efficiency is slightly higher). At II- and $\overline{\text{II}}$-PGTs self-locking occurs too, when the basic speed ratio i_0 is within the limits:

$$\eta_0 < i_0 < \frac{1}{\eta_0}. \tag{23.16}$$

This dependence is available if PGT works with one degree of freedom with output carrier [169].

Figure 23.8 shows a II-PGT working as a reducer with a positive speed ratio $i > 0$ and Figure 23.9—working as a reducer with a negative speed ratio $i < 0$.

Besides the above-presented PGTs with two internal meshings, there is also a third variant with two internal meshings (Figure 23.10) in which the only planet is with external and internal toothing. Its characteristic is the small axial gauge, which is an advantage in some cases.

As shown in Figure 23.11, II-PGT also distinguishes with small axial gauges. The only planet has two external toothings lying in one plane (one above the other), which determines the small axial gauge.

Finally, it is appropriate to define clearly the characteristics of the two types of PGT in question:

1. The correlation of ideal external torques T_1 and T_3 changes according to the basic speed ratio i_0 (Figure 23.5) as follows:
   For II-PGT, in which $i_0 \equiv i_{31(H)} > 1$, $T_3 < |T_1|$ is obtained.
   For $\overline{\text{II}}$-PGT, in which $i_0 \equiv i_{13(H)} > 1$, $|T_3| > T_1$ is obtained.

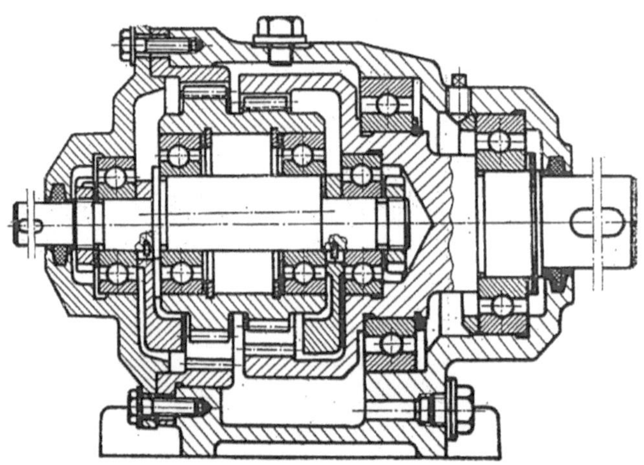

FIGURE 23.8
Arrangement of a II-PGT with one two-rim planet, working as a reducer with a positive speed ratio $i > 0$.

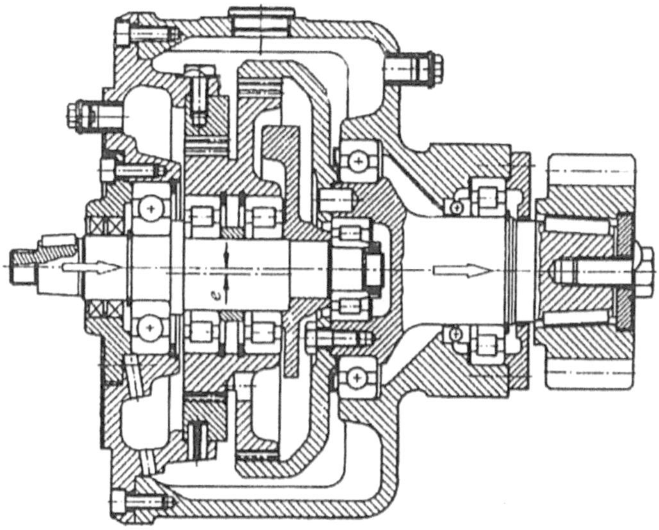

FIGURE 23.9
Arrangement of a II-PGT with one two-rim planet, working as a reducer with a negative speed ratio $i < 0$.

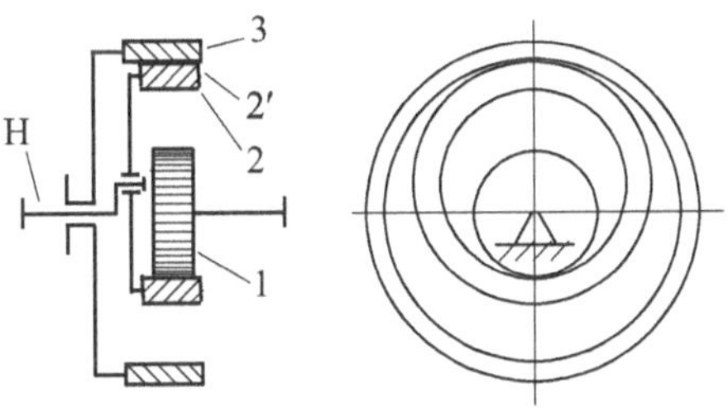

FIGURE 23.10
II-PGT with two different toothings on the only planet—external and internal, lying in one plane.

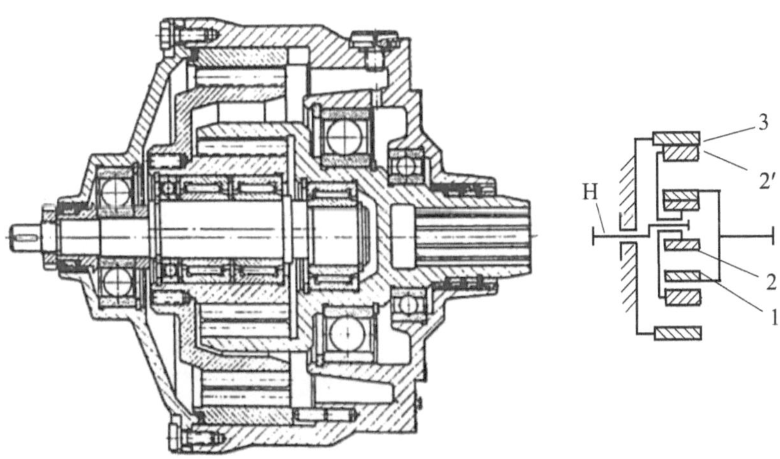

FIGURE 23.11
II-PGT with two external toothings on the only planet, lying in one plane.

2. In both cases, the torque T_H remains the lowest one (Figure 23.5).
3. In the two cases, however, the biggest torque is different (Figure 23.5)—T_1 in II-PGT and T_3 in $\overline{\text{II}}$-PGT.
4. Changing the torques correlation results in a change in the modified symbol of Wolf as shown in Figure 23.5.
5. In $\overline{\text{II}}$-PGT, in contrast to $\overline{\text{AA}}$-PGT, when proceeding from two- to one-rim planet, the direction of rotation of the output shaft changes, resp. speed ratio i changes its sign.
6. Although ring gear 3 is larger than ring gear 1, i.e., $z_3 > z_1$, when proceeding from two- to one-rim planet, the size of operating circles changes and they become $r_{w3} < r_{w1}$.

Advantages, disadvantages, and application of II- and $\overline{\text{II}}$-PGTs in comparison with AA- and $\overline{\text{AA}}$-PGTs are as follows:

Advantages

1. Very large kinematic capabilities—speed ratios of the order of $i = 1{,}000 \div 1{,}500$. Because of the low efficiency, however, usually $i = 30 \div 100$ is selected, with high values very rarely used.
2. Greater compactness due to internal meshing and planets positioning inside of central (ring) gears.
3. Manufacturing of $\overline{\text{II}}$-PGT with one-rim planet is substantially simplified.
4. Self-locking is possible, which in some cases is desirable.

Disadvantages

1. More complicated geometric calculations of the internal teeth and meshing than the external ones, especially in $\overline{\text{II}}$-PGT.
2. More complicated and low-productivity manufacturing of internal teeth (tooth cutting, not hobbing).
3. Harder manufacturing of two-rim planets of II-PGT (see Chapter 21). In these gear trains with number of planets $k > 1$, the problem arises with uneven load distribution between planets (mesh load factor K_{rmax}).
4. Low efficiency (although somewhat higher than of AA- and $\overline{\text{AA}}$-PGTs), which decreases as the speed ratio increases and results in heating, which makes these PGTs unsuitable for continuous operation.
5. The instantaneous speed ratio of these gears is not constant, too ($i \neq \text{const}$), but has significant variability leading, with very large speed ratios, even to the reverse rotation of the output shaft (Figure 22.6).
6. Self-locking in some cases is undesirable.
7. Increased noise due to high speed of the planets.

Application

1. Very limited, only as low-power, kinematic gear trains with significant and very large speed ratios, when low efficiency is tolerable or unimportant, even self-locking may be desirable.

2. Unavailability for continuous work due to heating as a result of low efficiency.
3. Mostly in metal-cutting machines and in instrumentation, where efficiency is irrelevant, and work is not continuous.

More details about kinematic calculations of $\overline{II}$-PGT can be found in [161] and for internal meshing with a small difference in the number of teeth—in [67, 68, 169, 212].

24

AAI- and $\overline{\text{AAI}}$-Planetary Gear Trains

Figure 24.1 gives a comparison of the negative AI- and $\overline{\text{AI}}$-planetary gear trains ($\overline{\text{AI}}$-PGTs), out of which the positive AAI- and $\overline{\text{AAI}}$-PGTs are obtained by adding another two- or one-rim planet to each of the power branches. Their basic speed ratio is positive:

$$i_0 = \frac{\omega_{1rel}}{\omega_{4rel}} = \frac{\omega_1 - \omega_H}{\omega_4 - \omega_H} = \left\{ \begin{array}{c} \dfrac{z_2}{z_1} \cdot \dfrac{z_3}{z_2'} \cdot \dfrac{z_4}{z_3'} \\ \dfrac{z_4}{z_1} \end{array} \right\} > 0. \tag{24.1}$$

The different gear trains shown in Figure 24.1 may be considered to have originated from one another as shown by arrows in the figure. This has been done in two ways—by proceeding from two- to one-rim planet or by adding another planet (two- or one-rim) to each power branch.

Figure 24.1 also shows external torques on all depicted PGTs and the differences that exist between them. As can be seen from the modified symbol of Wolf for the negative (AI and $\overline{\text{AI}}$) and positive (AAI and $\overline{\text{AAI}}$) PGTs, the torques T_4 and T_H on both ring 4 and carrier H have changed directions and magnitudes. Torques T_1 and T_H on sun gear 1 and carrier H, and not T_1 and T_4, become unidirectional.

In reality, though rarely, $\overline{\text{AAI}}$-PGT is the one that is mostly used. Its assembly condition is [161]

$$\boxed{\frac{z_4 - z_1}{k} = \text{an integer}}, \tag{24.2}$$

where k is the number of power branches, i.e., the number of planet pairs.

Of course, the adjacent condition must be observed so that the planets do not touch (see Section 4.3).

AAI-PGT can realize larger speed ratios (operating as a reducer) than the drive $\overline{\text{AAI}}$-PGT—approximately up to $i = 80 \div 100$. But it is very rarely used due to the high technological accuracy requirements and only as a reducer. $\overline{\text{AAI}}$-PGT is axially very compact, like $\overline{\text{AI}}$-PGT, and in some cases, the unidirectionality of the torques T_1 and T_H on sun gear 1 and carrier H leads to arrangement and design advantages. Figure 24.2 illustrates this advantage by an inter-axle differential of a vehicle. It is noteworthy that when the basic speed ratio is $i_0 = 2$, exactly $T_H = T_1$ is obtained. At $i_0 > 2$, $T_H > T_1$ is obtained, and if $i_0 < 2$, then $T_H < T_1$. Torque ratio t of this PGT is

$$\boxed{t = \frac{T_H}{T_1} = i_0 - 1 > +1}. \tag{24.3}$$

In this way, both symmetric and asymmetric differential can be obtained.

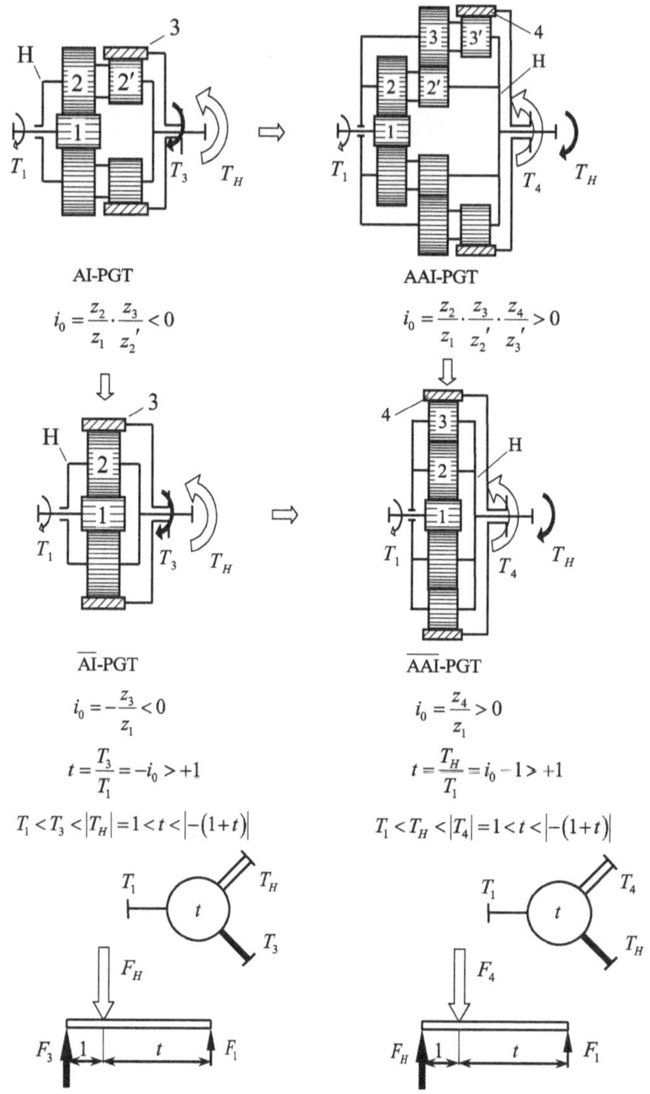

FIGURE 24.1
Comparing different PGTs and transforming them into one another.

Figure 24.2 [161] illustrates the more commonly used case when the basic speed ratio is $i_0 > 2$, for which the torque size correlation in the figure refers.

This gear is used in a vehicle, mostly as an inter-axle differential with $i_0 > 2$ and rarely as an inter-wheel differential with $i_0 = 2$.

The problem with the uneven distribution of the load among planets, especially in the case of the AAI-PGT, is much more acute than with the most commonly used $\overline{\text{AI}}$-PGT because of the higher number of negative factors, including the two-rim planets, and the high number of holes for the planet pins, which must be positioned very accurately on the carrier.

It is noteworthy that the $\overline{\text{AAI}}$-PGT considered here, just like the $\overline{\text{AI}}$-PGT (see Section 7.4.3) and $\overline{\text{AAA}}$-PGT (see Chapter 25), can work as a summation differential. The characteristic

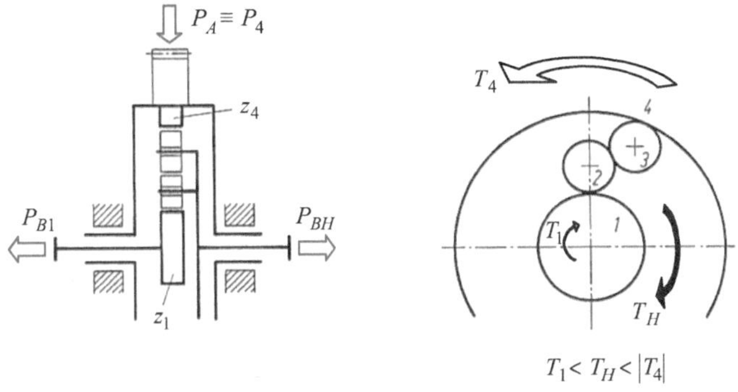

FIGURE 24.2
$\overline{\text{AAI}}$-PGT as an inter-axle differential in a vehicle.

feature here is that the driving elements (inputs) are sun gear 1 and carrier H (their external torques are unidirectional), and the output element (with the highest torque) is ring gear 4. In the case of twin-motor drive (summation differential), four speeds are possible, too, depending on how the motors work. The dependence of the output angular velocity ω_B as a function of the basic speed ratio i_0 is the same as on the diagram in Figure 7.8 for the $\overline{\text{AI}}$-PGT.

Advantages, disadvantages, and application of AAI- and $\overline{\text{AAI}}$-PGTs are as follows:

Advantages

1. Both PGTs can achieve greater speed ratios than AI and even greater than $\overline{\text{AI}}$-PGT—in particular, AAI-PGT speed ratio is in the order of $i = 80 \div 100$ and of $\overline{\text{AAI}}$-PGT—much lower.
2. $\overline{\text{AAI}}$-PGT is axially very compact, like $\overline{\text{AI}}$-PGT, which in some cases is an advantage.
3. The unidirectionality of sun gear and carrier torques T_1 and T_H in some cases may also be an advantage.

Disadvantages

1. More gears, which means higher complexity of the arrangement and increased production cost, also increase in the mass and the price of the gear train.
2. More complicated manufacturing, especially of AAI-PGT.
3. The problem with the uneven distribution of the load among planets is much more acute than in PGTs with one-rim planets and with a single planet in individual power flows.

Application

1. Of AAI-PGT—very limited and only as a reducer.
2. Of $\overline{\text{AAI}}$-PGT—not so limited mostly as an inter-axle differential in vehicles (Figure 24.2).

25

AAA-, $\overline{\text{AAA}}$-, IAI-, and $\overline{\text{IAI}}$-Planetary Gear Trains

The planetary gear trains (PGTs) shown in Figure 25.1 are from the group of negative-ratio PGTs (see Figure 2.4), as their basic speed ratio is negative, which is given by

$$i_0 = \frac{\omega_{1rel}}{\omega_{3rel}} = \frac{\omega_1 - \omega_H}{\omega_3 - \omega_H} < 0. \tag{25.1}$$

It can be assumed that all variants are derived from AAA- and IAI-PGTs by simplifying the planets—proceeding from two- to one-rim planet.

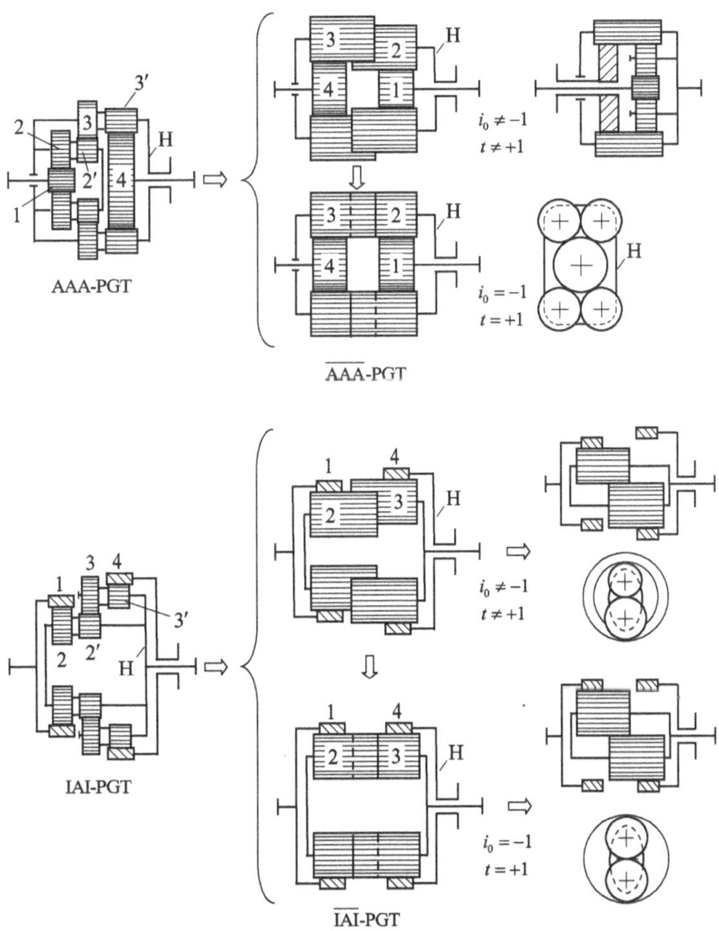

FIGURE 25.1
Different types of negative-ratio simple PGTs with three meshings.

Only $\overline{AAA}$- and $\overline{IAI}$-PGTs are used, mostly as a symmetric inter-wheel differential in the vehicles. In $\overline{AAA}$- and $\overline{IAI}$-PGTs, the carrier, the carrier H is the input element, the both output torques T_1 and T_4 are the same, since the basic speed ratio is $i_0 = -1$, and the torque ratio is $t = +1$.

$\overline{AAA}$- and $\overline{IAI}$-PGTs can also be used as asymmetric differentials when $i_0 \neq -1$ and $t \neq +1$, such as inter-axle differentials in vehicles. The drive is also performed from the carrier H, and the two output shafts leading to the front and rear axles are connected to the two central gears 1 and 4. The two output torques T_1 and T_4 are not equal, in which $T_1 < T_4$.

Assembly condition for $\overline{AAA}$- and $\overline{IAI}$-PGTs is [161]

$$\boxed{\frac{z_1 + z_4}{k} = \text{an integer}}, \tag{25.2}$$

where k is the number of planets.

Figure 25.2 shows symmetric inter-wheel $\overline{AAA}$-differential arrangements. The carrier is driven by a bevel or worm gear.

There is also the so-called double differential[1], which is used, though not very often, mostly in steering systems of continuous track vehicles (tractors, tugs, tanks) [79, 177].

For comparison, Figure 25.3 shows the two differentials—an ordinary one (Figure 25.3a) and a double one (Figure 25.3b). The difference is that in the double differential, on the shafts of the planets 2 and 3, there is one more gear on each—2a and 3a, respectively. They in turn mesh with additional sun gears 1a and 4a, which can be fixed by brakes.

In ordinary differential (Figure 25.3a), by stopping one of the sun gears (1 or 4), the other sun gear (4 or 1) starts to spin twice as fast and the vehicle (wheel or track) starts to turn very sharply.

In double differential (Figure 25.3b), in case of a fixed sun gear, e.g., 1a, the sun gear 1 on the same side decreases its speed but does not stop, and the other sun gear 4 starts to rotate faster, but not so fast as to cause a sharp turn (a concept known as regenerative, because during steering, there is no power lost in slipping the brakes [79]). Of course, the smoother turning takes place with a larger radius (by the way, the main disadvantage of this steering system is that there is only one turning radius).

More information about vehicle differentials can be found in [79].

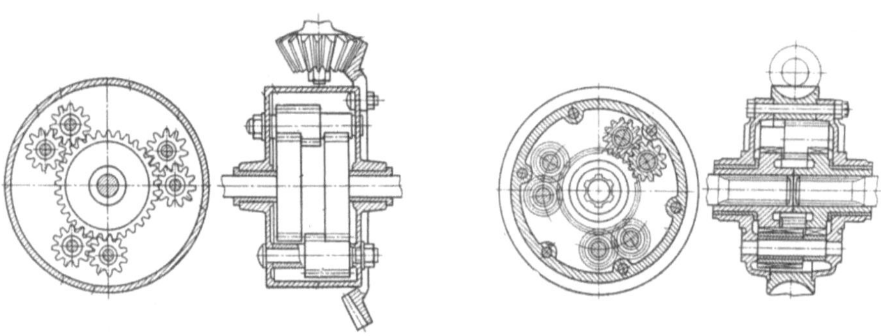

FIGURE 25.2
Arrangements of symmetric inter-wheel $\overline{AAA}$-differentials for vehicles.

[1] Invented by the Cleveland Tractor Co. (USA) in 1921 and still used, together with its variants [79].

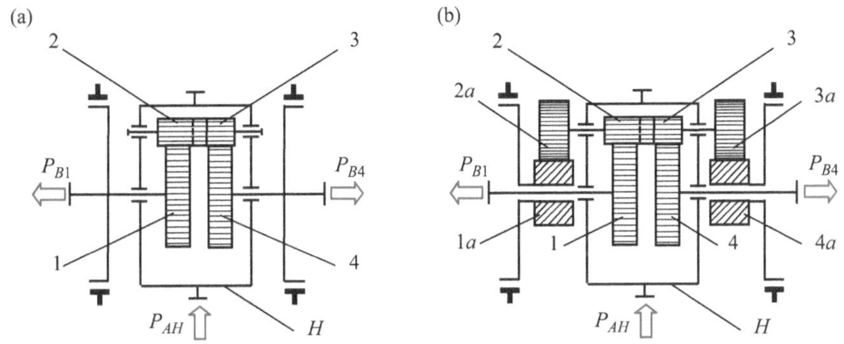

FIGURE 25.3
Symmetric inter-wheel differential—uniform (a) and double (b).

The above examples show the performance of differentials as division PGTs. Of course, they can also be used as a summation PGTs in twin-motor drives. Most commonly, the central gears 1 and 4 are the driving ones, at certain angular velocities ω_1 and ω_4. In symmetrical differential $\left(i_0 = -1, t = +1\right)$ for the carrier angular velocity $\omega_H = \frac{\omega_1 + \omega_4}{2}$ is obtained. Figure 25.4 shows the kinematic scheme of a twin-motor drive PGT used in tower cranes that can perform different rotation speeds of the rope drum of a hoisting mechanism. If $z_1 = z_3$ (as in Figure 25.4), when only one motor works, twice the speed ratio is obtained than when the two motors work. Thus, upon failure of one motor, the other one is able to complete the load handling at twice lower speed but at nominal power (without overloading). Figure 25.5 shows a drawing of the drum with a built-in PGT.

The same type of PGT, but when $z_1 \neq z_4$ (Figure 25.6) and the two motors work in turn one after the other, can perform four gear ratios as follows:

- At unidirectional rotation of both motors $i = 1$.
- At rotation only of the right motor and the large sun gear 4 $i > 1$.
- At rotation only of the left motor and the small sun gear 1 even bigger $i > 1$ is obtained.
- At opposite rotation of both motors, the highest speed ratio is obtained.

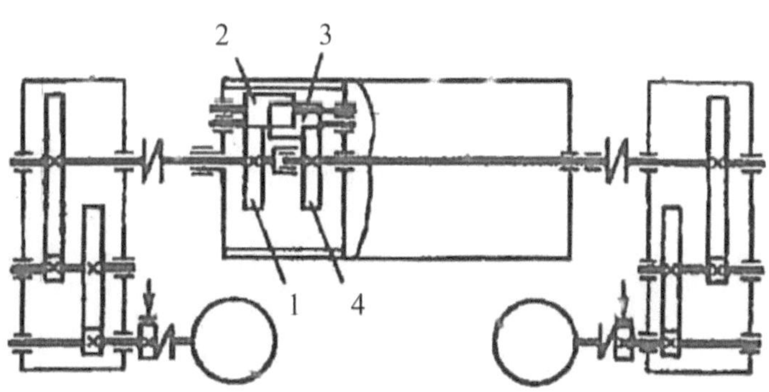

FIGURE 25.4
Kinematic scheme of a twin-motor drive $\overline{AAA}$-PGT for the hoisting mechanism of a tower crane ($z_1 = z_4$).

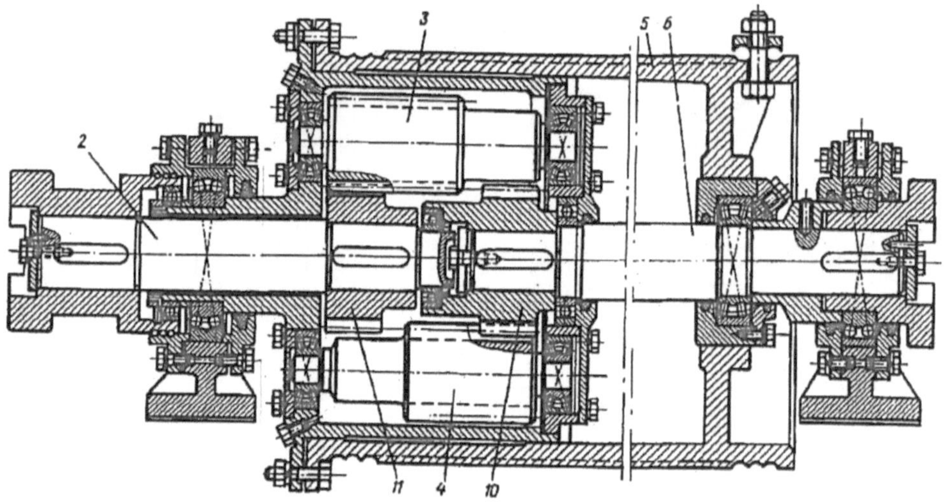

FIGURE 25.5
Drawing of the rope drum with the PGT from Figure 25.4.

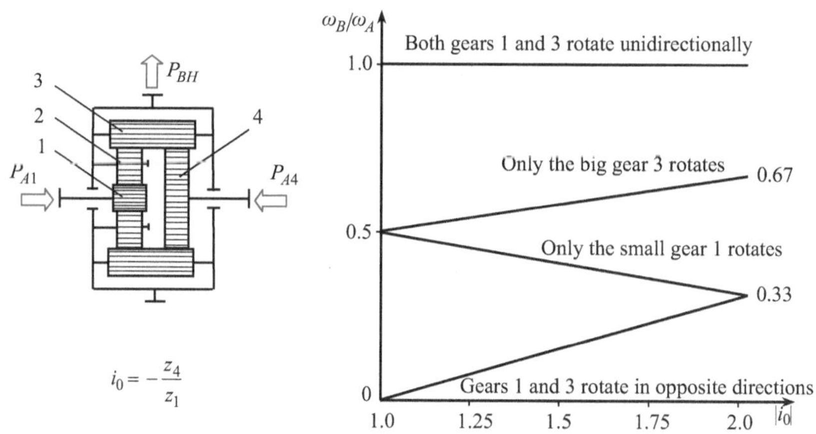

FIGURE 25.6
Kinematic scheme of a summation $\overline{\text{AAA}}$-PGT with a twin-motor drive, operating as a change-gear ($z_1 \neq z_4$).

Figure 25.6 shows the change of the output angular velocity (of the carrier H) $\omega_B \equiv \omega_H$, as a function of basic speed ratio i_0, at various cases of operation of the driving motors. By appropriately choosing the number of teeth, an appropriate (desired by the designer) correlation between these speed ratios can be obtained. By changing the motor speeds (e.g., by changing the number of pitch points), more speed ratios can also be obtained.

This case has much in common with the case of the $\overline{\text{AI}}$-PGT, operating as a twin-motor driven summation differential, considered in Section 7.4.3. Both PGTs are negative-ratio ones (Figure 2.4) with the following basic speed ratios:

$\overline{\text{AI}}$-PGT

$$i_0 = -\frac{z_3}{z_1}; \tag{25.3}$$

$\overline{\text{AAA}}$-PGT

$$i_0 = -\frac{z_4}{z_1}. \tag{25.4}$$

It can be seen that in $\overline{\text{AAA}}$-PGT, the large sun gear 4 plays the role of the ring gear 3 of the $\overline{\text{AI}}$-PGT.

Considering these differences, Figure 25.6 shows the valid dependence in this case

$$\omega_{\text{H}} = \omega_B = f\left(|i_0| = \frac{z_4}{z_1}\right). \tag{25.5}$$

Since, unlike the $\overline{\text{AI}}$-PGT, at $\overline{\text{AAA}}$-PGT, it is possible for the basic speed ratio to be $|i_0| = 1$, it is possible to obtain a speed ratio of the differential $|i_0| = 2$ and carrier angular velocity $\omega_H = \omega_A/2$, respectively.

For completeness, it must be noted that $\overline{\text{IAI}}$-PGT (Figures 2.4 and 25.1), which is only with two planets and central gears with internal teeth, can work as a differential.

Another interesting application of $\overline{\text{AAA}}$-PGT, also as a differential, is to reduce the load on the last stage of a heavily loaded transmission (Figure 25.7) [30]. $\overline{\text{AAA}}$-PGT operates as a division differential.

The kinematic scheme from Figure 25.7 is very suitable for driving of a rope drum of hoisting mechanisms. With some appropriate design solution, a very compact arrangement can be achieved (Figure 25.8). After learning Chapter 29, the reader could make a kinematic and power analysis of the two-carrier PGTs from the above two kinematic schemes.

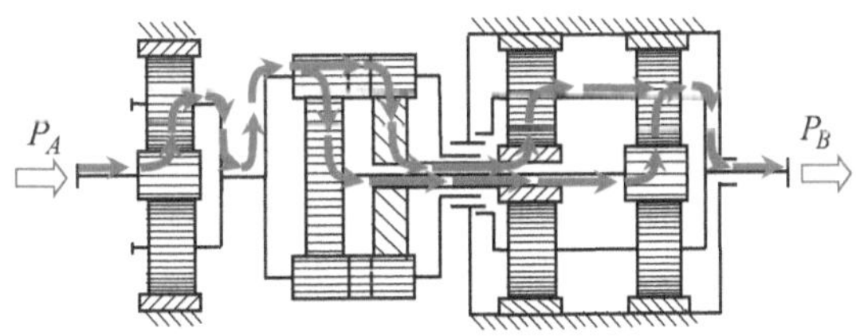

FIGURE 25.7
$\overline{\text{AAA}}$-PGT as a part of compound multi-carrier PGT.

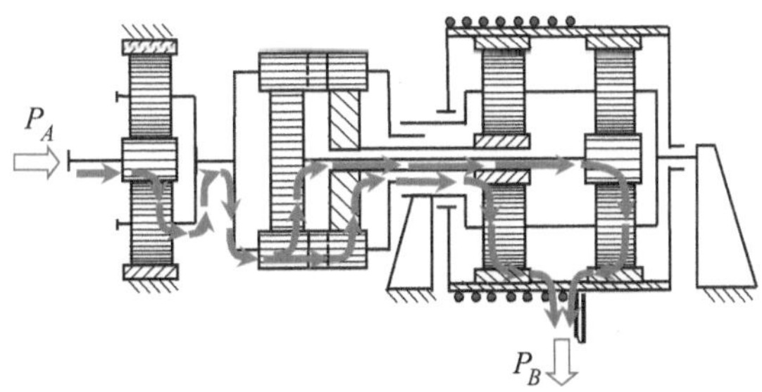

FIGURE 25.8
$\overline{\text{AAA}}$-PGT as a part of compound multi-carrier PGT for rope drum.

Advantages, disadvantages, and application of $\overline{AAA}$- and $\overline{IAI}$-PGTs are as follows:

Advantages

1. $\overline{AAA}$-PGT uses only external toothed gears which are easier for machining, but $\overline{IAI}$-PGT is more compact.
2. Unidirectional torques T_1 and T_4 on both central gears 1 and 4 make these PGTs appropriate for inter-wheel as well as for inter-axle differentials.

Disadvantages

1. More complicated manufacturing, especially of $\overline{IAI}$-PGT.
2. The problem with the uneven distribution of the load among planets is much more acute than in PGTs with one-rim planets and with a single planet in individual power flows.

Application

1. As division gear trains, mostly as symmetric inter-wheel and asymmetric inter-axle differentials.
2. As summation gear trains—in hoisting equipment with twin-motor driving (as symmetric or asymmetric differentials).

26

I-Planetary Gear Train (K-H-V Gear Train)

It is characteristic for this planetary gear train (PGT) that it has only one planet with one toothing, as $\overline{\text{II}}$-PGT (Figure 23.1), but in contrast, the planet meshes with only one gear. Figure 26.1 shows the two possible variants, where either planet 2 or ring gear 3 performs eccentric motion, the first option being used more often. By means of an eccentric (cranked) shaft playing the role of carrier H, the movement is transmitted through the planet 2 to the central element—shaft 1 or ring gear 3. There are also other ways of transmitting the movement between the eccentric and the central element—a cardan shaft (universal joint), a pin coupling, etc.

According to Prof. Kudryavtsev's classification [147], this PGT is referred to as K-H-V (wheel–carrier–shaft) gear train.

Both variants of this PGT are positive-ratio, as their basic speed ratio is clearly positive:

$$i_0 = \frac{\omega_{2rel}}{\omega_{3rel}} = \frac{\omega_2 - \omega_H}{\omega_3 - \omega_H} > 0. \tag{26.1}$$

Figure 26.2 shows an I-PGT with involute meshing.

The internal meshing is characterized by a small difference in the number of teeth of planet 2 and ring gear 3, which may even be $\Delta z_{32} = z_3 - z_2 = 1$ [67, 68, 212].

At a small difference in the number of teeth of the internal meshing, which is usually $\Delta z = z_3 - z_2 = 1 \div 4$, under load the gear teeth become deformed and more teeth mesh (Figure 26.3). This means distribution of the load on more teeth, which results in an increase in the load capacity of this gear train, which is its advantage [189]. As can be seen from the diagram in Figure 26.4, the number of loaded teeth n_{23} depends on the load T_H, on the difference in the number of teeth Δz and also varies depending on the number of teeth of

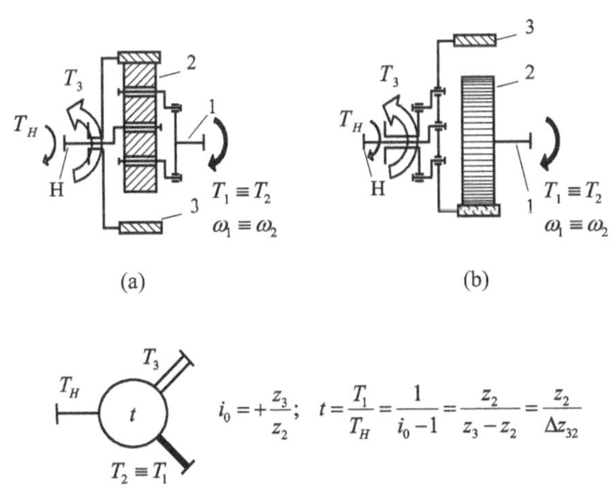

FIGURE 26.1
Both variants of I-PGT: (a) With eccentricity on planet 2; (b) with eccentricity on ring gear 3.

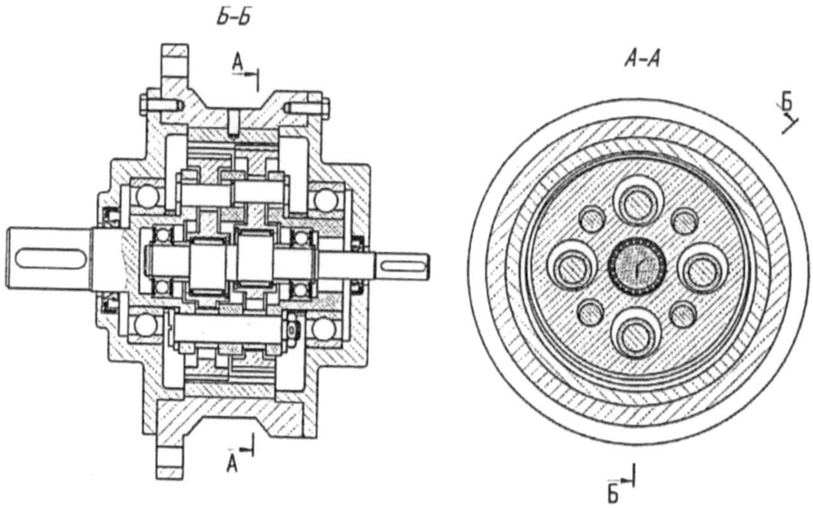

FIGURE 26.2
Arrangement of I-PGT with involute meshing.

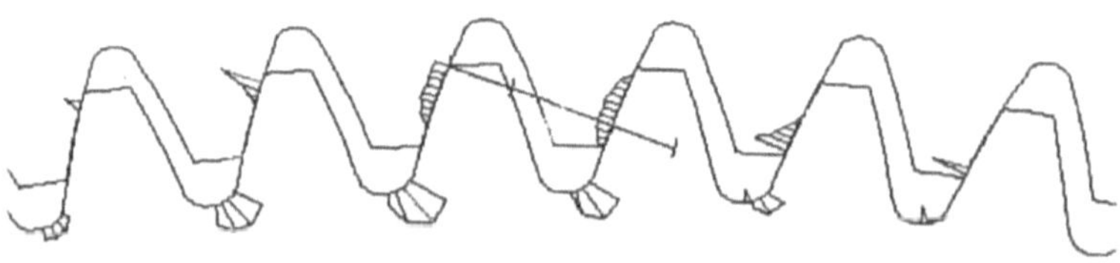

FIGURE 26.3
Increase in the number of meshed teeth due to deformations in internal meshing with a small difference in the number of teeth $\Delta z = 1 \div 4$. (Courtesy of POPOV, V. *Überdeckungsgrad und Tragfähigkeit von Exzenter-Getrieben mit Evolventenverzahnung*. Dissertation. Magdeburg: Otto-von-Guericke-Universität, 1996.)

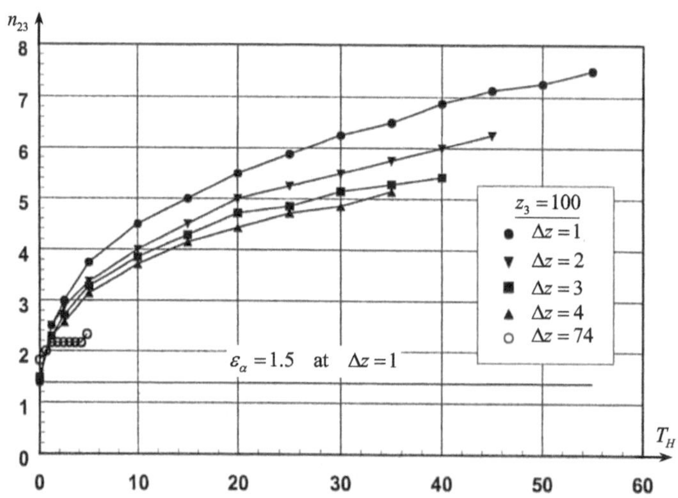

FIGURE 26.4
Number of teeth in mesh n_{23} depending on the carrier torque T_H and difference in the number of teeth Δz. (Courtesy of POPOV, V. *Überdeckungsgrad und Tragfähigkeit von Exzenter-Getrieben mit Evolventenverzahnung*. Dissertation. Magdeburg: Otto-von-Guericke-Universität, 1996.)

ring gear z_3. The largest number of loaded teeth is obtained at $\Delta z = 1$. This number increase with the increase in the load T_H and increase in the number of teeth of the ring gear z_3.

On the other hand, however, the inclusion of more teeth in mesh leads to an increase in the meshing loss. This is due to the fact that teeth meshed farther from the pitch point create greater losses due to greater friction.

It can be seen that in the internal meshing with a small difference in the number of teeth, as the load increases, some indicators improve, and others deteriorate.

Besides involute meshing (which is characterized by a large pressure angle [147]), these PGTs also use a pin meshing, which is a variation of the cycloid meshing [6, 70, 175, 234]. More details about the calculation of internal meshing with a small difference in the number of teeth can be found in [67, 68, 189, 212] and about the pin meshing in [69, 70, 163, 180, 185, 234]. These gear trains are known as cycloidal drives (Figure 26.5). Their efficiency is in the order of

$$\eta = 0.75 \div 0.90.$$

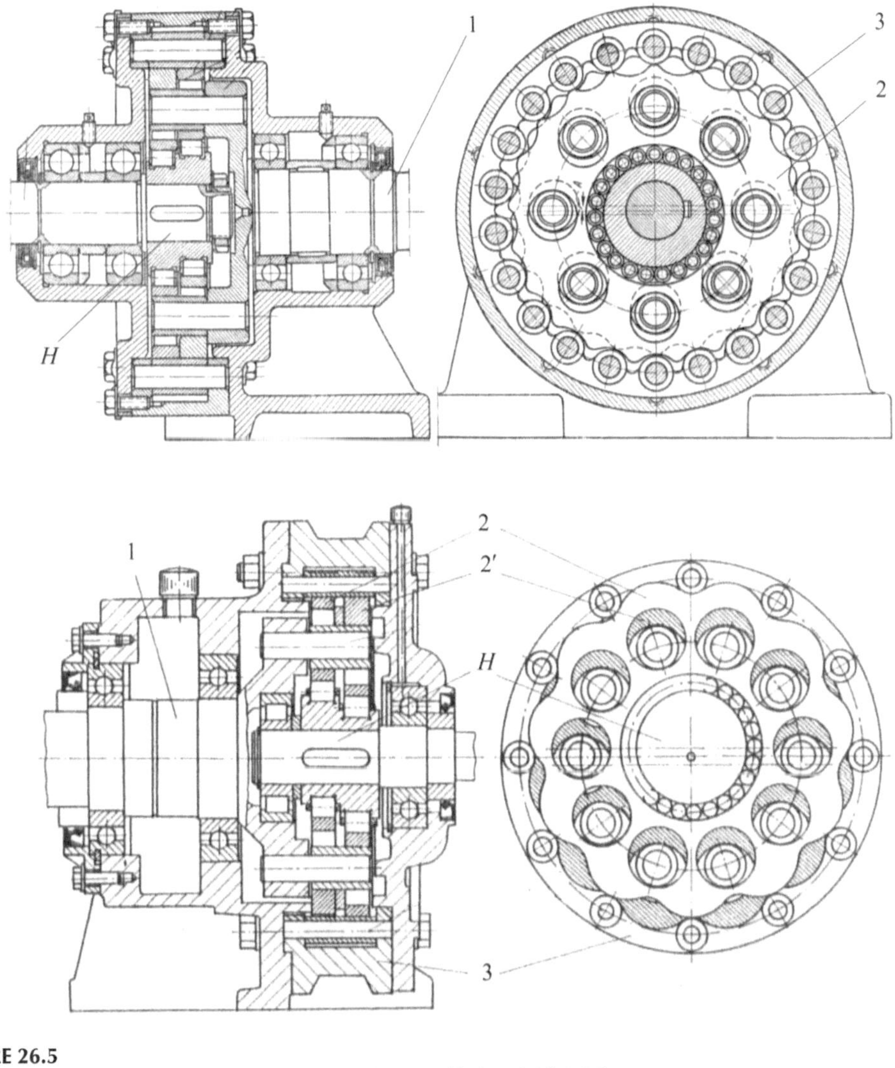

FIGURE 26.5
Arrangements of I-PGT with pin meshing, or the so-called cycloidal drive.

Efficiency of I-PGTs is not very high. It depends on the type of meshing—involute or pin (cycloidal) as well as on the manufacturing accuracy. Due to their not very high efficiency, these PGTs are not appropriate for continuous work.

I-PGT is used only as a reducer, and it is mainly the variant in Figure 26.1a with fixed ring gear 3 ($\omega_3 = 0$), as in Figure 26.5. Carrier H is input and shaft 1—output. Speed ratio is determined by the following formula [167]:

$$i_{H2(3)} \equiv i_{H1(3)} = -\frac{z_2}{z_3 - z_2} = -\frac{z_2}{\Delta z_{32}} < 0. \tag{26.2}$$

Figure 26.6 shows the diagram of peripheral velocity for this case.

The gear ratios that can usually be achieved by this PGT are up to $i \approx 100$.

Figure 26.7 shows the two variants of the PGT in question in their practical application in hoisting mechanisms. For both cases, the speed ratios are as follows [167]:

for the variant in Figure 26.7a

$$\boxed{i_{H3(2)} \equiv i_{H3(1)} = \frac{z_3}{z_3 - z_2} = \frac{z_3}{\Delta z_{32}} > 0}; \tag{26.3}$$

for the variant in Figure 26.7b

$$\boxed{i_{H2(3)} \equiv i_{H1(3)} = -\frac{z_2}{z_3 - z_2} = -\frac{z_2}{\Delta z_{32}} < 0}. \tag{26.4}$$

In Figure 26.8, the peripheral velocity diagrams for both cases are shown. The same formulae can also be obtained by the torque method—an appropriate task for the reader.

Advantages, disadvantages, and application of I-PGT are as follows:

Advantages

1. High compactness, especially in the axial direction due to internal meshing and only two gears.
2. Significant kinematic capabilities—speed ratios up to $i \approx 100$.
3. A small number of gears—just two.

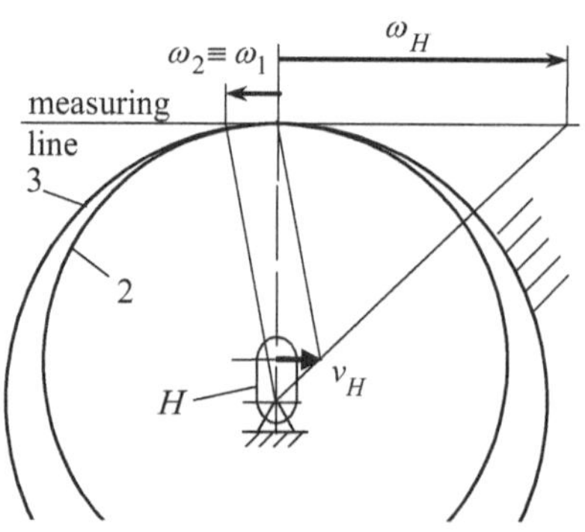

FIGURE 26.6
Peripheral velocity diagram of I-PGT with $F = 1$ degree of freedom as a reducer at $\omega_3 = 0$.

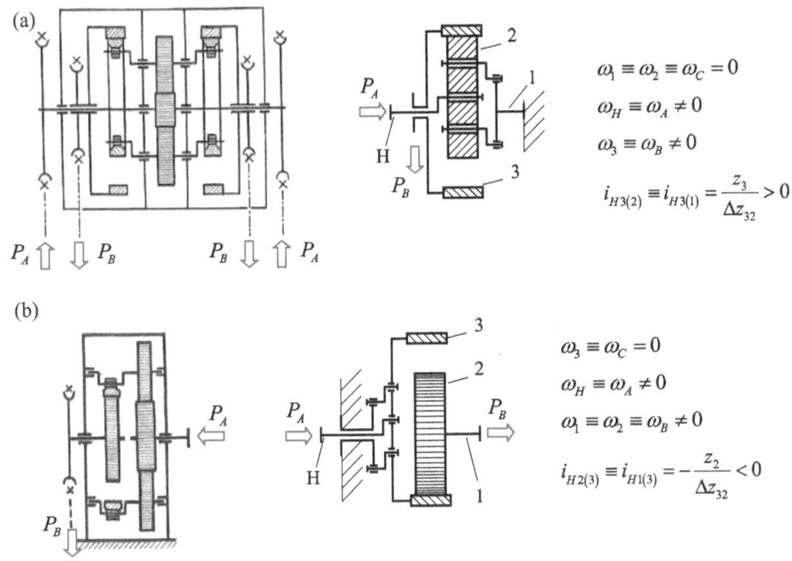

FIGURE 26.7
Hoisting mechanism with I-PGT: (a) With eccentric planet that performs circular translation; (b) with eccentric ring gear that performs circular translation.

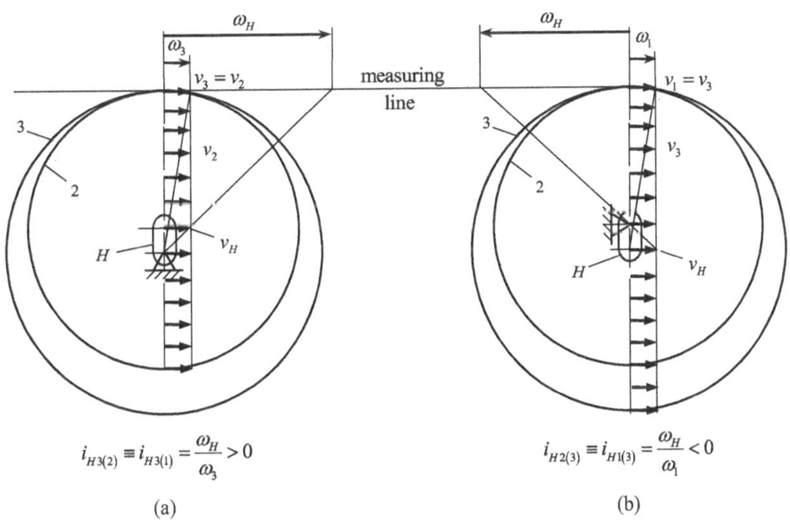

FIGURE 26.8
Peripheral velocity diagram of the I-PGT from Figure 26.7: (a) With eccentric planet 2 that performs circular translation and fixed shaft 1 $(\omega_1 \equiv \omega_2 = \omega_C = 0)$; (b) with eccentric ring gear 3 that performs circular translation $(\omega_3 = \omega_C = 0)$.

Disadvantages

1. Complex geometric calculations of shifted meshing with a small difference in the number of teeth reaching $\Delta z_{32} = z_3 - z_2 = 1$.
2. Complicated manufacturing of pin (cycloidal) meshing.

3. High load on the eccentric shaft bearing, which reduces its durability and makes it a weak point of the gear train [147].
4. Comparatively low efficiency, depending on the type of meshing (involute or cycloidal), which does not allow for continuous operation, as well as the transmission of large power.
5. Very high accuracy is required, which affects the price of the gear train.

Application

Above all, as low-power gear trains with narrow axial gauges and interrupted work in hoisting and handling machines (travel mechanisms), robots, etc.

27

Uncoaxial (Open) A- and I-Planetary Gear Trains

Unlike the different planetary gear trains (PGTs) considered in the previous chapters, A- and I-PGTs are uncoaxial gear trains which, according to [166, 167], are still called *open* PGTs. They do not have three, but only two coaxial shafts.

I-PGT shown in Figure 27.1 belongs to the category of positive-ratio gear trains because its basic speed ratio i_0 is positive:

$$i_0 = \frac{\omega_{2rel}}{\omega_{3rel}} = \frac{\omega_2 - \omega_H}{\omega_3 - \omega_H} = \frac{z_3}{z_2} > 0. \tag{27.1}$$

It is used, e.g., in textile machines (twining machines) and rope-twisting machines. Twine bobbins spin around their axes which revolve about the planet shafts which at the same time (with driven carrier H) revolve about the main axis. Ring gear 1 is fixed.

Figure 27.2 shows negative-ratio A-PGT with basic speed ratio given by

$$i_0 = \frac{\omega_{1rel}}{\omega_{2rel}} = \frac{\omega_2 - \omega_H}{\omega_2 - \omega_H} = -\frac{z_2}{z_1} < 0, \tag{27.2}$$

used to drive a mixer.

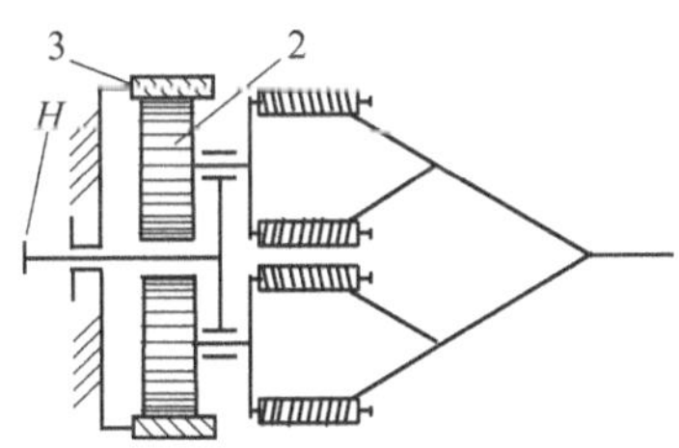

FIGURE 27.1
Positive uncoaxial I-PGT.

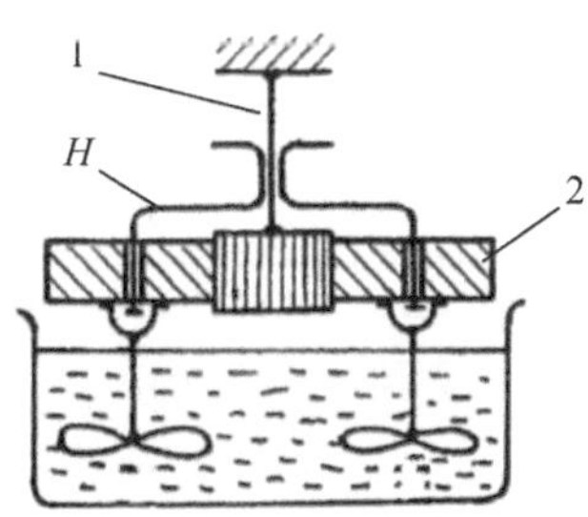

FIGURE 27.2
Negative-ratio A-PGT of a mixer.

Although the shown PGTs are with two central shafts (input and fixed), their planets are mounted on shafts that transmit the movement and power required to overcome the corresponding technological resistances (from mixing, twisting, etc.). Thus, the input and output shafts are not coaxial.

Various methods can be used to determine their kinematics (angular velocities ω_H and ω_2) (see Chapter 7). Especially, the open I-PGT has kinematics similar to the kinematics of I-PGT in Chapter 26 and Figure 26.6.

Unlike the coaxial PGT, in uncoaxial one, output power is transmitted by planet shafts, and their movement is part of a technological process. For these gears, therefore, it is difficult to talk about advantages and disadvantages, because in a sense, they are incomparable with the corresponding tri-shaft PGTs considered in previous chapters.

Besides the cases shown in Figures 27.1 and 27.2, the opposite arrangement is possible, in which (Figure 27.3) the planet 1 is the input and carrier H—output (with fixed ring gear 2 as in the previous examples). In both cases depicted in Figure 27.3, the motor is directly connected to the single planet 1. The rotation mechanisms shown in this figure are used in hoisting (handling) and dredging machines, excavators, turrets of tanks and battleships.

Interesting are two further examples shown in Figures 27.4 and 27.5.

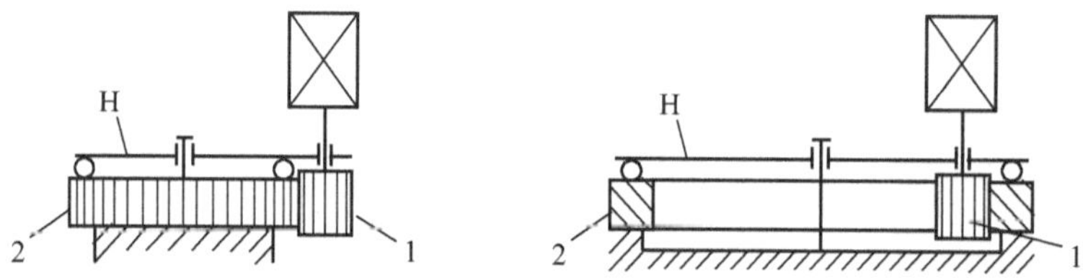

FIGURE 27.3
Uncoaxial A- and I-PGTs in rotation mechanisms of handling machines, excavators, etc.

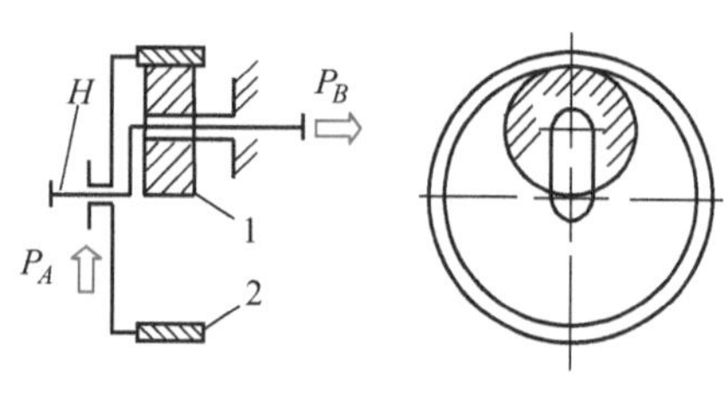

FIGURE 27.4
Kinematic scheme of Wankel engine.

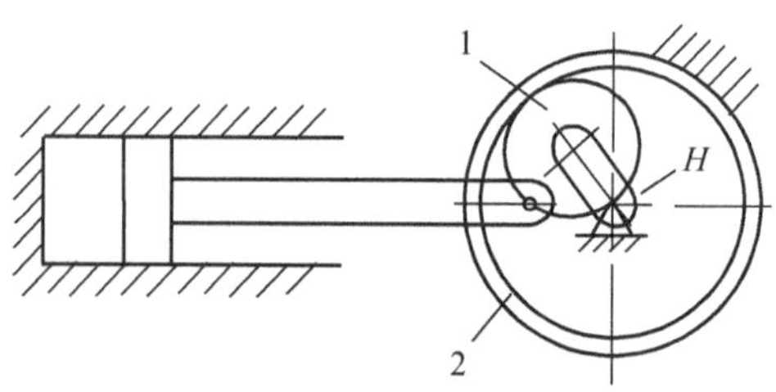

FIGURE 27.5
Conversion of the reciprocating movement of the piston into rotation by an I-PGT.

Figure 27.4 shows the kinematic scheme of the Wankel engine, in which the small gear 1 (of I-PGT) is fixed and ring gear 2 performs a pericyclic movement. The movement of the engine rotor connected with ring gear 2 is transmitted to the carrier H and to the output shaft.

Figure 27.5 shows a particular case of uncoaxial I-PGT where the planet 1 is twice smaller than the ring gear 2, i.e., $z_2/z_1 = 2$. In this particular case, the point Q lying on the operating circle of planet 1 in its movement describes a hypocycloid degenerated into a straight line. This allows for the reciprocating movement of the piston to be converted into a rotation. In this case, the carrier H acts as a crankshaft.

Uncoaxial (open) PGTs are used in textile, chemical, food, and other industries as part of the technology machines.

Part IV

Compound Planetary Gear Trains

28

Types of Compound Planetary Gear Trains

Compound planetary gear trains (PGTs) consist of a few simple (single-carrier) PGTs. As a building element of compound PGTs, AI-PGT is most commonly used, and hence, it was considered in a relatively detailed manner in Part II.

Considering the fact that simple PGTs have only one carrier and they are called *single-carrier*, the compound PGTs depending on the number of component simple PGTs are designated as follows [31]:

- *Two-carrier* compound PGT—consists of two component PGTs.
- *Three-carrier* compound PGT—consists of three component PGTs.
- *Four-carrier* compound PGT—consists of four component PGTs.
- *Multi-carrier* compound PGT—with more than four component PGTs.

According to [166, 167], two-carrier PGTs are designated as *simple* compound PGTs, and those with more than two component gear trains are designated as *higher* compound PGTs. On this basis, the terms *lower* and *higher* are adopted in the book (Figure 28.1).

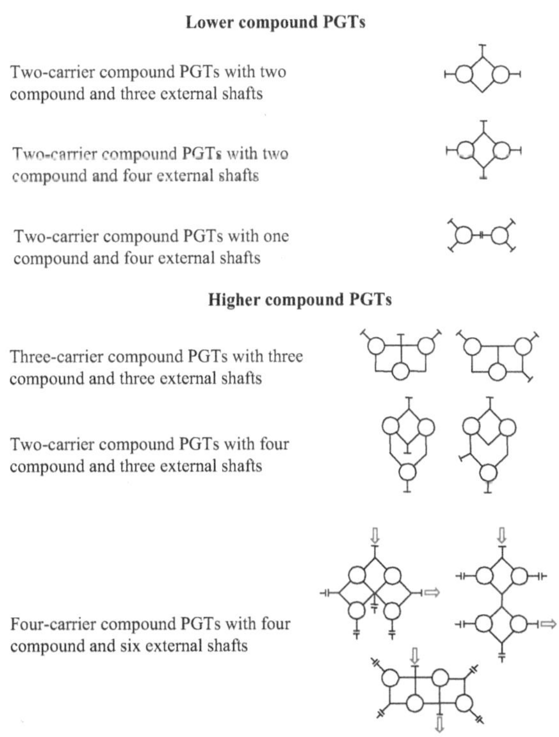

FIGURE 28.1
Types of compound PGTs and their structure depicted by the symbol of Wolf.

The area of compound PGTs is incredibly extensive. Possible combinations of component PGTs in a compound one are estimated at several thousand [202]. Therefore, the possible combinations shown in Figure 28.1 [25], especially in the three- and four-carrier PGTs, cannot claim to be exhaustive. They serve only as examples.

There is a particular category of compound PGTs, denoted as *reduced PGTs*. They are derived from two-carrier PGTs, both carriers of which have merged, i.e., are reduced to a single carrier. Such reduced gears are, e.g., Wolfrom gear train [255] and Ravigneaux gear train [161], which are discussed in Chapter 33. A crude error is to consider these PGTs to be simple (one-carrier) PGTs due to the fact that they are with only one carrier.

Compound PGTs similar to the simple ones can work with either $F = 1$ degree of freedom, i.e., as a reducer or a multiplier, or $F = 2$ degrees of freedom, as a summing or a division differential.

The extraordinary diversity of compound PGTs and the complexity of processes involved may impede the designer, and this requires developing both a clear and understandable system of these gear trains, as well as a simple, comprehensible, easy-to-use, and, above all, visual way for their analysis.

29

Two-Carrier Compound Planetary Gear Trains with Two Compound and Three External Shafts

29.1 Possible Ways of Connecting the Component Gear Trains—Types and Number

By using the modified symbol of Wolf (see Section 7.4.1), it is very easy to identify all possible ways of joining two simple (single-carrier) planetary gear trains (PGTs) in a compound PGT as shown in Table 29.1. Since the shafts of each of the component PGTs can occupy six different positions, the total possible number of compound gear trains naturally is $6 \times 6 = 36$.

TABLE 29.1

Possible Ways of Connecting the Two Component PGTs in a Compound PGT with Two Compound and Three External Shafts (in Frame are Given the Gears with Internal Division of Power, and the Rest are with Internal Circulation of Power) [25, 130]

	...1 t_{II}	...2 t_{II}	...3 t_{II}	...4 t_{II}	...5 t_{II}	...6 t_{II}
1... t_I	11	12	13	14	15	16
2... t_I	21 ≡ 12	22	23	24	25	26
3... t_I	31 ≡ 13	32 ≡ 23	33	34	35	36
4... t_I	41 ≡ 14	42 ≡ 24	43 ≡ 34	44	45	46
5... t_I	51 ≡ 15	52 ≡ 25	53 ≡ 35	54 ≡ 45	55	56
6... t_I	61 ≡ 16	62 ≡ 26	63 ≡ 36	64 ≡ 46	65 ≡ 56	66

21 ≡ 12   62 ≡ 26

With regard to the compound PGTs given in Table 29.1 [18, 25, 130, 133], can be done the following:

Findings and comments

1. The number of different compound PGTs is not 36, as it appears at first glance, and as it is indeed accepted by some authors [155], but it is 21 due to the existing isomorphism. The PGTs at the bottom left half of the table are isomorphic to the top right half PGTs. The table diagonal is the dividing line between the two sets of gear trains. This is clearly visible in the two examples below the table where the left and right compound gear trains are simply swapped.
2. Further, using the rule of algebraic signs (see Section 29.3.2), the delicate problem of internal division and internal circulation of power in two-carrier PGTs has been addressed. In the table, the PGTs with internal power division are framed. The other two-carrier PGTs are with internal power circulation. It is very important that this question is clear to the designer in advance, as it affects both the size of efficiency with its consequences for gear train heating and the required lubrication system, and the magnitude of load on the individual elements of compound gear train.
3. The compound gear trains in the table also have three external shafts as the simple (single-carrier) PGTs; i.e., they are also tri-shafts gear trains. And the torques of two of the shafts are unidirectional—$T_{D\,min}$ (smaller) and $T_{D\,max}$ (larger), and the torque of the third shaft T_Σ is opposite, the biggest one by absolute value and equal to the sum of the other two torques. In other words, the situation is completely analogous to the situation in the simple (single-carrier) PGTs, so the dependencies are in force:

$$\boxed{\sum T_i = T_{D\,min} + T_{D\,max} + T_\Sigma = 0} \text{ and } \boxed{T_{D\,min} < T_{D\,max} < |T_\Sigma|}. \tag{29.1}$$

Because each of the unidirectional torques is equal to the difference of the other two torques:

$$T_{D\,min} = T_\Sigma - T_{D\,max} \text{ and } T_{D\,max} = T_\Sigma - T_{D\,min}, \tag{29.2}$$

they are called *differential* and are denoted by index D, and the third one, which is equal to the sum of the other two

$$T_\Sigma = T_{D\,max} + T_{D\,min}, \tag{29.3}$$

is called *summation* and is denoted by an index Σ (sum). In the display of the compound PGTs with the modified symbol of Wolf (Figure 29.1), as in the simple AI-PGT, the shafts with unidirectional torques $T_{D\,min}$ and $T_{D\,max}$ are marked by different single lines depending on the torque size, and the shaft with the summation torque T_Σ—with a double line.

4. Similar to simple (single-carrier) PGTs, and for compound PGTs, an *aligned (reduced) torque ratio* t_{red} can be defined, which is the ratio of unidirectional torques $T_{D\,min}$ and $T_{D\,max}$, and which depends on the torque ratios t_I and t_{II} of the component PGTs I and II.

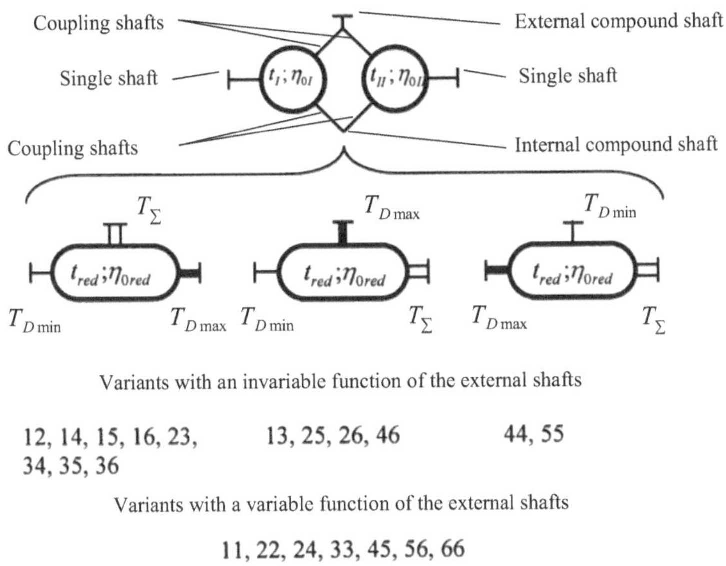

FIGURE 29.1
Functions of the three external shafts of a two-carrier PGT with two compound and three external shafts.

$$\boxed{t_{red} = \frac{T_{D\,max}}{T_{D\,min}} = f\left(t_I, t_{II}\right) > +1}. \tag{29.4}$$

By means of this aligned (reduced) torque ratio t_{red}, similar to simple PGTs (see Section 7.4.1), the ratio of the three ideal external torques (without taking into account the losses, i.e., $\eta_0 = 1$ and $\eta = 1$) of the compound PGTs can be expressed:

$$\boxed{T_{D\,min} : T_{D\,max} : T_{\Sigma} = T_{D\,min} : t_{red} \cdot T_{D\,min} : -\left(1 + t_{red}\right)T_{D\,min} = +1 : +t_{red} : -\left(1 + t_{red}\right)}, \tag{29.5}$$

which is invariable irrespective of the following:

- With how many degrees of freedom the gear operates, with $F = 1$ or $F = 2$.
- Which element of the PGT is fixed at $F = 1$ degree of freedom.
- How a component PGT works—as a reducer or a multiplier at $F = 1$, or as a summation or division differential at $F = 2$, respectively.

5. On the basis of the aligned torque ratio t_{red}, with the compound PGTs, a lever analogy exists (Figure 29.2), as well as an *aligned basic efficiency* η_{0red} (when the PGT works with a fixed summation shaft). This makes the calculations of the compound PGT analogous to those of the simple one [25, 130].
6. For displaying of the two-carrier PGT here, it is accepted that Wolf's modified symbol should be elongated rather than circular (Figure 29.1).
7. Figure 29.1 shows the names of the individual shafts of compound PGTs—external and internal. It also indicates the function that the three external shafts perform at various compound gear trains in Table 29.1. It can be seen that in some of the compound PGTs, their shafts have an invariable function and, in other compound PGTs, a changing function depending on the value of the torque ratios t_I and t_{II} of

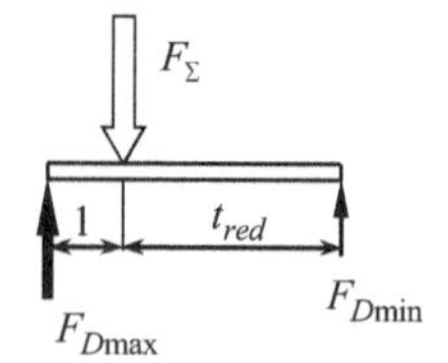

FIGURE 29.2
Lever analogy of a compound PGT with three external shafts.

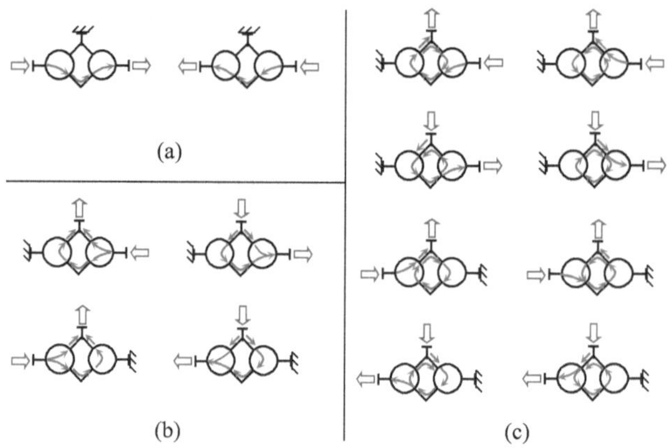

FIGURE 29.3
Working modes of two-carrier PGTs with two compound and three external shafts: (a) Sequentially (in series) coupled component gear trains; (b) closed differential with internal power division; (c) closed differential with internal power circulation.

the component gear trains. It is seen, e.g., that the external compound shaft can be with different torques—with the smallest $T_{D\,min}$, the medium $T_{D\,max}$, or the largest T_Σ [25, 130].

8. As mentioned above, compound PGTs, like the simple ones, can operate with either $F = 1$ degrees of freedom, such as reducers or multipliers, or $F = 2$ degrees of freedom, such as summation or division differentials. Specifically for two-carrier PGTs, this is discussed in more detail in Sections 29.3 and 29.4.
9. As set forth in Section 29.3, at $F = 1$ degree of freedom, if the compound PGT is operated with a fixed external compound shaft, then there is a sequential (in series) connection of the component PGTs. When one of the single shafts is fixed, there is a *closed-loop differential* gear train. In such gear train, two cases are possible—with internal power division and internal power circulation (see Sections 29.3.3 and 29.3.4). All these different working modes are illustrated in Figure 29.3.

29.2 Torque Method Analysis of Compound PGTs and Its Application

Compound PGTs are complex devices whose analysis is considerably more complex than the analysis of simple ones. In general, different methods of PGTs analysis exist. Two, however, are the most commonly used: one is the analytical method of Willis, and the other is the graphical method of Kutzbach discussed in Sections 7.1 and 7.2. Both methods have proven their worth over the course of many years. Willis's method has existed for nearly 2 centuries since 1841 when his book was published [248], and Kutzbach's method has existed for nearly a century—since 1927 [151].

A comparison of the advantages and disadvantages of both methods is given in Table 29.2. It can be seen that even at complex compound PGTs, both methods lose their advantages.

In contrast to these classical methods, the torque method [13, 18, 31] is better suited for the analysis of PGTs in general and especially for complex compound PGTs, because it sets more goals than just determining speed ratios as in Willis's and Kutzbach's methods in question:

1. Determination of speed ratio of both simple (single-carrier) and compound multi-carrier PGTs; i.e., the method possesses universality.
2. Determination of internal power flows in compound PGTs by size and direction as a prerequisite for determination of efficiency and loading of their elements. A particular advantage is the preliminary clarification of whether there is an internal division or an internal circulation of power.
3. Determination of load spectrum of variable loaded elements of planetary change-gears—gears, shafts, bearings—as a prerequisite for their adequate load capacity calculation.
4. Combining a number of favorable features, such as precision, simplicity, ease of use, easy verification of results, and, above all, maximum clarity and visibility, which is extremely important for the engineer.

The essence of the torque method is already presented in Section 7.4.1 for the purpose of simple $\overline{\text{AI}}$-PGT analysis. Therefore, only its particularities related to its application in compound PGTs will be considered here in a concentrated form.

The torque method is based on the following:

Well-known principles of mechanics

1. Equilibrium of the three ideal external torques (without taking into account the losses in the gear train, i.e., at efficiency $\eta_0 = 1$ and $\eta = 1$)

$$\sum T_i = T_{D\,min} + T_{D\,max} + T_\Sigma = 0, \tag{29.6}$$

TABLE 29.2

Characteristics of the Analytical and Graphical Method for PGT Analysis

Analytical method (Willis, 1841)	
Advantages	**Disadvantages**
• Precision • Universality	• Lack of any clarity (visibility) • Increasing volume of calculations, especially with compound PGTs • Increasing risk of error • Determination of speed ratio only
Graphical method (Kutzbach, 1927)	
Advantages	**Disadvantages**
• Clarity (visibility), especially with simple PGTs	• Losing clarity with complex compound PGTs • Precision of graphical methods is generally limited • Determination of speed ratio only

both of which $T_{D\,min}$ and $T_{D\,max}$ are unidirectional and the third T_Σ is equal to their sum (see Section 29.1).

2. Equilibrium of the three real external torques (taking into account the losses in the gear train, i.e., at efficiency $\eta_0 < 1$ and $\eta < 1$)

$$\sum T_i' = T_{D\,min}' + T_{D\,max}' + T_\Sigma' = 0. \tag{29.7}$$

3. Sum of ideal powers ($\eta_0 = 1$ and $\eta = 1$)

$$\sum P_i = P_A + P_B = T_A \cdot \omega_A + T_B \cdot \omega_B = 0, \tag{29.8}$$

from where the *speed ratio of the compound gear train* is determined, which is given by

$$\boxed{i_k = \frac{\omega_A}{\omega_B} = -\frac{T_B}{T_A}}. \tag{29.9}$$

In these formulae, P_A, T_A, and ω_A are input: power, torque, and angular velocity, and P_B, T_B, and ω_B—output: power, torque, and angular velocity.

4. Sum of real powers ($\eta_0 < 1$ and $\eta < 1$)

$$\sum P_i = \eta \cdot P_A + P_B = \eta \cdot T_A' \cdot \omega_A + T_B' \cdot \omega_B = 0, \tag{29.10}$$

from where the efficiency of the compound gear train is determined, which is given by

$$\boxed{\eta = -\frac{T_B'/T_A'}{\omega_A/\omega_B} = -\frac{i_T}{i_k}} \tag{29.11}$$

or $\eta = T_B'/T_B$, when $T_A' = T_A$, resp. $\eta = T_A/T_A'$, when $T_B' = T_B$ (see Section 12.3).

In these formulae, i_T is the ratio of real external torques T_A' and T_B' on the input A and output B shaft, so-called *torque transmit ratio* or *torque transformation* (see Section 7.4.1). The other ratio i_k is the usual speed ratio, and, unlike above, the index k (kinematic) can be used, if necessary.

Along with the abovementioned known principles of mechanics, in the method are used, as already explained in Section 7.4.1 for $\overline{\text{AI}}$-PGT, three small but practically very useful *innovations*:

1. The symbol of Wolf is used, but modified (Figure 7.4).
2. A new torque ratio is defined in compound PGT, called *aligned (reduced) torque ratio* $t_{red} = T_{D\,max}/T_{D\,min} > +1$.
3. The lever analogy is used (Figures 7.4 and 29.2).

Using the aligned (reduced) torque ratio t_{red}, the following dependencies (29.1) and (29.5), similar to those for $\overline{\text{AI}}$-PGT (7.27) and (7.28), are valid for the compound PGT:

$$\sum T_i = T_{D\,min} + T_{D\,max} + T_{\Sigma} = 0 \text{ and } T_{D\,min} < T_{D\,max} < |T_{\Sigma}|;$$

$$T_{D\,min} : T_{D\,max} : T_{\Sigma} = T_{D\,min} : t_{red} \cdot T_{D\,min} : -(1+t_{red})T_{D\,min} = 1 : t_{red} : -(1+r_{red}).$$

The torque method is used for the analysis of compound PGTs in the following *sequence*:

1. Determination of torque ratios t_I and t_{II} of the component PGTs I and II.
2. Determination of basic efficiencies η_{0I} and η_{0II} of the component PGTs I and II.
3. Determination of ideal torques T_i on all the shafts of the compound PGT.
4. Checking the equilibrium condition of the ideal external torques according to formula (29.6).
5. Determination of speed ratio i_k of the compound PGT according to formula (29.9).
6. Clarifying the presence of internal division or circulation of power, as well as the direction of relative power in the component PGTs (see Section 29.3.2).
7. Determination of real torques T_i' on all the shafts.
8. Checking the equilibrium condition of the real external torques according to (29.7).
9. Determination of torque transmit ratio (torque transformation) i_T of the compound PGT.
10. Determination of efficiency η of the compound PGT according to formula (29.11).

This sequence will be clarified in the examination of individual examples.

29.3 Work of Two-Carrier Compound PGTs with $F = 1$ Degree of Freedom

29.3.1 Sequentially (in Series) Coupled Two-Carrier Compound PGTs

This is the simplest and most commonly used compound PGT of this kind. Its kinematic and structural schemes, its analysis, and lever analogy are shown in Figure 29.4. This initial example, as well as the next two, will be considered in detail for understandable reasons. Further, the examinations will be shorter.

Following the procedure set out in Section 29.2, the torque ratios t_I and t_{II}, and basic efficiencies η_{0I} and η_{0II} of the component simple (single-carrier) $\overline{\text{AI}}$-PGTs are determined in the ways described in Section 11.5 as follows:

$$t_I = \frac{z_3}{z_1} \text{ and } t_{II} = \frac{z_6}{z_4}; \tag{29.12}$$

$$\eta_{0I} = 1 - \psi_{0I} \approx 1 - \left[0.15\left(\frac{1}{z_1} + \frac{1}{z_2}\right) + 0.2\left(\frac{1}{z_2} - \frac{1}{z_3}\right) + \Delta\psi\right]; \tag{29.13}$$

$$\eta_{0II} = 1 - \psi_{0II} \approx 1 - \left[0.15\left(\frac{1}{z_4} + \frac{1}{z_5}\right) + 0.2\left(\frac{1}{z_5} - \frac{1}{z_6}\right) + \Delta\psi\right]. \tag{29.14}$$

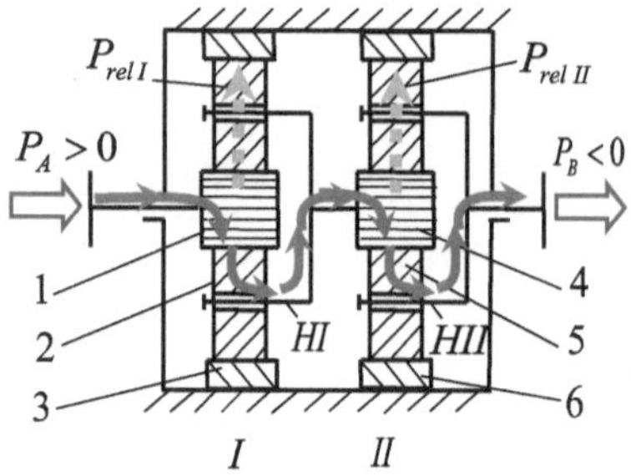

$$t_I = \frac{z_3}{z_1} = 4; \quad t_{II} = \frac{z_6}{z_4} = 3$$

$$\eta_{0I} = \eta_{0II} = 0.97$$

$$T_A \equiv T_{D\min} = +1; \quad T_B \equiv T_\Sigma = -20; \quad T_C \equiv T_{D\max} = +19$$

$$\sum T_i = T_A + T_B + T_C = +1 - 20 + 19 = 0$$

$$t_{red} = \frac{T_{D\max}}{T_{D\min}} = \frac{+19}{+1} = +19$$

$$i_k = \frac{\omega_A}{\omega_B} = \begin{cases} -\dfrac{T_B}{T_A} = -\dfrac{-20}{+1} = +20 \\ -\dfrac{T_\Sigma}{T_{D\min}} = -\dfrac{-(1+t_{red})}{+1} = +\dfrac{1+19}{1} = +20 \end{cases}$$

$$T'_A \equiv T'_{D\min} = +1; \quad T'_B \equiv T'_\Sigma = -19.08; \quad T'_C \equiv T'_{D\max} = +18.08$$

$$\sum T'_i = T'_A + T'_B + T'_C = +1 - 19.08 + 18.08 = 0$$

$$i_T = \frac{T'_B}{T'_A} = \frac{-19.08}{+1} = -19.08$$

$$\eta = \begin{Bmatrix} -\dfrac{i_T}{i_k} = -\dfrac{-19.08}{+20} \\ \dfrac{T'_B}{T_B} = \dfrac{-19.08}{-20} \end{Bmatrix} = +0.954 = 95.4\%$$

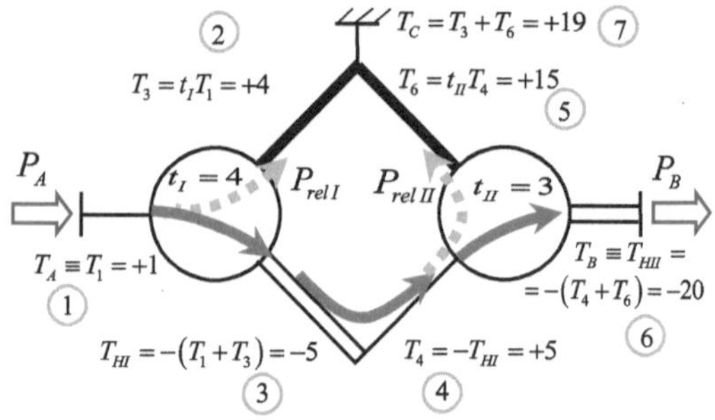

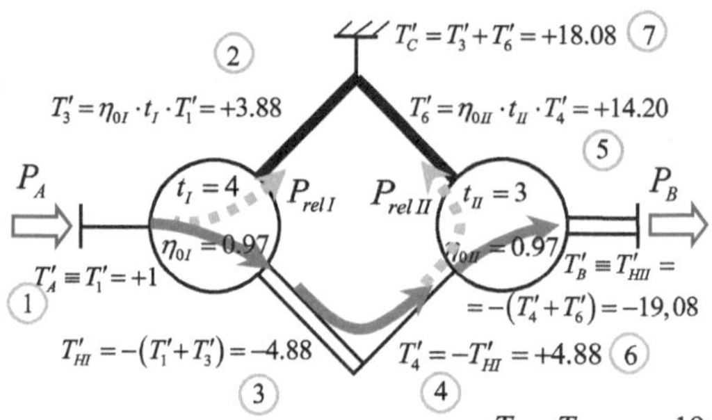

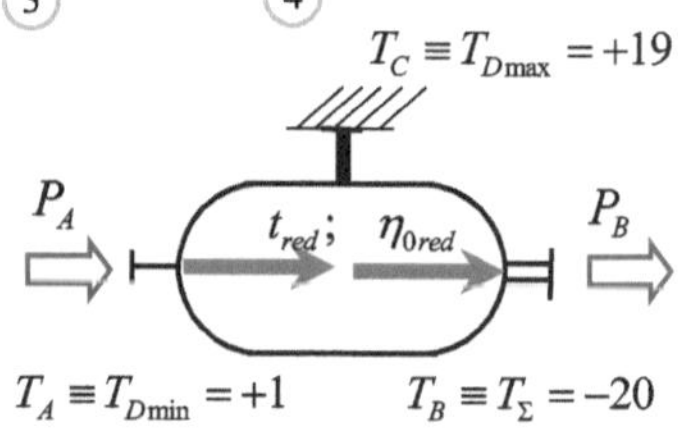

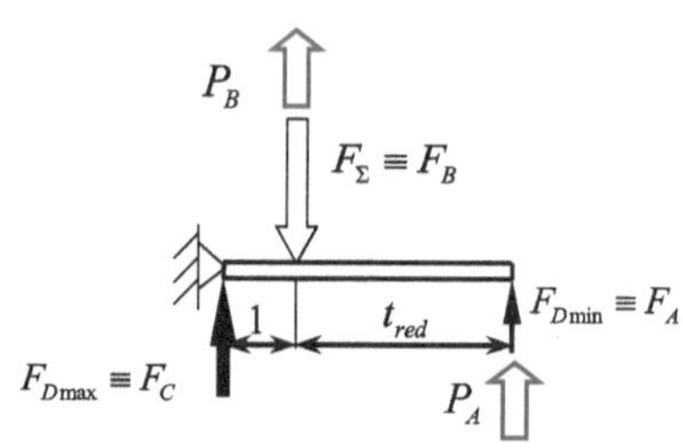

FIGURE 29.4
Analysis of in series coupled two-carrier PGT (variant 26 ≡ 62 from Table 29.1).

Since this is an example illustrating the method of analysis, directly are accepted

$$t_I = 4 \text{ and } t_{II} = 3$$

$\eta_{0I} = \eta_{0II} = 0.97$ (accepted, not calculated).

When calculating real PGTs, efficiency can be determined as shown in Chapter 11.

The sequence of the individual torques determination is indicated in Figure 29.4 with numbers surrounded by circles.

For convenience, not because it is mandatory, +1 is set to the smallest torque in the compound PGT. In this case, this is the shaft of the sun gear 1 of the first stage.

At the component PGTs, however, determination of the shaft with the smallest torque is no difficulty. Furthermore, it is necessary to start neither with the shaft of one of the sun gears nor with a numerical value of +1. It is entirely possible to start with the torque of any shaft, with arbitrary value, and without limitation in the algebraic sign "+" or "–".

The three ideal external torques of each of the component PGTs, which are in invariable proportions (Section 29.1), are determined by the respective torque ratios t_I and t_{II}. As can be seen from Figure 29.4, both edges of the internal shaft act equal in size, but different in direction torques (with different algebraic signs). The torque on the external compound shaft is defined as the algebraic sum of both torques acting on the coupling shafts (Figure 29.4).

Checking the correctness of calculations for ideal external torques is made by the condition (29.6) $\sum T_i = 0$, as shown in the figure. Also, in the figure, the kind of the external torques (input T_A, output T_B, and reactive T_C) is indicated according to their size, which is given by

$$T_A \equiv T_{D\,min} = +1 < T_C \equiv T_{D\,max} = +19 < |T_B| \equiv |T_\Sigma| = +20.$$

Obviously, the aligned torque ratio is

$$t_{red} = \frac{T_{D\,max}}{T_{D\,min}} = \frac{+19}{+1} = +19.$$

The speed ratio i_k can be determined in two ways—by the ideal output T_B and input T_A torques or by the aligned torque ratio t_{red}, given by

$$i_k = \frac{\omega_A}{\omega_B} = \begin{cases} -\dfrac{T_B}{T_A} = -\dfrac{-20}{+1} = +20 \\ -\dfrac{T_\Sigma}{T_{D\,min}} = -\dfrac{-(1+t_{red})}{+1} = \dfrac{1+19}{+1} = +20 \end{cases}.$$

In the compound PGT in question, there is no internal division or circulation of power, so this issue is not considered. Power is transmitted sequentially through the both component PGTs.

In determining the real torques, the direction of relative (rolling) power $P_{rel\,I}$ and $P_{rel\,II}$, and the value of basic efficiency η_{0I} and η_{0II} of the component PGTs are taken into account.

In the investigated compound gear train, it can be considered without any difficulty that the angular velocities of sun gears 1 and 4 are higher than the angular velocities of the respective carriers HI and HII, so in both cases, the relative power in the component PGTs is transmitted from the sun gears 1 and 4 to the ring gears 3 and 6 (in the figure, the direction of relative power is indicated with dotted arrows and the direction of transmitted power—with solid arrows). In this situation, the real torques of ring gears 3 and 6 are calculated by the following formulae:

$$T_3' = \eta_{0I} \cdot t_I \cdot T_1' \text{ and } T_6' = \eta_{0II} \cdot t_{II} \cdot T_4'.$$

The sequence of determination of the individual real external torques is the same as in the determination of the ideal ones. Similarly, the correctness of the calculations is checked by the condition (29.7) $\sum T_i' = 0$.

The compound gear train efficiency is determined according to formula (29.11) through the torque i_T and speed i_k ratios:

$$\left.\begin{aligned} i_T &= \frac{T'_B}{T'_A} = \frac{-19.08}{+1} = -19.08 \\ i_k &= +20 \end{aligned}\right\} \eta = -\frac{i_T}{i_k} = -\frac{-19.08}{+20} = +0.954 = 95.4\%.$$

Since in this case, the input torques, ideal T_A and real T'_A, are the same ($T_A = T'_A = +1$), efficiency can also be determined by the ratio of both output torques—ideal T_B and real T'_B, as shown in the figure (see Section 12.3 for the case $\omega_3 = 0$):

$$\eta = \frac{T'_B}{T_B} = \frac{-19.08}{-20} = +0.954 = 95.4\%.$$

In this most simple case of a compound PGT, its speed ratio, of course, can also be determined by the method of Willis, since there is a sequential coupling of two simple (single-carrier) AI-PGTs. Speed ratios of each of them at fixed ring gears 3 and 6 are determined by the known formulae, and the speed ratio of the compound gear train is equal to their product, i.e.,

$$i_{1HII(3,6)} = i_{1HI(3)} \cdot i_{4HII(6)} = \left(1 + \frac{z_3}{z_1}\right)\left(1 + \frac{z_6}{z_4}\right).$$

In Figure 29.4, as well as in the following Figures 29.6, 29.8, and 29.9 for the other two cases, the analysis is given in more detail, which is dictated by the need for a comprehensive and comprehensible explanation. Provided that the torque method has already been utilized, analysis can be much shorter, e.g., as in Figures 29.10 and 29.11.

29.3.2 Determining the Presence of Internal Division or Internal Circulation of Power in Two-Carrier Compound PGTs

This question is not only delicate but also a real trap for the designer. Therefore, the following two examples (Sections 29.3.3 and 29.3.4) are intended to illustrate this problem and elucidate it. They are more interesting than the first simple example, since they are closed differential gear trains, one of which is with internal power division (Section 29.3.3) and the other one with internal power circulation (Section 29.3.4), which is very unfavorable.

From the ways offered in the literature to determine the presence of division or circulation of power, some of which are rather complex [145], the simplest one is the rule of algebraic signs [13, 23, 25, 31, 133], illustrated in Figure 29.5.

Rule of algebraic signs to determine the presence of internal division or internal circulation of power

When the algebraic signs of the ideal torques of the two shafts forming the external compound shaft are the same, there is an internal power division (see Table 29.1).

With different algebraic signs, there is an internal power circulation.

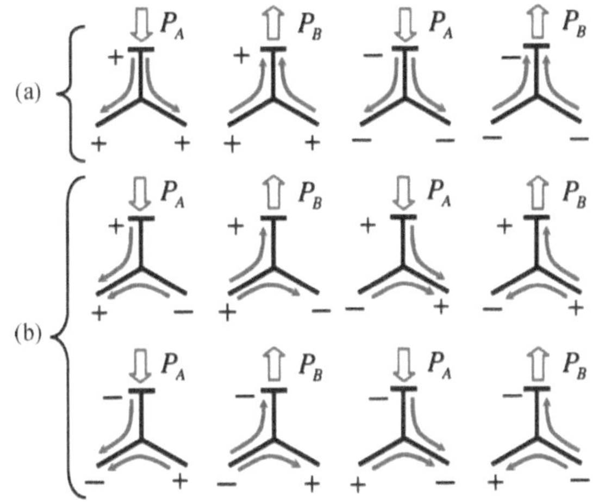

FIGURE 29.5
Rule of algebraic signs in determining internal power flows: (a) Internal power division; (b) internal power circulation.

This rule is applied to compound PGTs given in Table 29.1. It is also illustrated by the examples in Sections 29.3.3 and 29.3.4, where the direction and magnitude of the circulating power P_{circ} are determined, depending on torque ratios t_I and t_{II} of the component PGTs. The direction of circulating power is determined by the following:

Rule of algebraic signs to determine the direction of circulating power

The direction of the circulating power P_{circ} coincides with the direction of the input P_A or output P_B power of this one of the two coupling shafts (forming the external compound shaft) whose algebraic sign of its torque coincides with the algebraic sign of the torque on the external compound shaft (T_A or T_B) [31].

When in compound two- or three-carrier PGTs difficulties are encountered in determining the direction of relative power P_{rel} in the component PGTs, the method of samples [209], illustrated by the examples in Figures 29.11 and 34.9, may be used.

29.3.3 Closed Differential PGT with Internal Power Division

Figure 29.6 shows kinematic and structural schemes, analysis, and lever analogy of such a gear train. The same procedure for the analysis as in the previous example (Figure 29.4) is followed. The sequence of individual torque determination is indicated with numbers surrounded by circles. Identical algebraic signs of the torques on the shafts forming the external compound shaft (T_{HI} and T_6) indicate the presence of an internal power division. And here, before the calculation of the real external torques, the direction of relative power $P_{rel\,I}$ and $P_{rel\,II}$ in the component PGTs must be determined, too. As in the previous example, the alternative way of determining the efficiency is also used here by means of the output torques—ideal T_B and real T_B', but on condition that input torques—ideal T_A and real T_A', are equal ($T_A = T_A'$).

Figure 29.7 shows a very important fact that under the same structural scheme of a compound PGT, its kinematic schemes may be different.

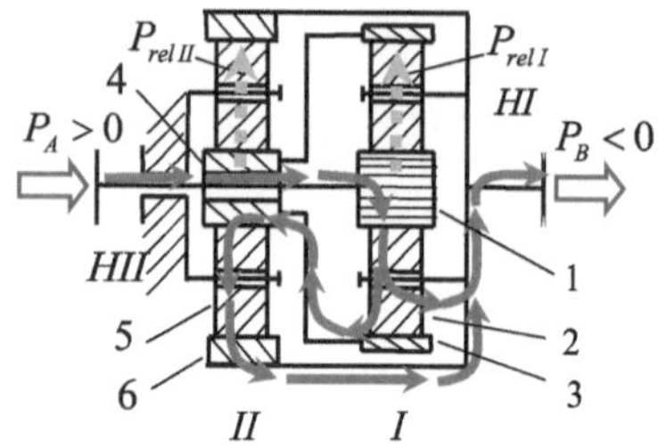

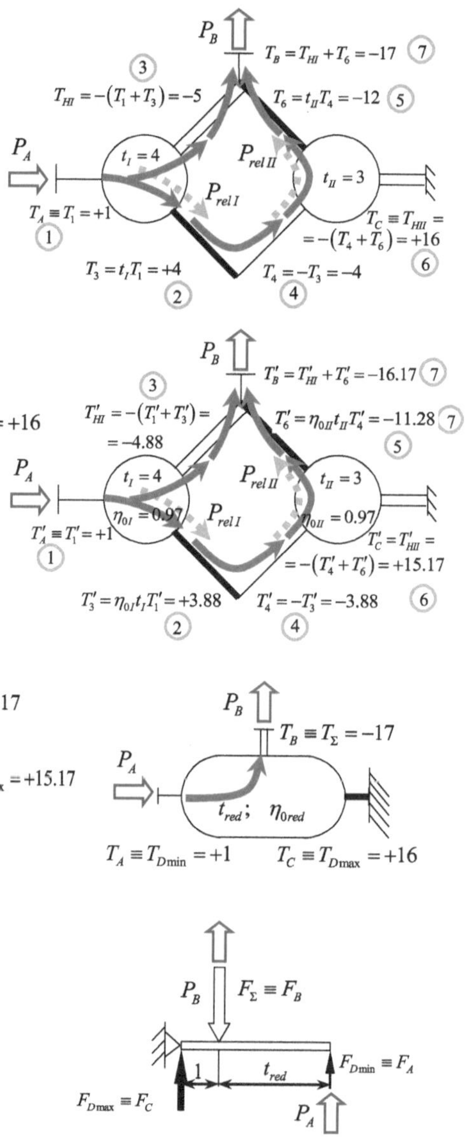

$$t_I = \frac{z_3}{z_1} = 4; \quad t_{II} = \frac{z_6}{z_4} = 3$$

$$\eta_{0I} = \eta_{0II} = 0.97$$

$$T_A \equiv T_{D\min} = +1; \quad T_B \equiv T_\Sigma = -17; \quad T_C \equiv T_{D\max} = +16$$

$$\sum T_i = T_A + T_B + T_C = +1 - 17 + 16 = 0$$

$$t_{red} = \frac{T_{D\max}}{T_{D\min}} = \frac{+16}{+1} = +16$$

$$i_k = \frac{\omega_A}{\omega_B} = \begin{cases} -\dfrac{T_B}{T_A} = -\dfrac{-17}{+1} = +17 \\ -\dfrac{T_\Sigma}{T_{D\min}} = -\dfrac{-(1+t_{red})}{+1} = +\dfrac{1+16}{1} = +17 \end{cases}$$

$$T'_A \equiv T'_{D\min} = +1; \quad T'_B \equiv T'_\Sigma \equiv T'_{D\min} = -16.17; \quad T'_C = T'_{D\max} = +15.17$$

$$\sum T'_i = T'_A + T'_B + T'_C = +1 - 16.17 + 15.17 = 0$$

$$i_T = \frac{T'_B}{T'_A} = \frac{-16.17}{+1} = -16.17$$

$$\eta = \begin{Bmatrix} -\dfrac{i_T}{i_k} = -\dfrac{-16.17}{+17} \\ \dfrac{T'_B}{T_B} = \dfrac{-16.17}{-17} \end{Bmatrix} = +0.951 = 95.1\%$$

FIGURE 29.6
Analysis of closed differential PGT with internal power division (variant 16 ≡ 61 from Table 29.1).

29.3.4 Closed Differential PGT with Internal Power Circulation

This third example is the most complicated and interesting one, and hides the most dangers. Figure 29.8 shows kinematic and structural schemes, analysis, and lever analogy of such a gear train. Here, again the procedure described in Section 29.2 is followed, starting from the shaft of sun gear 4 of II component gear train ($T_4 = +1$). Of course, it could be started from the sun gear 1 of I component train ($T_1 = 1$). The different algebraic signs of the torques on the shafts forming the external compound shaft (T_1 and T_4) indicate the presence of an internal power circulation, whose direction is determined by the rule of algebraic signs (Section 29.3.2).

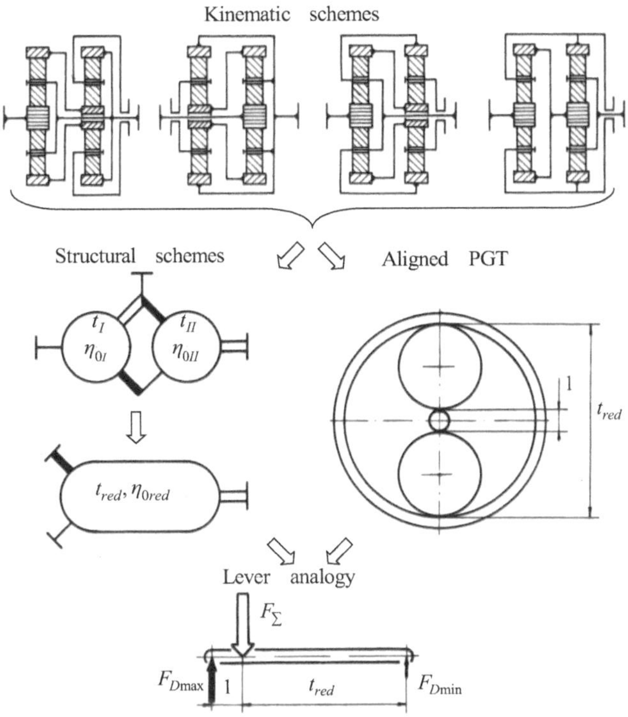

FIGURE 29.7
An example of the same structural scheme for different kinematic schemes of two-carrier PGTs (variant 16 ≡ 61 from Table 29.1).

Unlike the previous two examples, the alternative way of efficiency determining (by means of the output torques T_B and T'_B) cannot be used here, because input torques—ideal $T_A = +0.25$ and real $T'_A = +0.2943$, are not equal.

Figure 29.8 shows that the circulating power P_{circ} is approximately 2.4 times bigger than the input power, which is given by

$$\frac{T'_1}{T'_A} = \left|\frac{-0.7057}{+0.2943}\right| \approx 2.4.$$

Moreover, power magnitude changes along the way because of losses. Power circulation also results in a significantly lower efficiency of this gear train compared to the previous two ones (83 versus 95%), although their speed ratios i are close (16, 17, and 20), and the same basic efficiency η_{0I} and η_{0II} of the component PGTs is assumed.

29.3.5 Closed Differential PGT with Self-Locking

This gear train (Figure 29.9) is interesting in that it is self-locking, although it is composed of two negative-ratio $\overline{\text{AI}}$-PGTs which are not of the category of self-locking gear trains, such as positive-ratio PGTs (Section 2.1 and Figure 2.3). Self-locking is obtained at close values of torque ratios t_I and t_{II} of the component PGTs when the compound PGT works as a multiplier. Moreover, this PGT is with internal power circulation. The direction of the circulating power P_{circ} (as well as of the relative powers $P_{rel\,I}$ and $P_{rel\,II}$) varies depending on

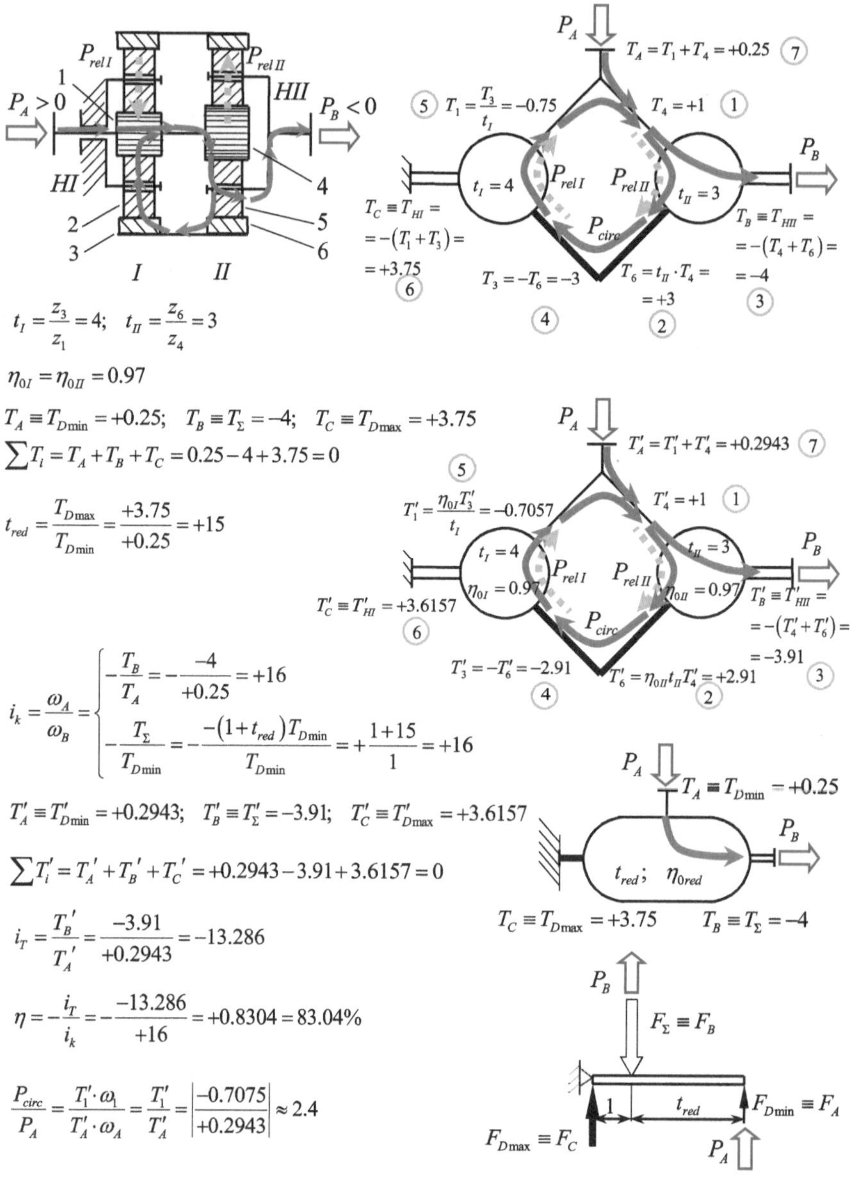

FIGURE 29.8
Analysis of closed differential PGT with internal power circulation (variant 55 from Table 29.1).

which one of the torque ratios t_I and t_{II} is higher. The torque method helps for easy orientation in this complex situation.

It can be seen from Figure 29.9 that the circulating power P_{circ} is much bigger than the input one—approximately 25 times (!) and that the efficiency is negative ($\eta = -24.2\%$), so self-locking occurs. The magnitude of the circulating power is determined by the real powers, and by the real torques $T_{HI}' = -5.124$ and $T_A' = +4.124$, respectively. The first torque T_{HI}' acts on the carrier HI of the first (I) component PGT through which only the circulating power P_{circ} passes. The second torque T_A' acts on the input shaft through which only

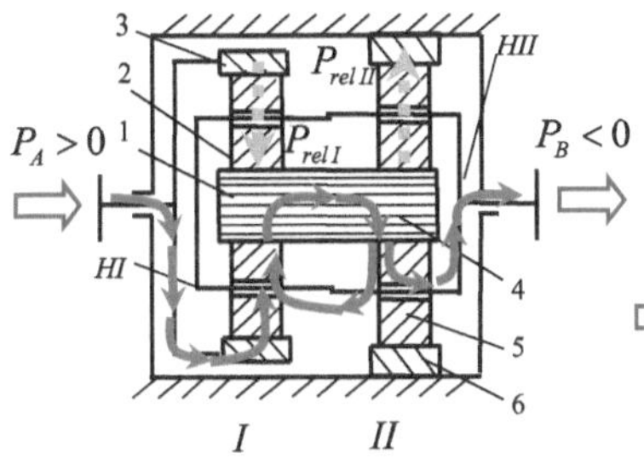

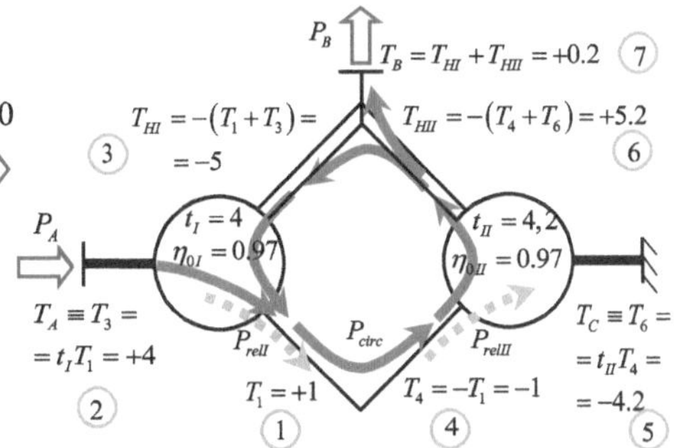

$$t_I = \frac{z_3}{z_1} = 4; \quad t_{II} = \frac{z_6}{z_4} = 4.2$$

$$\eta_{0I} = \eta_{0II} = 0.97$$

$$T_A \equiv T_{D\max} = +4; \quad T_B \equiv T_{D\min} = +0.2; \quad T_C \equiv T_\Sigma = -4.2$$

$$\sum T_i = T_A + T_B + T_C = 4 + 0.2 - 4.2 = 0$$

$$t_{red} = \frac{T_{D\max}}{T_{D\min}} = \frac{+4}{+0.2} = +20$$

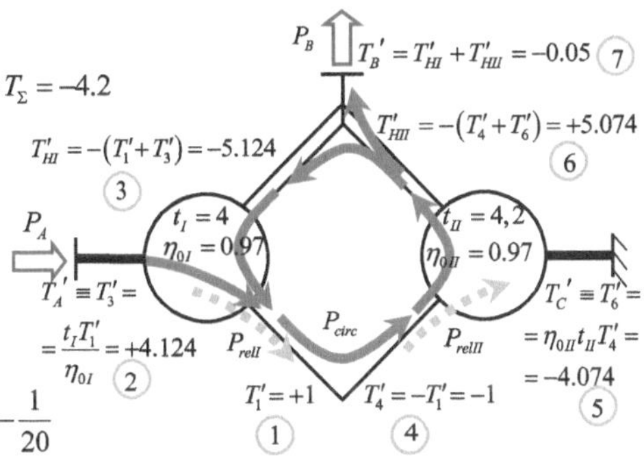

$$i_k = \frac{\omega_A}{\omega_B} = \begin{cases} -\dfrac{T_B}{T_A} = -\dfrac{+0.2}{+4} = -0.05 = -\dfrac{1}{20} \\ -\dfrac{T_{D\min}}{T_{D\max}} = -\dfrac{T_{D\min}}{t_{red} \cdot T_{D\min}} = -\dfrac{1}{t_{red}} = -\dfrac{1}{20} \end{cases}$$

$$T'_A \equiv T'_\Sigma = +4.124; \quad T'_B \equiv T'_{D\min} = -0.05; \quad T'_C \equiv T'_{D\max} = -4.074$$

$$\sum T'_i = T'_A + T'_B + T'_C = +4.124 - 0.05 - 4.074 = 0$$

$$i_T = \frac{T'_B}{T'_A} = \frac{-0.05}{+4.124} = -0.0121$$

$$\eta = -\frac{i_T}{i_k} = -\frac{-0.0121}{-0.05} = -0.242 = -24.2\% < 0!$$

$$\frac{P_{circ}}{P'_A} = \left|\frac{T'_{HI} \cdot \omega_{HI}}{T'_A \cdot \omega_A}\right| = \left|\frac{T'_{HI}}{T'_A} \cdot \frac{1}{i_k}\right| = \left|\frac{-5.124}{+4.124 \cdot 0.05}\right| \approx 25$$

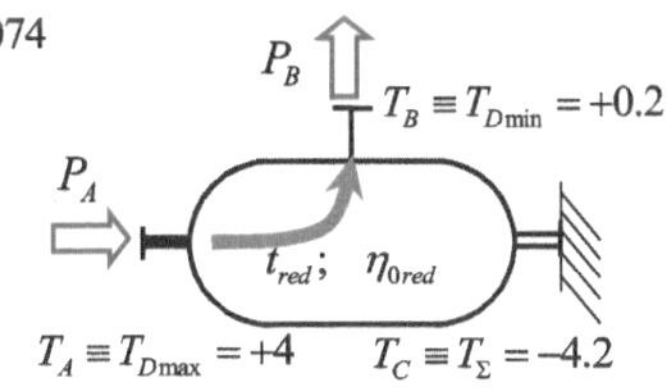

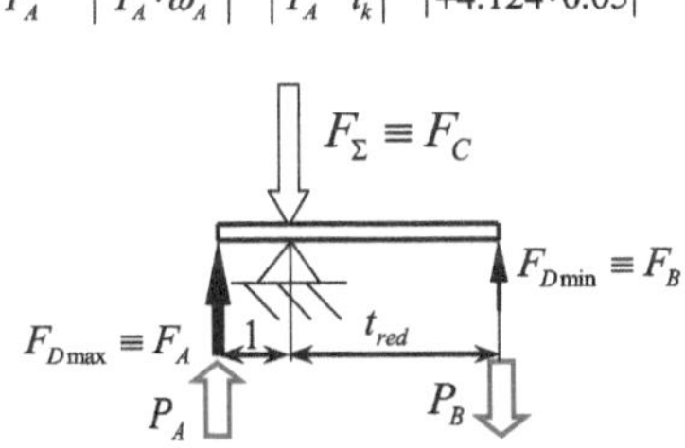

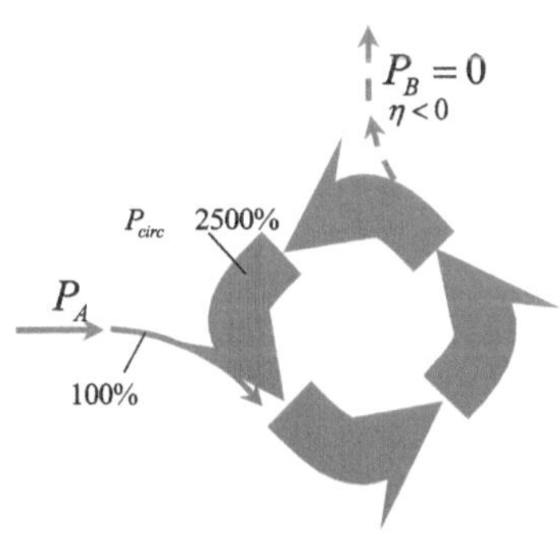

FIGURE 29.9
Analysis of closed differential PGT with self-locking as a multiplier (variant 33 from Table 29.1).

the input power P_A passes. The ratio of the angular velocities of the respective elements is taken into account, too. The angular velocity ω_{HI} of the carrier HI is identical to the output angular velocity ω_B ($\omega_{HI} \equiv \omega_B$), and for the ratio in question is obtained

$$\frac{\omega_{HI}}{\omega_A} = \frac{\omega_B}{\omega_A} = \frac{1}{i_k} = \frac{1}{-\dfrac{1}{20}} = -20.$$

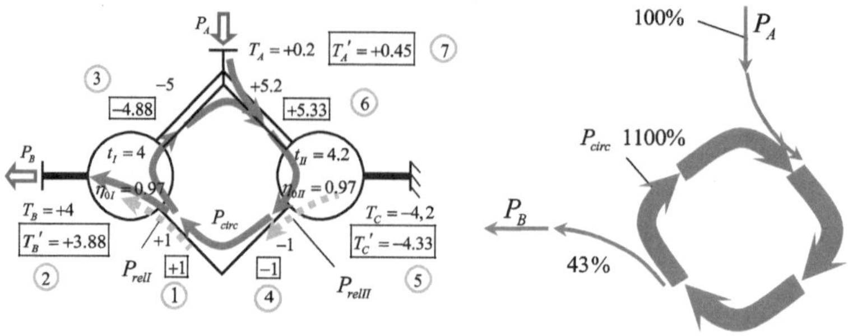

FIGURE 29.10
Torques and power flows of the PGT in Figure 29.9, working as a reducer.

Hence, the ratio of powers is obtained:

$$\frac{P_{circ}}{P'_A} = \left|\frac{T'_{HI}\cdot\omega_{HI}}{T'_A\cdot\omega_A}\right| = \left|\frac{T'_{HI}}{T'_A}\frac{1}{i_k}\right| = \left|\frac{-5.124}{+4.124}(-20)\right| \approx 25.$$

As noted in Section 29.3.4, the magnitude of circulating power P_{circ} changes along the way due to losses.

Of course, at a self-locking gear train which cannot rotate, as in this case, the term "circulating power" should be understood only conditionally.

Figure 29.10 shows the operating of the same PGT as a reducer. Unlike the previous examples, for brevity, in one structural scheme here the ideal and real torques (in a frame) and directions of power flows are shown. The size of the ideal torque remains the same. What changes is the direction of the power flows, absolute and relative, and hence the size of the real torque changes, too. With the numerical values for the ideal and real torques shown in the figure, both conditions $\sum T_i = 0$ and $\sum T'_i = 0$ are checked, the speed ratio $i_k = -20$ is calculated, and the torque transmit ratio $i_T = +8.0222$ and ultimately the efficiency $\eta = 43\% > 0$ are determined. Although very low, this efficiency is not negative, as in the case when the PGT works as a multiplier. In this case, the ratio of the circulating power to the real input power is more favorable and is determined similarly to the previous case by the real torques $T'_H = -4.88$ and $T'_A = +0.45$ at $\omega_{HI} \equiv \omega_A$, which is given by

$$\frac{P_{circ}}{P'_A} = \left|\frac{T'_{HI}\cdot\omega_{HI}}{T'_A\cdot\omega_A}\right| = \left|\frac{T'_{HI}}{T'_A}\right| = \left|\frac{-4.88}{+0.45}\right| \approx 11.$$

It can be seen that when compound PGT works as a reducer, the circulating power P_{circ} is more than two times lower (25:11 ≈ 2.3) than when PGT works as a multiplier. Hence, the losses in the gear train are smaller and efficiency—higher.

29.4 Work of Two-Carrier Compound PGTs with $F = 2$ Degrees of Freedom

Figure 29.11 shows a compound PGT working with $F = 2$ degrees of freedom as a summation differential with internal power division. However, for the reader, it is not difficult to

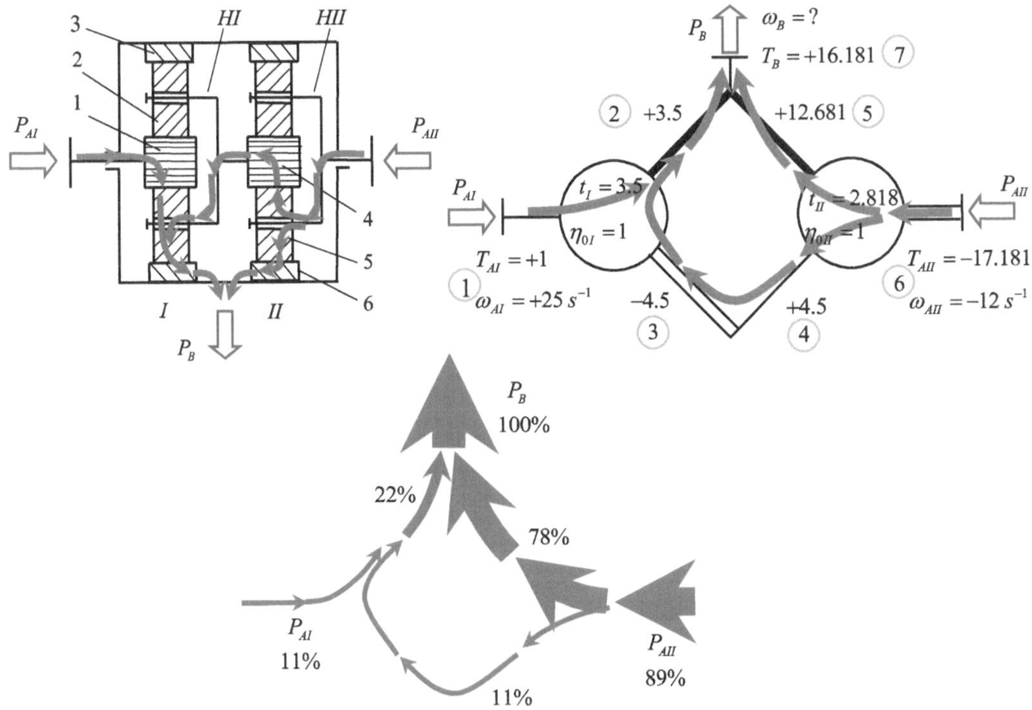

FIGURE 29.11
Two-carrier PGT with $F = 2$ degrees of freedom (twin-motor drive), its ideal torques, and power flows (variant 26 ≡ 62 from Table 29.1).

find out that the PGT in Figure 29.11 has the same structure as the sequentially coupled PGT in Figure 29.4. The difference is that the former output shaft B has become a second A_{II} input shaft, and the fixed shaft C has become an output shaft B.

The task is to determine torques and angular velocities (absolute and relative) on all shafts, and the efficiency of the gear train.

Below are the ideal external torques and angular velocities of all the shafts of the compound PGT at given:

Torque ratios $t_I = 3.5$ and $t_{II} = 2.818$ of both component PGTs

Angular velocities $\omega_{AI} \equiv \omega_1 = 25\ \text{s}^{-1}$ and $\omega_{AII} \equiv \omega_{H\,II} = -12\,\text{s}^{-1}$ of both input shafts—the sun gear 1 and carrier HII.

The determination of ideal external torques is done in the known sequence (Section 29.2), which is shown in the figure with numbers surrounded by circles. Check $\sum T_i = 0$ shows the calculation correctness

$$\sum T_i = T_{AI} + T_{AII} + T_B = 1 - 17.181 + 16.181 = 0.$$

By assuming the input torque of the first (I) component PGT $T_{AI} = +1$, from the equilibrium of ideal powers of the compound gear train as a whole

$$\sum P_i = P_{AI} + P_{AII} + P_B = T_{AI} \cdot \omega_{AI} + T_{AII} \cdot \omega_{AII} + T_B \cdot \omega_B$$

$$= 1 \cdot 25 + (-17.181)(-12) + 16.181 \cdot \omega_B = 0$$

its output angular velocity is determined:

$$\omega_B \equiv \omega_3 \equiv \omega_6 = -\frac{25 + 206.172}{16.181} = -14.29 \text{ s}^{-1}.$$

From the equilibrium of ideal powers only of the first (I) component PGT

$$\sum P_i = P_1 + P_3 + P_{HI} = T_1 \cdot \omega_1 + T_3 \cdot \omega_3 + T_{HI} \cdot \omega_{HI}$$

$$= 1 \cdot 25 + 3.5(-14.29) + (-4.5)\omega_{HI} = 0$$

the angular velocity of internal compound shaft is determined:

$$\omega_{HI} \equiv \omega_4 = -\frac{25 - 50.51}{4.5} = -5.559 \text{ s}^{-1}.$$

The two identical negative signs of angular velocity $\omega_{HI} = -5.559 \text{ s}^{-1}$ and torque $T_{HI} = -4.5$ on the shaft of the carrier HI of the first (I) component PGT indicate that this is an input shaft for the internal power flow (11%) of this component gear train $(P_{HI} = T_{HI} \cdot \omega_{HI} > 0!)$.

Found angular velocity ω_{HI} is required to determine relative angular velocities of the individual gears as follows:

$$\omega_{1rel} = \omega_1 - \omega_{HI} = +25 - (-5.559) = +30.559 \text{s}^{-1}$$

$$\omega_{2rel} = \omega_{1rel} \frac{z_1}{z_2}$$

$$\omega_{3rel} = \omega_3 - \omega_{HI} = -14.29 - (-5.559) = -8.731 \text{s}^{-1}$$

$$\omega_{4rel} = \omega_4 - \omega_{HII} = -5.559 - (-12) = +6.441 \text{s}^{-1}$$

$$\omega_{5rel} = \omega_{4rel} \frac{z_4}{z_5}$$

$$\omega_{6rel} = \omega_6 - \omega_{HII} = \omega_B - \omega_{AII} = -14.29 - (-12) = +2.29 \text{s}^{-1}.$$

These angular velocities are important not only for calculating load capacity of the gears but also of the shafts and bearings.

Determination of efficiency of this compound PGT (Figure 29.12), which works with $F = 2$ degrees of freedom, has some features, unlike the previous examples (working with $F = 1$), in which the determination of direction of both relative powers $P_{rel\,I}$ and $P_{rel\,II}$ is not a problem.

Determination of direction of the relative power in the first (I) component PGT is not difficult. Since the angular velocity of the output shaft $\omega_B \equiv \omega_3$ (coupled to ring gear 3) and the angular velocity ω_{HI} of the carrier HI (of the first (I) component PGT) are lower than the angular velocity of the input shaft $\omega_A \equiv \omega_1$ (coupled to sun gear 1), the relative power $P_{rel\,I}$ is undoubtedly transmitted from sun gear 1 to ring gear 3.

A particular situation is with the second (II) component PGT in which the input power $P_{A\,II}$ is divided into two branches—to sun gear 4 and to ring gear 6. The direction of relative power $P_{rel\,II}$ is not obvious. In such cases, the *method of samples* is very successful [209]. The procedure is as follows. There are two possible directions of relative power $P_{rel\,II}$, as shown in Figure 29.12. In both cases, for the second (II) component PGT, a lowered basic efficiency is assumed (e.g., $\eta_{0II} = 0.8$), whereas for the first (I) component PGT, whose direction of relative power is clear, $\eta_{0I} = 1$ is taken. Then, the real external torques are determined, and

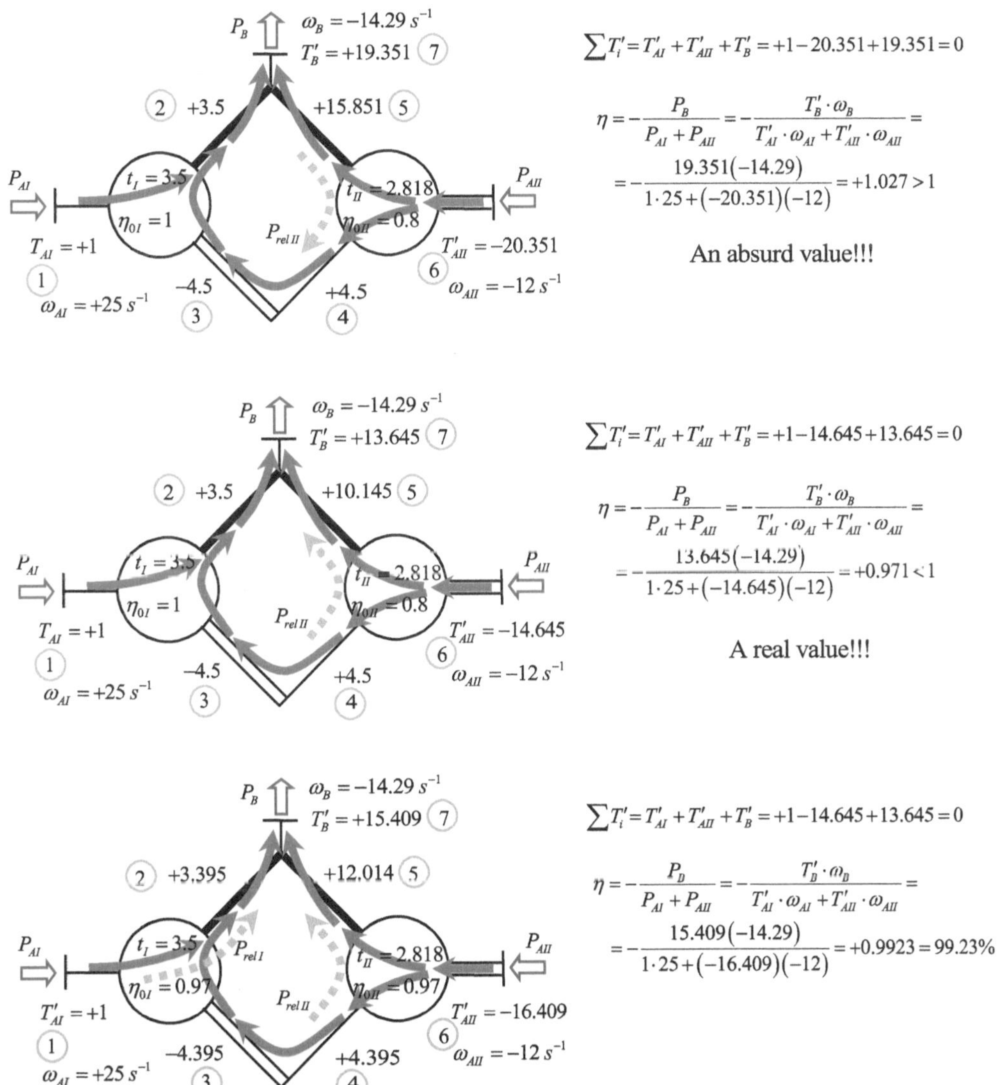

FIGURE 29.12
Determination of efficiency of the PGT from Figure 29.11 after determining the direction of the relative power $P_{rel\,II}$ in the second (II) component gear train by the method of samples.

it is found that in one of the cases for the efficiency of compound PGT, an absurd result is obtained ($\eta > 1$!) and in the other case—realistic ($\eta < 1$). So, the right direction of relative power $P_{rel\,II}$ is determined.

Knowing the correct directions of the two relative powers $P_{rel\,I}$ and $P_{rel\,II}$ allows for determining the real external torques in the known way and through them—the efficiency of the compound gear train.

It is not difficult for a designer to estimate how simple and easy to apply in practice is the method of samples compared to, e.g., the method of partial derivatives [145]. It is very helpful and advisable for the reader to see how to determine the direction of relative power in even more complex PGTs—the three-carrier PGTs (Figures 34.10 and 34.11).

30

Two-Carrier Compound Planetary Gear Trains with Two Compound and Four External Shafts

30.1 Possible Ways of Connecting the Component PGTs —Types and Numbers

Planetary gear trains (PGTs) with *two* compound and *three* external shafts discussed in the previous chapter can be converted into PGTs with *two* compound and *four* external shafts if both compound shafts become external. In this case, only 12 different structural schemes remain (Table 30.1) [23, 31], since the schemes with interchanged places of upper and lower compound shafts are isomorphic, as demonstrated by the two examples below the table. Of the twelve structural schemes, three are without a "twin": 12 ≡ 21, 34 ≡ 43 and 56 ≡ 65.

TABLE 30.1

Possible Variants of Two-Carrier PGTs with Two Compound and Four External Shafts

	...1 t_{II}	...2 t_{II}	...3 t_{II}	...4 t_{II}	...5 t_{II}	...6 t_{II}
1... t_I	11	12 ≡ 21	13 ≡ 31	14 ≡ 41	15 ≡ 51	16 ≡ 61
2... t_I	21 ≡ 12	22	23 ≡ 32	24 ≡ 42	25 ≡ 52	26 ≡ 62
3... t_I	31 ≡ 13	32 ≡ 23	33	34 ≡ 43	35 ≡ 53	36 ≡ 63
4... t_I	41 ≡ 14	42 ≡ 24	43 ≡ 34	44	45 ≡ 54	46 ≡ 64
5... t_I	51 ≡ 15	52 ≡ 25	53 ≡ 35	54 ≡ 45	55	56 ≡ 65
6... t_I	61 ≡ 16	62 ≡ 26	63 ≡ 36	64 ≡ 46	65 ≡ 56	66

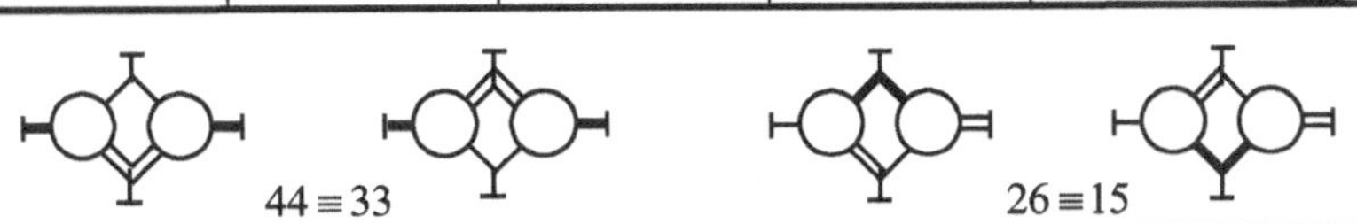

44 ≡ 33   26 ≡ 15

30.2 Working Modes

Two-carrier PGTs under consideration can be used as change-gears to realize two speed ratios [25, 31, 134, 230, 232]. Figure 30.1 shows the three possible locations of both brakes as well as the resulting power flows. Possible brake locations are:

1. Brakes on both single shafts—case (a)
2. Brakes on both compound shafts—case (b)
3. One brake on a single shaft, the other brake on a compound shaft—case (c)

In case (a), when one of the brakes is locking, the power flow passes only through the component PGT with locked shaft. The other component PGT is idling.

In case (b) described in Section 29.3.1, a sequentially (in series) coupled compound PGT is obtained, and the power flow passes sequentially through the component PGTs.

In case (c), depending on the locked brake, either a closed differential with internal division or circulation of power is obtained, or only one of the component PGTs works.

PGTs of the type in question are very convenient for reversible mechanisms with one speed in each direction. Figure 30.2 shows such a PGT with brakes on both single shafts (Figure 30.1a), which is mainly used in vessels. Figure 30.3 shows a case with brakes on both compound shafts (Figure 30.1b) and Figure 30.4—with brakes on a single and a compound shaft (Figure 30.1c). Different speed ratios in both directions are convenient for technology machines with slow-moving and a rapidly returning working body. At the PGT in Figure 30.4 the direction of rotation of output shaft at the second gear depends on the magnitude of the torque ratios t_I and t_{II} of the component PGTs. Reversing of rotation direction

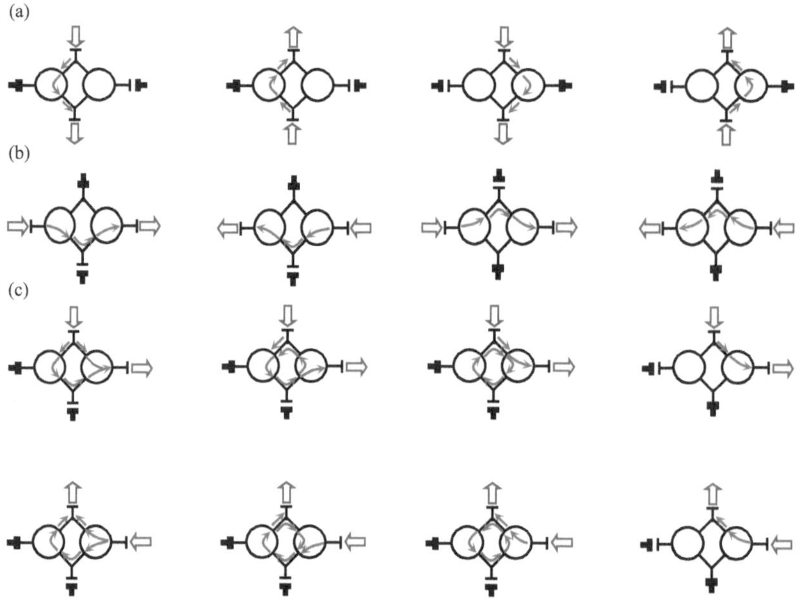

FIGURE 30.1
Possible locations of both brakes at two-carrier PGTs with two compound and four external shafts: (a) Brakes on both single shafts; (b) brakes on both compound shafts; (c) brakes on a single and on a compound shaft.

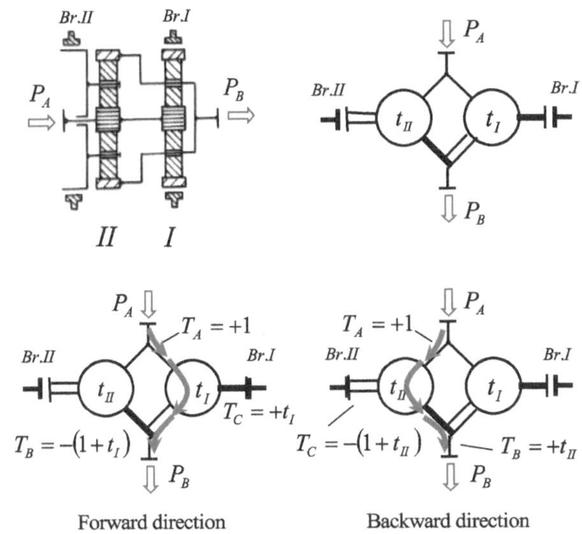

$$i_{Br.I} = -\frac{T_B}{T_A} = +(1+t_I) > 0 \qquad i_{Br.II} = -\frac{T_B}{T_A} = -t_{II} < 0$$

$$i_{Br.I} = |i_{Br.II}| \quad \text{when} \quad t_{II} = 1 + t_I$$

FIGURE 30.2
Reversible PGT (variant 36 ≡ 45, resp. 36 ≡ 63 and 45 ≡ 54 from Table 30.1).

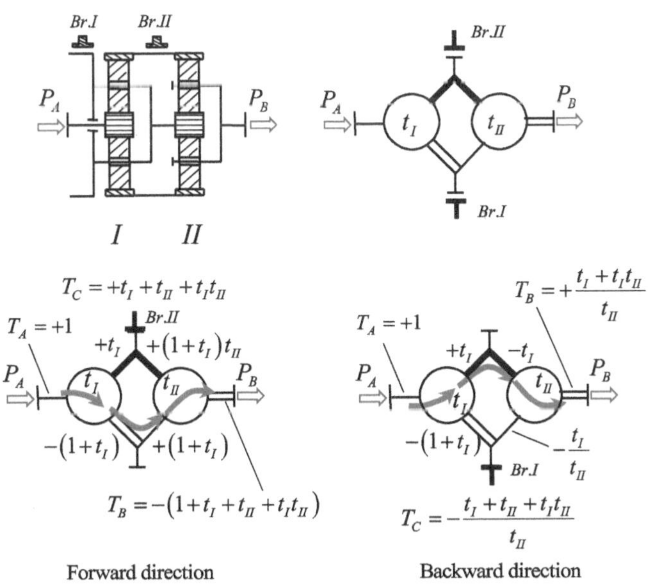

$$i_{Br.II} = -\frac{T_B}{T_A} = +(1+t_I+t_{II}+t_I t_{II}) > 0 \qquad i_{Br.I} = -\frac{T_B}{T_A} = -\frac{t_I + t_I t_{II}}{t_{II}} < 0$$

FIGURE 30.3
Reversible PGT (variant 26 ≡ 15, resp. 26 ≡ 62 and 15 ≡ 51 from Table 30.1).

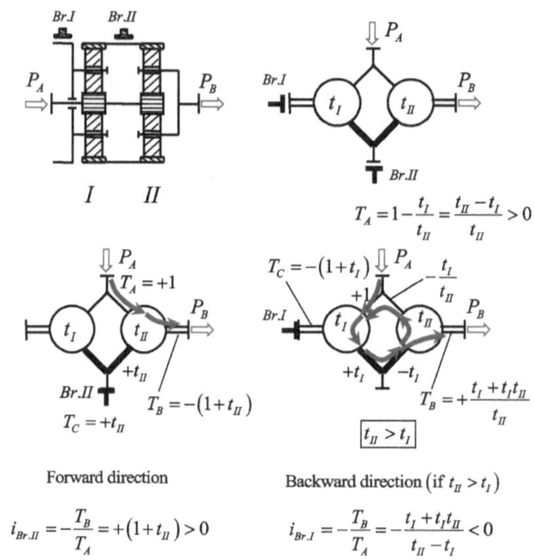

FIGURE 30.4
Reversible PGT (variant 55 ≡ 66 from Table 30.1).

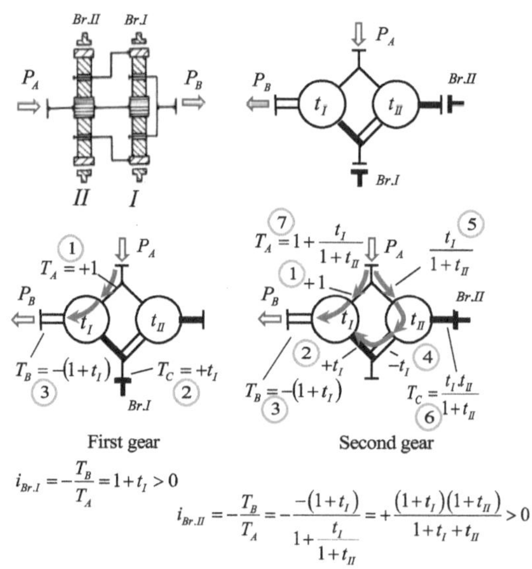

FIGURE 30.5
Two-speed change-gear (variant 45 ≡ 36, resp. 45 ≡ 54 and 36 ≡ 63 from Table 30.1).

is obtained at $t_{II} > t_I$. Otherwise at both gears input and output shafts will rotate in the same direction. Depending on the ratio of t_I and t_{II} the direction of the circulating power is changed. The direction shown in the figure is for the case when $t_{II} > t_I$. These phenomena are investigated in detail in Refs. [130, 230].

Figure 30.5 shows a change-gear with two speeds in one direction, suitable for two-speed drives for handling machines, technological machines or as part of a complex transmission.

Figure 30.6 shows the kinematic and structural schemes of a two-carrier change-gear (transmission gearbox) of a Mercedes-Benz® automobile [161]. This structural scheme corresponds to variant 36 ≡ 63 from Table 30.1. Although the structural scheme is the same as of the PGT in Figure 30.5, the kinematic schemes of these gear trains are quite different. Figure 30.6 also presents a table showing which of the gear elements—brakes *Br.I* and *Br.II* or couplings *Cp.I* and *Cp.II*, are locked in the different speed ratios (gears). In forward direction, by coupling *Cp.I*, the input shaft is connected to ring gear 3 and, in reverse direction, by *Cp.II*,—to the coupled sun gears 1 and 4. The output shaft is always connected to the carrier HI and ring gear 6. In various cases (gears), either carrier HII, or coupled sun gears 1 and 4 are fixed. In Figure 30.7, speed ratios for the four gears are determined.

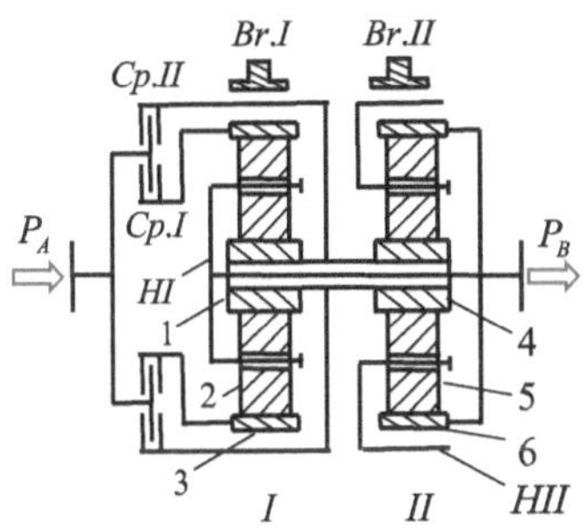

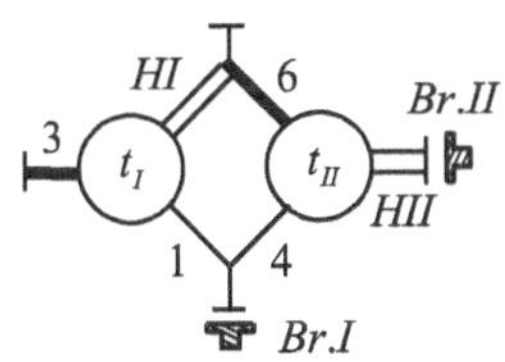

Gear	Cp.I	Cp.II	Br.I	Br.II
1	•			•
2	•		•	
3	•	•		
R		•		•

FIGURE 30.6
Four-speed two-carrier change-gear with two brakes (*Br.I* and *Br.II*) and two couplings (*Cp.I* and *Cp.II*).

First gear:

$\omega_3 = \omega_A;\ \omega_{HI} = \omega_6 = \omega_B;\ \omega_{HII} = 0$

$\sum T_i = T_A + T_B + T_C = +t_I - 1 - t_I - t_{II} + 1 + t_{II} = 0$

$i_1 = \frac{\omega_A}{\omega_B} = -\frac{T_B}{T_A} = -\frac{-(1+t_I+t_{II})}{+t_I} = +\frac{1+t_I+t_{II}}{+t_I}$

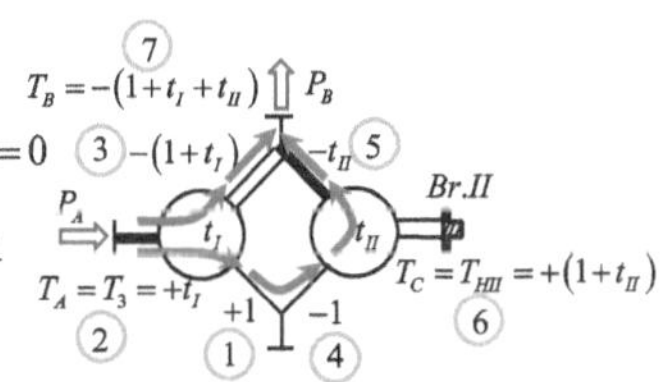

Second gear:

$\omega_3 = \omega_A;\ \omega_{HI} = \omega_6 = \omega_B;\ \omega_1 = \omega_4 = 0$

$\sum T_i = T_A + T_B + T_C = +t_I - 1 - t_I + 1 = 0$

$i_2 = \frac{\omega_A}{\omega_B} = -\frac{T_B}{T_A} = -\frac{-(1+t_I)}{+t_I} = +\frac{1+t_I}{+t_I}$

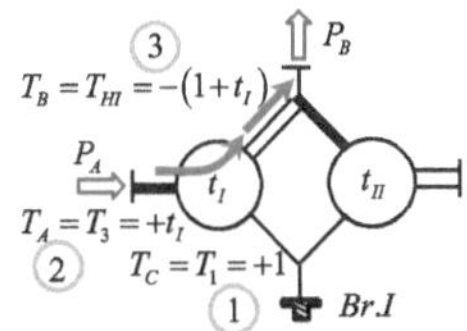

Third gear:

$\omega_3 = \omega_A = \omega_{HI} = \omega_6 = \omega_B$. PGT rotates as coupling.

$i_3 = +1$

Reverse gear:

$\omega_1 = \omega_4 = \omega_A;\ \omega_{HI} = \omega_6 = \omega_B;\ \omega_{HII} = 0$

$\sum T_i = T_A + T_B + T_C = +1 + t_{II} - 1 - t_{II} = 0$

$i_R = \frac{\omega_A}{\omega_B} = -\frac{T_B}{T_A} = -\frac{+t_{II}}{+1} = -t_{II}$

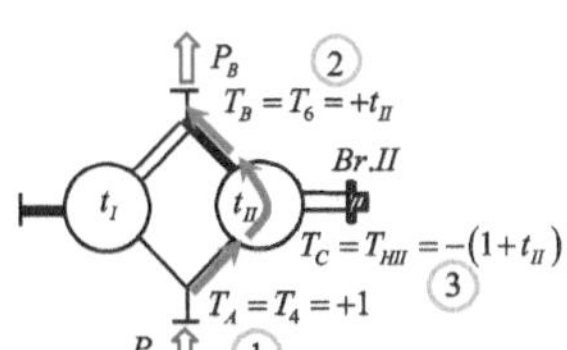

FIGURE 30.7
Determination of speed ratios of the change-gear from Figure 30.6.

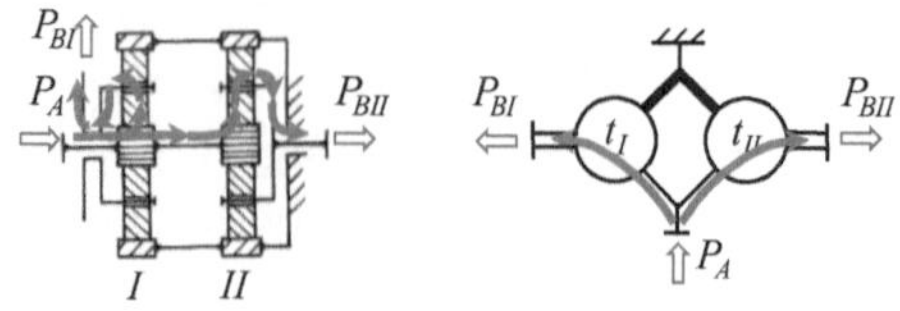

FIGURE 30.8
Two parallel coupled component PGTs (variant 55 ≡ 66 from Table 30.1).

Unlike the PGT in Figure 30.5, this one in Figure 30.6 except two braces has two couplings thanks to which performs more speed ratios, including in reverse direction. Only the ratios of the PGT are considered here. Actually, the car can work with more gears (speeds) by changing the input torque through a hydrotransformer, for example.

Torque method allows for direct determination of load on the individual gears. If, from experience, the operating time of the gear under given load is known (e.g., for cars), it is possible to determine load spectrum of the individual gears and, consequently, their equivalent loads (equivalent stress) and, ultimately, their safety factors S_F and S_H (see Figure 34.18). This possibility is a very significant advantage of the torque method.

For completeness, it can still be noted that the PGTs under consideration can also operate as parallel coupled power splitting (to two driven machines) gear trains, as shown in Figure 30.8.

31

Two-Carrier Compound Planetary Gear Trains with One Compound and Four External Shafts

31.1 Possible Ways of Connecting the Component PGTs—Types and Numbers

In these compound planetary gear trains (PGTs), the four single shafts of the component gear trains are external; i.e., they are four-shaft PGTs [25, 31]. Here too, it is important for a designer to know the number of possible variants. As can be seen from Table 31.1, of the nine possible combinations, only six are different, and the other three are isomorphic, as shown by the examples.

Since these gear trains have only one compound shaft, there is no power loop and the power circulation problem does not exist.

31.2 Working Modes

The following three figures illustrate the work of the PGTs in question:

Figure 31.1 shows a PGT driving two counterrotating airplane propellers. Both component PGTs have the same torque ratios $t_I = t_{II}$, as a result of which the output ideal torques

TABLE 31.1

Possible Variants of Two-Carrier PGTs with One Compound and Four External Shafts

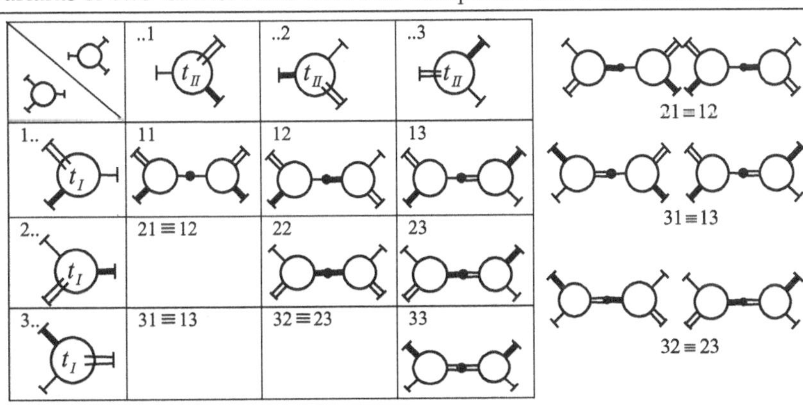

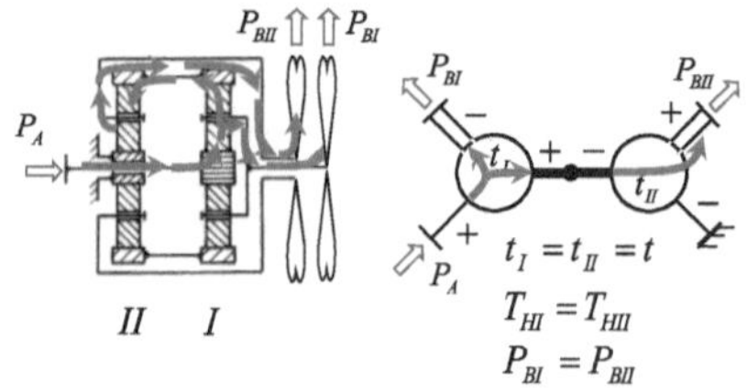

FIGURE 31.1
Single-motor drive of two counterrotating airplane propellers (a division differential, variant 22 from Table 31.1).

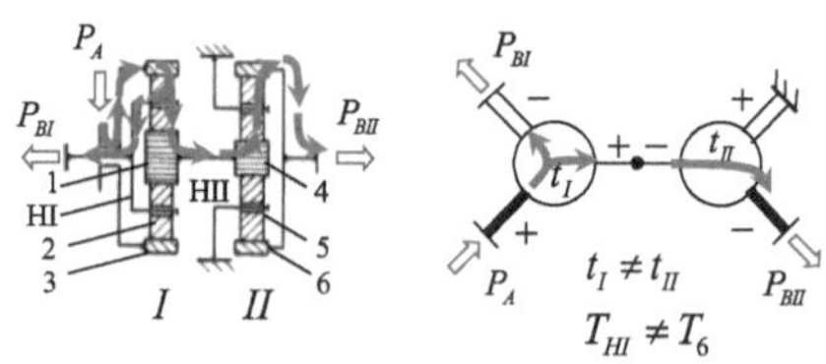

FIGURE 31.2
Single-motor drive of an asymmetric inter-axle differential of a heavy vehicle (a division differential, variant 11 from Table 31.1).

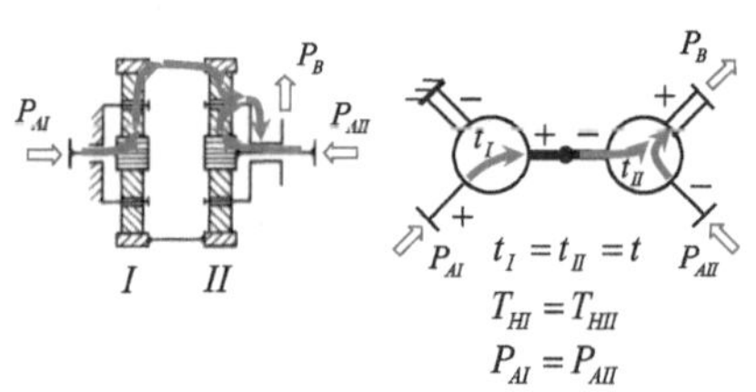

FIGURE 31.3
Twin-motor drive of an airplane spoiler (summation differential, variant 11 from Table 31.1).

T_{HI} and T_{HII} acting on the carriers HI and HII are equal. Since the two propeller blades have the same geometry and the same mechanical properties $T_{BI} = T_{BII} = f(\omega_B)$, their angular velocities will be also equal $\omega_{BI} = \omega_{BII}$. This means equal output powers $P_{BI} = P_{BII}$. Obviously, the PGT works as a division differential.

Figure 31.2 also shows a division PGT with a single-motor drive. Unlike the previous case, however, the two component PGTs are with different torque ratios $t_I \neq t_{II}$. The torques on the output shafts HI and 6 are not equal $(T_{HI} \neq T_6)$, except for the case of $t_{II} = 1 + t_I$. This gear train is used as an asymmetric inter-axle differential in heavy vehicles to transmit different torques to the front and rear axles.

PGT shown in Figure 31.3 consists of two identical component PGTs with equal torque ratios $t_I = t_{II}$ [161]. It works as a summation differential in a twin-motor driven airplane spoiler. A twin-motor drive is intentionally provided to enable the second motor to operate the spoiler, though at a reduced speed, in the event of a failure of the first motor. This type of duplication is often used to increase the security of responsible mechanisms—aircraft industry, metallurgy, etc.

32

Load Distribution between the Planets in Two-Carrier Compound Planetary Gear Trains, Its Unevenness, and Equalization

This problem, specifically for $\overline{\text{AI}}$-planetary gear train ($\overline{\text{AI}}$-PGT), is discussed in detail in Chapter 8. The exposure outlined there applies to a very large extent to compound PGTs. In order to achieve maximum equalization of load distribution in compound PGTs, it is necessary to observe the recommendations given in Section 8.6.6, namely,

- Use of a kinematic equalizing (distribution) device with a possible lower mass of the floating element.
- Achieving technically the highest and economically feasible accuracy of gear train elements.
- Increasing the pliability of certain gear train elements.
- Equal radial clearance of planet bearings.
- Use of bearings with increased clearance for central elements of the gear train.
- Sinphase montage of planets.

As the most suitable for a floating element, it is perfectly logical to choose the internal compound shaft that has no outlet. Figure 32.1 illustrates possible ways of joining two individual elements into a common floating element.

As can be seen from the figure, the joints can be either fixed or movable with one or two (better!) gear couplings. Since during the equalizing movement, the floating element is tilted, it is advisable that the planets have only one self-aligning bearing. Thanks to such a bearing, the unevenness of the load on the tooth face width (face load factor K_β!) is reduced. For factors K_β and K_γ, the values recommended in Chapter 8 can be used. Moreover, for greater sensitivity ξ of the equalizing device (see Section 8.6.5 and Figure 8.17), it is more advantageous to have the coupling sleeve longer.

Figure 32.2 graphically shows the results of the experimental studies of a two-stage planetary gear reducer of a Bulgarian hoist, carried out at the Technical University of Gabrovo [35]. The assembly of the carrier of the first stage (I) with the sun gear of the second stage (II) is a floating element. An original issue in these studies was to carry away the electrical signal from the strain gauges on the planet pins of the first stage, which are difficult to access as opposed to those of the second stage.

Based on own and on other authors' experimental studies [49], the following tentative values may be considered for the mesh load factor at $k=3$ planets:

- For the first (high-speed) stage $K_{\gamma\,max} \approx 1.6 \div 1.75$
- For the second (low-speed) stage $K_{\gamma\,max} \approx 1.15 \div 1.25$

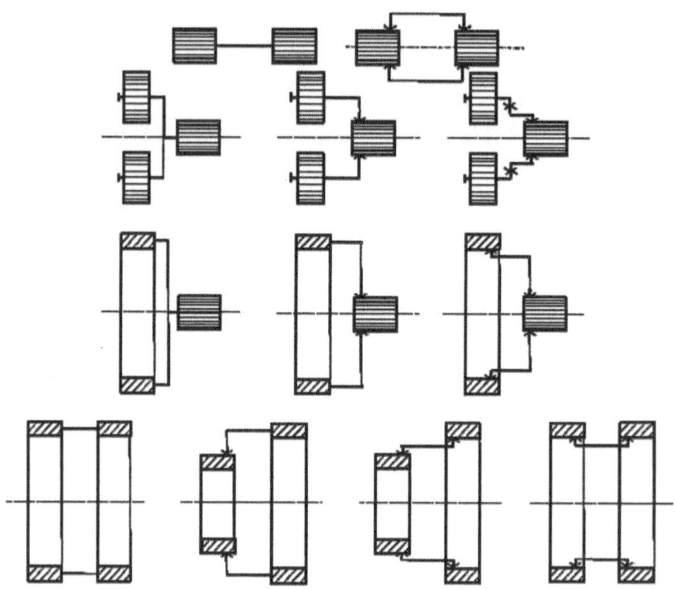

FIGURE 32.1
Possible ways of joining two central elements into a common floating element.

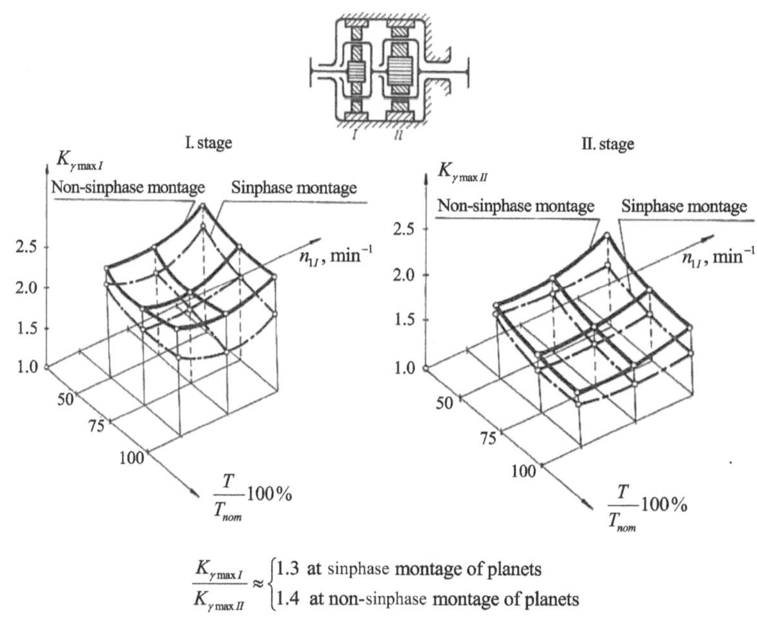

FIGURE 32.2
Experimentally determined mesh load factors $K_{\gamma\ max\ I}$ and $K_{\gamma\ max\ II}$ for both stages of hoist reducer at various input speed $n_{1,I}$ and input torque T as a percent of the nominal one T_{nom} [35].

As for factors K_v and K_β, the following can be noted:

In the first stage, which is naturally with narrower gears and usually with one bearing, influence of the dynamic factor K_v (considering the internal dynamics) is stronger than influence of the face load factor K_β due to the self-adjusting of the planet bearings.

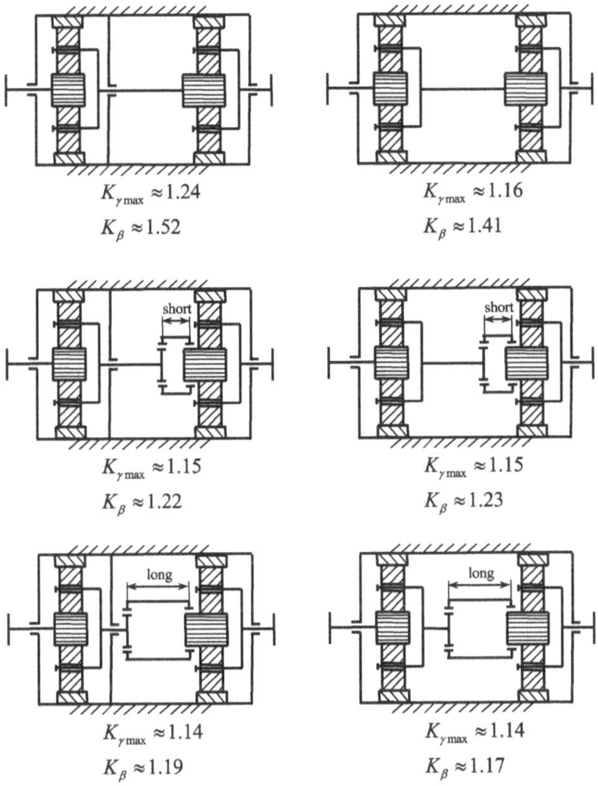

FIGURE 32.3
Experimental results for K_γ and K_β factors at various equalizing devices [49].

In the low-speed second stage, influence of K_υ decreases, and K_β (taking into account the load unevenness on the tooth face width) comes to the fore.

The problem of uneven load distribution and its solution are even more complicated in two-carrier PGTs than in single-carrier ones. Therefore, it is no wonder that experimental results are even less [35, 49, 53, 91, 92].

Other interesting and varied studies of two-carrier PGTs are contained in [49]. Two different arrangements of the investigated gear unit have been studied:

- Each component PGT is self-dependent due to the intermediate support between the two stages.
- Both component PGTs are coupled by a common floating element—the carrier of the first stage is joined to the sun gear of the second stage.

The common floating element itself is in three variants: as a rigid joint, and single and double-articulated gear coupling.

The study (determination of K_γ and K_β factors) was done only for the second stage. In order to have a basis for comparison of the experimental results for all variants, before each test, all gears are mounted in the same starting position. In Figure 32.3, the results of these studies are shown. The experiments were made with different bearings—single-row ball bearings (one or two) or double-row self-aligning (spherical) ball bearings. Quite expectedly, with self-aligning bearings, mesh load factor K_γ is the lowest.

33

Reduced Planetary Gear Trains

33.1 Wolfrom Gear Train (3K-Gear Train)

Wolfrom planetary gear train (PGT) [255], which according to Prof. Kudryavtsev's classification [147] is referred to as 3K-gear train, is a compound PGT consisting of two sequentially connected simple PGTs (Figure 33.1). As shown in the figure, one of them is a negative-ratio $\overline{AI}$-PGT, and the other one is a positive-ratio II-PGT, discussed in Chapters about $\overline{AI}$-, II- and $\overline{I}I$-PGTs, respectively. Wolfrom gear train is a reduced PGT because the carriers of the component PGTs are merged (reduced) into a single carrier. This single carrier has no output shaft. Its only purpose is to carry planets. Some authors classify it as a simple PGT because it has one carrier only, which is completely wrong.

The initial AII-variant with two-rim planets shown in Figure 33.1 is not very appropriate from a technological and operational point of view [124]. A more favorable one is $\overline{AI}$I-variant with one-rim planets shown in Figure 33.2, which is why it is discussed here in greater detail.

Thanks to the one-rim planet $(z_2' = z_2)$, the formula for the speed ratio $i_{14(3)}$ given in Figure 33.1 is simplified to

$$i_{14(3)} = \left(1+\frac{z_3}{z_1}\right)\frac{1}{1-\dfrac{z_3}{z_4}} = \left(1+\frac{z_3}{z_1}\right)\frac{z_4}{z_4-z_3} = \left(1+\frac{z_3}{z_1}\right)\frac{z_4}{\Delta z_{43}} = f\left(z_1, z_3, z_4\right). \tag{33.1}$$

Absolutely, the same formula is obtained if the torque method is used. Figure 32.2 also shows the structural scheme of the gear train as a compound two-carrier PGT (variant 26 from Table 29.1) with the corresponding torque ratios of the two component PGTs.

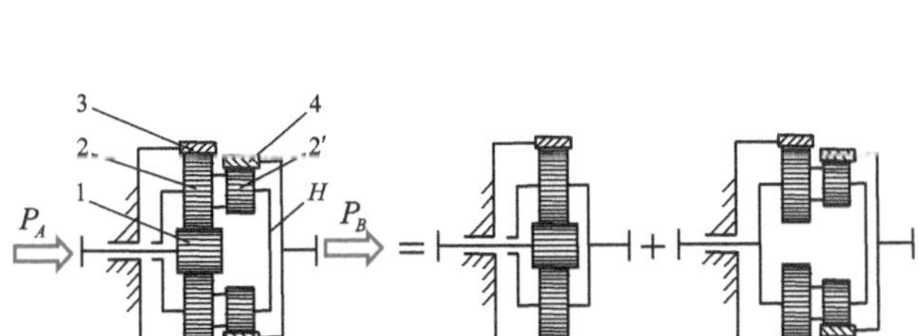

$$i_{14(3)} = \left(1+\frac{z_3}{z_1}\right)\frac{1}{1-\dfrac{z_2'}{z_4}\cdot\dfrac{z_3}{z_2}} \qquad i_{1H(3)} = 1+\frac{z_3}{z_1} \qquad i_{H4(3)} = \frac{1}{1-\dfrac{z_2'}{z_4}\cdot\dfrac{z_3}{z_2}}$$

FIGURE 33.1
Wolfrom gear train as a compound PGT.

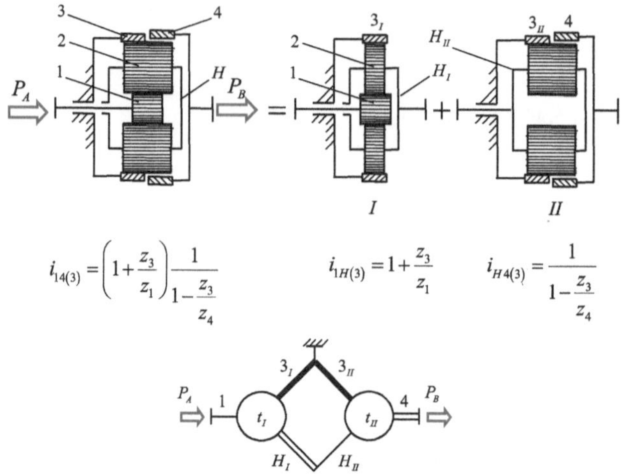

FIGURE 33.2
Wolfrom gear train with one-rim planet as compound PGT and its structural scheme (variant 26 from Table 29.1).

Torque ratios of the component PGTs are determined (see Section 7.4 and Chapter 23), taking into account the element designations adopted in Figure 33.2, wherein $z_{3I} \equiv z_{3II} = z_3 \cdot 2$:

Torque ratio of the first (I) component $\overline{\text{AI}}$-PGT is

$$t_I = \frac{T_3}{T_1} = \frac{z_3}{z_1}. \tag{33.2}$$

From equilibrium of torques in the second (II) component $\overline{\overline{\text{II}}}$-PGT

$$\sum T_i = T_3 + T_4 + T_{HII} = 0, \tag{33.3}$$

where

$T_{HII} \equiv T_{HI} = T_H$ and

$T_4 = -\frac{z_4}{z_3} T_3$, i.e.,

$$\sum T_i = T_3 - \frac{z_4}{z_3} T_3 + T_H = \left(1 - \frac{z_4}{z_3}\right) T_3 + T_H = 0,$$

carrier torque is determined

$$T_H = -\left(1 - \frac{z_4}{z_3}\right) T_3 = \left(\frac{z_4}{z_3} - 1\right) T_3, \tag{33.4}$$

As well as torque ratio of the second component PGT:

$$t_{II} = \frac{T_3}{T_H} = \frac{1}{\frac{z_4}{z_3} - 1} = \frac{z_3}{z_4 - z_3}. \tag{33.5}$$

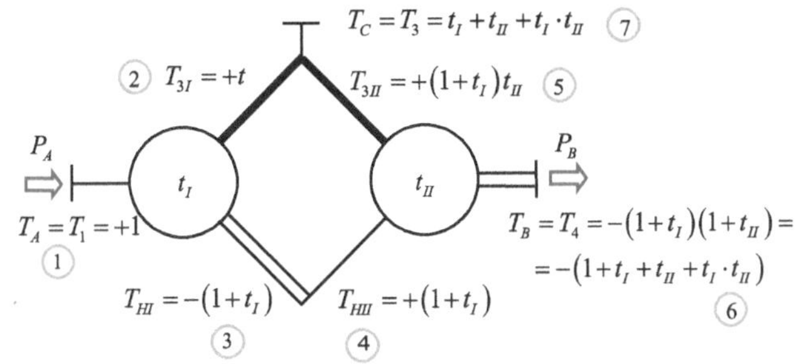

FIGURE 33.3
Determination of speed ratio of Wolfrom gear train through the torque method.

Figure 33.3 shows ideal torques determination. After check by formula (29.1)

$$\sum T_i = T_A + T_B + T_C = +1 - (1 + t_I + t_{II} + t_I \cdot t_{II}) + t_I + t_{II} + t_I \cdot t_{II} = 0, \tag{33.6}$$

speed ratio by formula (29.9) is determined:

$$\begin{aligned} i_k &= -\frac{T_B}{T_A} = 1 + t_I + t_{II} + t_I \cdot t_{II} = 1 + \frac{z_3}{z_1} + \frac{z_3}{z_4 - z_3} + \frac{z_3}{z_1} \cdot \frac{z_3}{z_4 - z_3} \\ &= \left(1 + \frac{z_3}{z_1}\right)\frac{z_4}{z_4 - z_3} = \left(1 + \frac{z_3}{z_1}\right)\frac{z_4}{\Delta z_{43}}. \end{aligned} \tag{33.7}$$

In the case of a one-rim planet Wolfrom gear train (Figure 33.2), the *following should be considered*:

- At a positive difference in the number of teeth of ring gears 4 and 3, i.e., $\Delta z_{43} = z_4 - z_3 > 0$ (as in Figure 33.2), the speed ratio $i_{14(3)}$ is positive ($i_{14(3)} > 0$), i.e., directions of rotation of the input and output shaft coincide.
- At a negative difference in the number of teeth of ring gears 4 and 3, i.e., $\Delta z_{43} = z_4 - z_3 < 0$, the speed ratio $i_{14(3)}$ is negative $\left(i_{14(3)} < 0\right)$, and directions of rotation of the input and output shaft are opposite.

The explanation of why $\Delta z_{43} = z_4 - z_3 > 0$ results in $i_{14(3)} > 0$, and $\Delta z_{43} = z_4 - z_3 < 0$ results in $i_{14(3)} < 0$ is discussed later in relation to Figure 33.6 and profile shift of the gears (also see Figure 23.6 and the explanations for it).

The one-rim planet variant $\left(\overline{\text{AII}}\right)$ does not allow so close values of speed ratios as two-rim planet variant (AII), which in some sense is its disadvantage. However, the difference is often very small.

In addition, it should be borne in mind that the difference in the number of teeth Δz_{43} should be equal to the number of planets k, since otherwise the gear train cannot be assembled, i.e., it is necessary that

$$\Delta z_{43} = k, \tag{33.8}$$

as most often $k = 3$, and the sum of the number of teeth z_1, z_3, and z_4 of center gears should be multiple of the number of planets k:

$$\boxed{\frac{z_1 + z_3}{k} = \text{an integer}} \text{ and } \boxed{\frac{z_1 + z_4}{k} = \text{an integer}}. \tag{33.9}$$

The determination of the number of teeth at a given speed ratio $i_{14(3)}$, which, as shown in Figure 33.1, depends on z_1, z_3, and z_4, and on Δz_{43}, respectively, is done in the following *practical way* [27, 131]:

1. Number of teeth of sun gear 1 is chosen, as is customary, in the range $z_1 = 12 \div 21$ [257]. Once chosen, this number is considered as $z_1 = \text{const}$. A special case where $z_1 = 12 = \text{const}$ is discussed in [257, 258]. However, this value is not mandatory.
2. Difference in the number of teeth of ring gears 4 and 3 according to (33.2) is chosen

$$\Delta z_{43} = k, \tag{33.10}$$

as most often the number of planets is $k = 3$. Very rarely $k = 2$ or $k = 4$. This dimension can also be considered further as $\Delta z_{43} = \text{const}$. Thus, e.g., if $z_1 = 15$ and $\Delta z_{43} = 3$, only z_3 remains unknown; i.e., the formula for speed ratio $i_{14(3)}$ is a function of only one variable quantity:

$$i_{14(3)} = \left(1 + \frac{z_3}{z_1}\right)\frac{z_4}{z_4 - z_3} = \left(1 + \frac{z_3}{z_1}\right)\frac{z_3 + \Delta z_{43}}{\Delta z_{43}} = \left(1 + \frac{z_3}{15}\right)\frac{z_3 + 3}{3} = f(z_3). \tag{33.11}$$

3. Although, in this case, formula (33.11) is a square equation:

$$a \cdot z_3^2 + b \cdot z_3 + c = 0, \tag{33.12}$$

solved by a known mathematical way, the method of samples proves to be more suitable for the practice. By varying the number of teeth z_3, the speed ratio $i_{14(3)}$ which is closest to the desired one is obtained.

4. Planet teeth number z_2 is determined as follows [257, 258]:
   - If the difference in teeth number of the larger ring gear 4 ($z_4 > z_3$) and the sun gear 1 $\Delta z_{41} = z_4 - z_1$ (not $\Delta z_{43} = z_4 - z_3$!) is *an even number*, z_2 is determined by the following formula:

$$z_2 = \frac{z_4 - z_1}{2} - 1 = \frac{\Delta z_{41}}{2} - 1. \tag{33.13}$$

   - If the difference Δz_{41} is *an odd number*, z_2 is determined by the following formula:

$$z_2 = \frac{z_4 - z_1}{2} - 0.5 = \frac{\Delta z_{41}}{2} - 0.5. \tag{33.14}$$

One-rim planet Wolfrom gear train is the most complex case of tooth geometry (Section 5.3). Each planet 2 participates in three meshings—with sun gear 1 and ring gears 3 and 4.

Moreover, two of the meshings are internal, which are generally much more complex than the external one. These three meshes are in mutual connection, and their parameters cannot be chosen independently of one another. This requires a specific approach to reach acceptable, balanced results for all gears.

The approach is as follows:

1. Initially, an operating center distance a_w is chosen, which lies within the boundaries defined by the three reference center distances given by

$$a_{12} = m\frac{z_1 + z_2}{2},\ a_{23} = m\frac{z_3 - z_2}{2},\ \text{and } a_{24} = m\frac{z_4 - z_2}{2}. \tag{33.15}$$

2. The three operating pressure angles are determined at chosen operating a_w and reference a_{12}, a_{23}, and a_{24} center distances given by

$$\begin{aligned}\alpha_{w12} &= \arccos\left(\frac{a_{12}}{a_w}\cos\alpha\right) = f(a_w);\\ \alpha_{w23} &= \arccos\left(\frac{a_{23}}{a_w}\cos\alpha\right) = f(a_w);\\ \alpha_{w24} &= \arccos\left(\frac{a_{24}}{a_w}\cos\alpha\right) = f(a_w).\end{aligned} \tag{33.16}$$

   An exemplary variation of the three pressure angles as a function of a_w is shown in Figure 33.4.

3. The sum of profile shift coefficients $(x_1 + x_2)$ of external meshing and the differences $(x_3 - x_2)$ and $(x_4 - x_2)$ of internal meshings are determined. They determine the individual shift coefficients x_1, x_3, and x_4, as a function of the planet profile shift coefficient x_2, which is given by

$$\begin{aligned}x_1 &= \frac{z_1 + z_2}{2}\cdot\frac{\text{inv}\alpha_{w12} - \text{inv}\alpha}{\tan\alpha} - x_2 = \text{const} - x_2 = f(\alpha_{w12}, x_2) = f(a_w, x_2)\\ x_3 &= \frac{z_3 - z_2}{2}\cdot\frac{\text{inv}\alpha_{w23} - \text{inv}\alpha}{\tan\alpha} + x_2 = \text{const} + x_2 = f(\alpha_{w23}, x_2) = f(a_w, x_2)\\ x_4 &= \frac{z_4 - z_2}{2}\cdot\frac{\text{inv}\alpha_{w24} - \text{inv}\alpha}{\tan\alpha} + x_2 = \text{const} + x_2 = f(\alpha_{w24}, x_2) = f(a_w, x_2).\end{aligned} \tag{33.17}$$

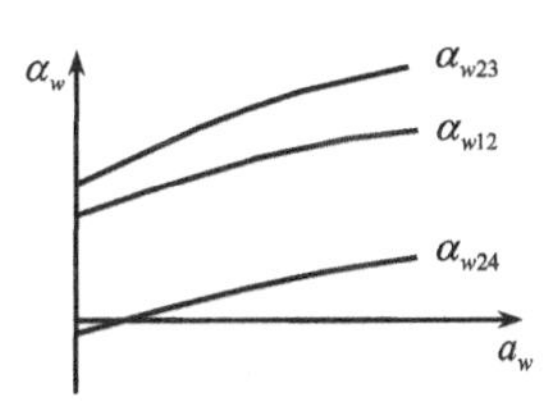

FIGURE 33.4
Exemplary variation of the three operating pressure angles α_w as a function of the operating center distance a_w.

Dependencies are obviously linear, as shown in Figure 33.5. After choosing the planet profile shift coefficient x_2, other shift coefficients x_1, x_3, and x_4 are determined by (33.8).

4. Checking whether the resulting pressure angles and profile shift coefficients meet certain *requirements*, concerning
   - Operating pressure angles
   - Profile shift coefficients
   - Transverse contact ratios
   - Interference

The requirements are as follows:

1. Values of pressure angles α_{w12}, α_{w23}, and α_{w24} in line with experience [147, 157, 158] must be approximately within the following ranges:
   - For external meshing of sun gear 1 with planets 2: $\alpha_{w12} \approx 20° \div 27°$
   - For internal meshing of planets 2 with ring gear 3: $\alpha_{w23} \approx 23° \div 27°$
   - For internal meshing of planets 2 with ring gear 4: $\alpha_{w24} \approx 15° \div 18°$

   These values, of course, are approximate. If the chosen operating center distance a_w does not obtain acceptable operating pressure angles α_{w12}, α_{w23}, and α_{w24}, another center distance is chosen, based on Figure 33.4.
2. Profile shift coefficients x_1, x_2, x_3, and x_4 must have acceptable values in view of undercutting and tooth tip pointing (see Figure 5.1). As it is known (see Sections 5.1 and 5.5), moderate values for profile shift coefficients are generally recommended [175]

$$x = \pm(0.5 \div 0.65),$$

   but which cannot always be complied with. For example, for the smaller ring gear 3, large positive values for the coefficient are sometimes obtained $x_3 = +1.3 \div +1.5$, and for the larger ring gear 4—smaller negative values for $x_4 = -0.2 \div -0.7$.
3. Transverse contact ratios $\varepsilon_{\alpha 12}$, $\varepsilon_{\alpha 23}$, and $\varepsilon_{\alpha 24}$ must be high enough. Recommended values are [175]

$$\varepsilon_{\alpha 12} \geq 1.2,\ \varepsilon_{\alpha 23} \geq 1.4, \text{ and } \varepsilon_{\alpha 24} \geq 1.4.$$

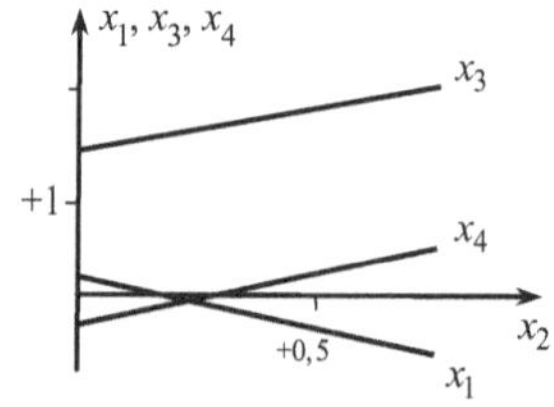

FIGURE 33.5
Exemplary variation of the profile shift coefficients x_1, x_3, and x_4 as a function of the planet profile shift coefficient x_2.

These values are only approximate, too. Moreover, the contact ratio ε_α affects the loss factor ψ_z in the three meshings (see Section 11.1), which are the dominant losses in the gear train and are important when determining its efficiency (see Section 12.1).

Tentative efficiency can be determined from the diagrams in Figure 33.6. Two diagrams of two sources are given purposely. According to the first diagram [147], efficiency $\eta_{14(3)}$ is determined depending on the speed ratio $i_{14(3)}$ and on the number of teeth z_2 of planet 2 which is given by

$$\eta_{14(3)} = f\left(i_{14(3)}, z_2\right).$$

According to the second diagram [155], efficiency $\eta_{14(3)}$ is also determined depending on the speed ratio $i_{14(3)}$, but instead on the number of teeth, it is dependent on the gear ratio of the external meshing $u_{12} = z_2/z_1$

$$\eta_{14(3)} = f\left(i_{14(3)}; u_{12}\right).$$

A designer may, of course, use the formulae for meshing loss and basic efficiency (ψ_0 and η_0) defined in Section 11.1, and calculate these for the specific case. In doing so, the designer can determine efficiency $\eta_{14(3)}$ depending on the variation of the coefficient of friction in teeth meshing μ_z, using the following formula [161]:

$$\eta_{14(3)} = \frac{\dfrac{1 + \dfrac{z_3}{z_1}\eta_{013}}{1 - \dfrac{z_3}{z_4}\eta_{034}}}{\dfrac{1 + \dfrac{z_3}{z_1}}{1 - \dfrac{z_3}{z_4}}} = f\left(\mu_z\right), \tag{33.18}$$

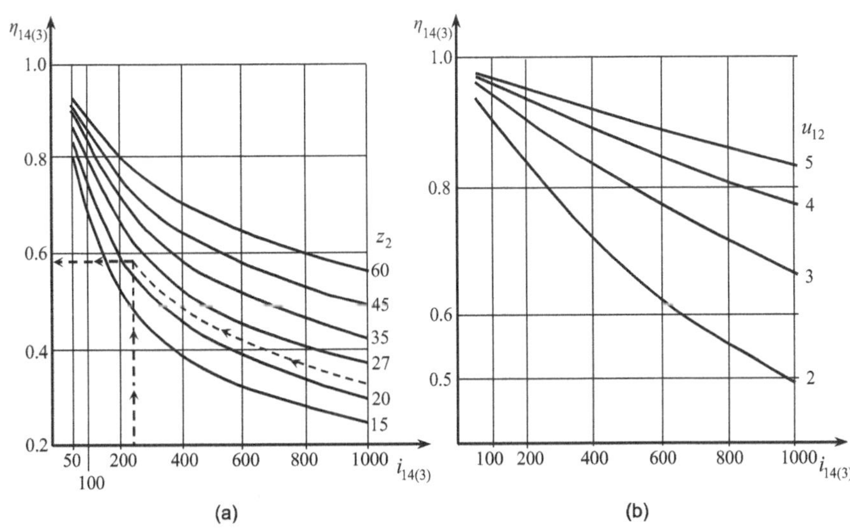

FIGURE 33.6
Tentative determination of efficiency of Wolfrom gear train according to various sources: (a) $\eta_{14(3)} = f\left(i_{14(3)}; z_2\right)$ according to [147]; (b) $\eta_{14(3)} = f\left(i_{14(3)}; u_{12}\right)$ according to [155]. (Redrawn from the mentioned sources.)

so that based on μ_z the efficiency scattering can be determined. This third way to determine the efficiency, unlike the other two ways, has the advantage of taking into account the specific tooth geometry (profile shift, contact ratio) and scattering of the friction coefficient μ_z, so that the result is as reliable as possible. It should be borne in mind that the losses in the external meshing $\psi_{z12(H)}$ are considerably greater than the losses in the internal meshings $\psi_{z23(H)}$ and $\psi_{z24(H)}$. These losses can be made as small as possible if in-front-of-pitch-point and beyond-of-pitch-point contact ratios $\varepsilon_{\alpha 1}$ and $\varepsilon_{\alpha 2(1)}$ (see Section 11.1) are approximately equal, i.e.,

$$\varepsilon_{\alpha 1} \approx \varepsilon_{\alpha 2(1)},$$

respectively, the segments $\overline{AC} \approx \overline{CE}$ in Figure 11.2.

Figure 33.6 shows that Wolfrom gear train efficiency is not very high. This is due to the fact that it is a hybrid between a high-efficiency negative-ratio $\overline{\text{AI}}$-PGT and a positive-ratio II-PGT with low efficiency. However, its efficiency is not quite as low as at AA-, $\overline{\text{AA}}$-, II-, and $\overline{\text{II}}$-PGTs, and therefore, in some cases, it is used. Since, as can be seen from Figure 33.6, efficiency $\eta_{14(3)}$ depends not only on the speed ratio but also on the number of teeth of planets z_2, respectively, of the gear ratio of the external gears u_{12}, which increases with the increase in speed ratios, ultimately, $\eta_{14(3)}$ slightly depends on the speed ratio $i_{14(3)}$ [257, 258]. It is usually $\eta_{14(3)} \approx 0.7 \div 0.8$. The influence of quality of the tooth surfaces is stronger, i.e., whether they are only milled, further shaved, or ground. It is also interesting that in trains with $k = 3$ planets, the efficiency is higher than with $k = 2$ planets, and this is experimentally determined [257, 258]. The ways for increasing the efficiency of Wolfrom gear train are proposed in [161].

4. It is necessary to avoid interference, the danger of which is greater in internal meshing (see Section 5.5, Chapter 23, and Figure 23.4).

Determination of ideal external torques T_1, T_3, and T_4 on all three gears and tangential forces $F_{t12} = F_{t21}$, $F_{t23} = F_{t32}$, and $F_{t24} = F_{t42}$ in all three meshings is as follows:

It begins from the input torque $T_A \equiv T_1$ on the sun gear 1. Output torque $T_B \equiv T_4$ on ring gear 4 is determined by the speed ratio $i_{14(3)}$:

$$T_B \equiv T_4 = -T_1 \cdot i_{14(3)}. \tag{33.19}$$

Reactive torque $T_C \equiv T_3$ on the fixed ring gear 3 is determined by the equilibrium of the three torques $\sum T_i = 0$:

$$T_C \equiv T_3 = T_B - T_A = T_4 - T_1 = \left(i_{14(3)} - 1\right)T_1. \tag{33.20}$$

As noted above, unlike other PGTs, at the Wolfrom gear train the carrier is not loaded with torque but only carries the planets.

Nominal tangential forces in the three meshings are determined by the following formulae (Section 6.1):

$$F_{t12} = F_{t21} = 2,000\frac{T_1}{k \cdot d_1},$$

$$F_{t23} = F_{t32} = 2,000\frac{T_3}{k \cdot d_3}, \tag{33.21}$$

$$F_{t24} = F_{t42} = 2,000\frac{T_4}{k \cdot d_4},$$

where F_t is the nominal tangential force in corresponding meshing, N, T_1, T_3, and T_4 are the torques on corresponding shafts, Nm, and d_1, d_3, and d_4 are the reference diameters of corresponding gears, mm.

Wolfram gear train has the same feature, characteristic of $\overline{\text{II}}$-PGT (Chapter 23 and Figure 23.6). Although the number of teeth of ring gear 4 is greater than the number of teeth of ring gear 3 ($z_4 > z_3$), the diameter of the operating circle $d_{w2(4)}$ of planet 2 in mesh with ring gear 4 is smaller than the diameter of its operating circle $d_{w2(3)}$ in mesh with ring gear 3, i.e.,

$d_{w2(4)} < d_{w2(3)}$, although $z_4 > z_3$.

This is clearly visible in Figure 33.7 (see Figure 23.6 too). It is due to the fact that the three operating pressure angles are different:

$\alpha_{w12} \neq \alpha_{w23} \neq \alpha_{w24}$, in that $\alpha_{w23} > \alpha_{w24}$,

resulting in three different operating circles of planet 2, one for each meshing, given by

$$d_{w2(1)} \neq d_{w2(3)} \neq d_{w2(4)}.$$

The fact that $d_{w2(4)} < d_{w2(3)}$ leads to $d_{w4} < d_{w3}$ and to $i_{14(3)} > 0$.

As can be seen from Figure 33.7, peripheral forces F_{t32} and F_{t42} in the internal meshing of ring gear 3 and 4 with planets 2 are several times greater than the force in the external meshing of planets 2 with sun gear 1. Because the forces F_{t32} and F_{t42} act in two different parallel planes, they create a moment that tilts the planets within their bearing clearances. This leads to nonuniform distribution of load over the tooth face, which is considered in load capacity calculation according to face load factor K_β. This factor depends, of course, very heavily on the accuracy of machining, clearances, and pliability of gear train elements, more in particular the bearings. As a mean value in the load capacity calculations, tentatively it can be accepted:

$$K_\beta \approx 1.3.$$

In order to reduce this nonuniformity, resp. K_β, planets are made with a groove in the middle (Figure 33.8). This groove increases teeth pliability where the maximum load is, and the load decreases.

Because of the inevitable manufacturing anomalies within the tolerances of elements, in Wolfrom gear train, planet loading is unevenly distributed, too. The question of uneven load distribution between the planets in $\overline{\text{AI}}$-PGT is discussed in detail in Chapter 8. The points outlined there are generally applicable to Wolfrom gear train. In it, however, the picture of load distribution is more complex. At a floating sun gear 1 or carrier H in the load capacity calculations, mesh load factor can be taken (quite tentatively) $K_{\gamma max} = 1.2 \div 1.4$ [258].

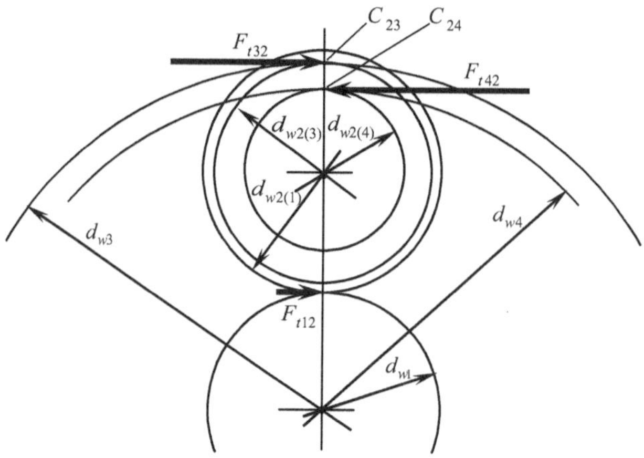

FIGURE 33.7
Forces in Wolfrom gear train meshings (also see Figure 23.6).

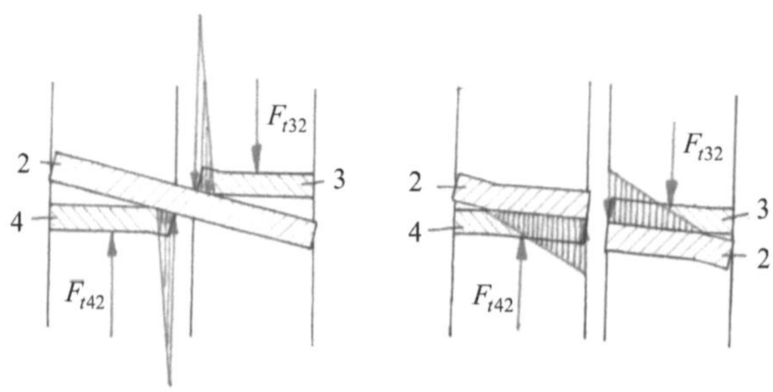

FIGURE 33.8
Groove in the middle of planet to reduce nonuniform distribution of load over the tooth face (face load factor K_β!) in gears 2, 3, and 4.

Of course, the same recommendations for a more even load distribution can be given to Wolffrom gear train, namely, accuracy, pliability (including floating) of central elements, and equal clearance of planet bearings (see Section 8.6.6).

Wolfrom gear train may also be *self-locking* if it is necessary, due to the fact that one of the component trains ($\overline{\text{II}}$-PGT) may be self-locking. This occurs (see Chapter 23) when its basic speed ratio i_0 is within the limits defined by its basic efficiency η_0 which is given by

$$\boxed{\eta_0 < i_0 < \frac{1}{\eta_0}}. \tag{33.22}$$

Figure 33.9 shows a Wolfram gear train with one-rim planet, and Figures 33.10 and 33.11—the so-called carrier-less Wolfrom gear trains.

In the gear train in Figure 33.10, the large radial forces in internal meshings strongly press the planets to the sun gear, where friction increases, and efficiency decreases, respectively. It may even reach to self-locking. Increased friction also occurs on the faces of the planets because of the large tangential forces in the internal meshings which act in parallel planes and thus create the moment. This gear train, in particular, is embedded in an

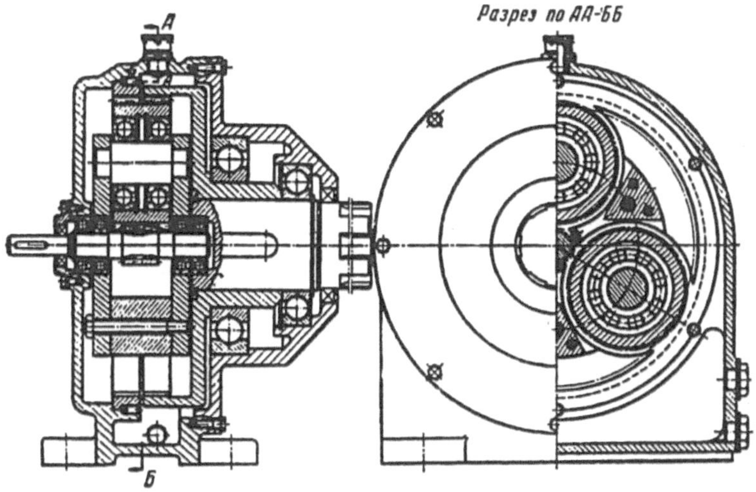

FIGURE 33.9
Wolfrom gear train with one-rim planet ($\overline{\text{AI}}$-PGT).

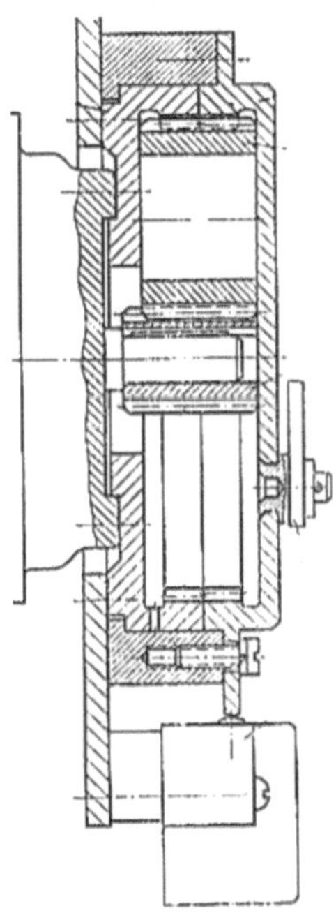

FIGURE 33.10
Carrier-less Wolfrom gear train, embedded in an eccentric control mechanism.

eccentric control mechanism, and the load is low enough so that the increased friction is not relevant for its operation.

In the gear of Figure 33.11, the radial forces are borne by the two rings, so there is not so much friction between the teeth as in the previous case. This makes it suitable to embed in the wheels of travel mechanisms of railroad handling machines (the shown case).

Kinematic capabilities and assembly condition of different types of Wolfrom gear train are considered in [124].

Wolfrom gear train has the following advantages, disadvantages, and application:

Advantages

1. Large kinematic capabilities—starting with speed ratios of about 50, reaching up to about 1,000, but with decreasing efficiency with speed ratio increase.
2. Very large compactness due to internal meshing and planet positioning inside of ring gears. The relative weight and volume are three to six times smaller than non-PGTs [258].
3. Simplified manufacturing of one-rim planet. Especially simple is the arrangement of the carrier-less variant.
4. Self-locking is possible, which in some cases is desirable.
5. Kilogram cost is not much higher (only 20 ÷ 30%) than that of non-PGTs.

Disadvantages

1. Complicated geometric calculations, since here we have the most complex case of tooth geometry—three mutual connected meshings, two of which are internal.
2. More complicated and low-productivity manufacturing of internal teeth (by tooth cutting, not hobbing).
3. Comparatively low efficiency decreasing as the speed ratio increases, resulting in heating and making the gear train unsuitable for continuous operation. However, it is not as low as of AA-, $\overline{AA}$-, II-, and $\overline{II}$-PGTs. At high speed ratios, it is comparable to the efficiency of multistage non-planetary cylindrical gear trains.
4. Self-locking in some cases is undesirable.

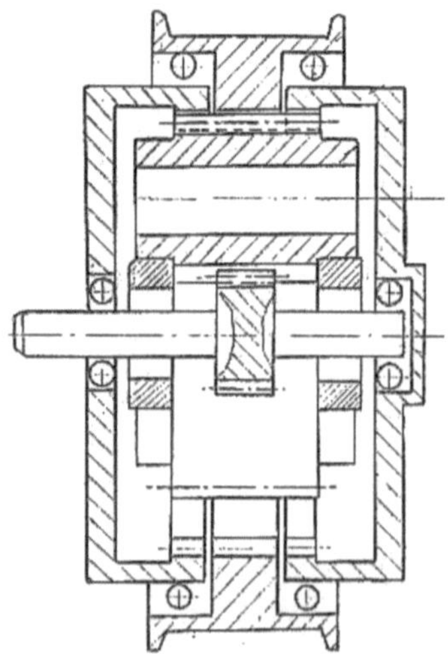

FIGURE 33.11
Carrier-less Wolfrom gear train, embedded in a wheel of the travel mechanisms of a railroad handling machine.

Application

Relatively limited, though not as much as the application of the AA-, $\overline{AA}$-, II-, and $\overline{II}$-PGTs. It is very suitable for the cases of small power (10÷15 kW) and noncontinuous operation, and in cases where low efficiency is tolerable or does not matter, respectively, even self-locking may be desirable. Wolfrom gear train can successfully replace the three-carrier PGTs when it comes to noncontinuous work and the arrangement should be simplified, e.g., in travel mechanisms of cranes, hoists, etc. In [108], a low-loss Wolfrom gear train for wind turbine is presented.

33.2 Ravigneaux Gear Train

Ravigneaux gear train, similar to Wolfrom gear train, is a compound PGT consisting of two sequentially connected simple PGTs (Figure 33.12) [161], one of which is a positive-ratio AAI-PGT (discussed in Chapter 24) and the other—negative-ratio AAA-PGT (discussed in Chapter 25). Ravigneaux gear train is a reduced PGT because carriers of the component PGTs are merged (reduced) into a single carrier. It can also be with two- or one-rim planet (Figure 33.13). The figure shows the structural scheme of the reduced PGT (which corresponds to variant 45 ≡ 54 from Table 30.1) operating as a change-gear and a table showing which of the elements (brakes *Br.1* and *Br.2* or couplings *Cp.1*, *Cp.2*, and *Cp.3*) are locked in the different speed ratios (gears). By the three couplings, the input shaft can be connected to sun gear 1 (most often), sun gear 5, or carrier H. The output shaft is always connected to ring gear 4.

Figure 33.14 shows speed ratio determination for all operation cases of the change-gear from Figure 33.13 through the torque method. The example is solved with specific values for the number of gear teeth ($z_1 = 28$, $z_4 = 88$, and $z_5 = 36$) which are the same as in the example solved in the German standard VDI 2157 [237], so that not only the results but also the volume and complexity of the calculations can be compared. It is no surprise that the torque method produces the same values as VDI 2157, but in a shorter, simpler, and more visible way.

Ravigneaux gear train is widely used in automotive automatic transmissions (change-gears). It is also used in continuous track vehicles, including tanks. More about this interesting gear train can be found in [161].

For completeness, it should be noted that both reduced gear trains, Wolfrom and Ravigneaux, can be represented differently as a compound of two component PGTs. In kinematic terms, i.e., to determine the speed ratio, this is irrelevant, but not when the

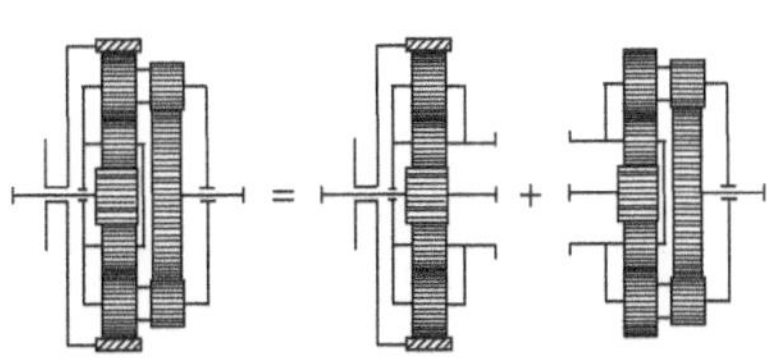

FIGURE 33.12
Ravigneaux gear train as compound PGT.

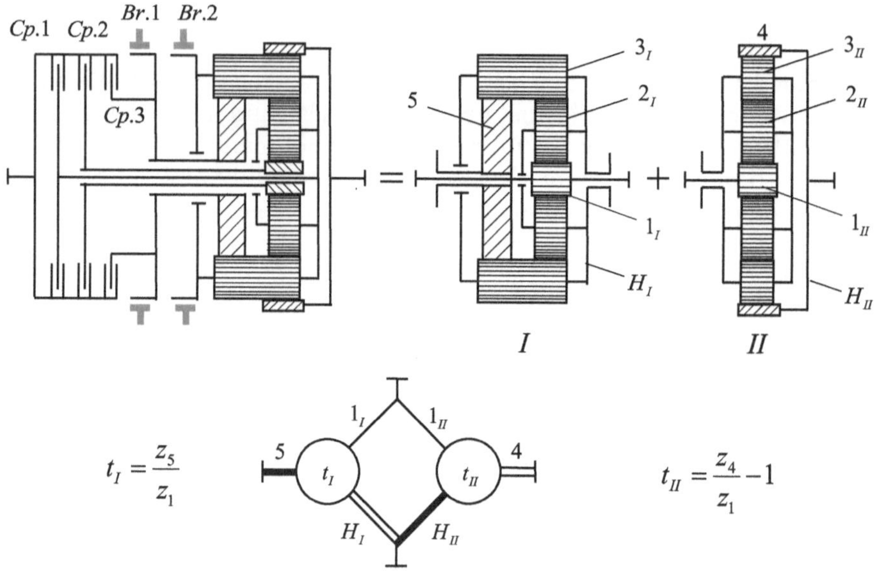

Gear	Cp.1	Cp.2	Cp.3	Br.1	Br.2
1		•			•
2		•		•	
3	•	•			
4	•			•	
R			•		•

FIGURE 33.13
Simplified variant of Ravigneaux gear train as change-gear and its structural scheme (variant 45 ≡ 54 from Table 30.1).

internal losses and efficiency are determined. More details on this and other problems can be found in [161, 232].

The torque method allows for a direct determination of torques on the individual gears. With known or accepted (e.g., from normative documents) duration of action of a given load, it is possible to determine load spectrum of the individual gears and, consequently, their equivalent loads (equivalent stress) and, ultimately, their safety factors S_F and S_H (see Figure 34.18).

First gear: Locked Cp.3 and Br.1

$$\omega_1 = \omega_A;\ \omega_4 = \omega_B;\ \omega_H = 0$$

$$\sum T_i = T_A + T_B + T_C = +1 - 3.143 + 2.143 = 0$$

$$i_1 = \frac{\omega_A}{\omega_B} = -\frac{T_B}{T_A} = -\frac{-3.143}{+1} = +3.143$$

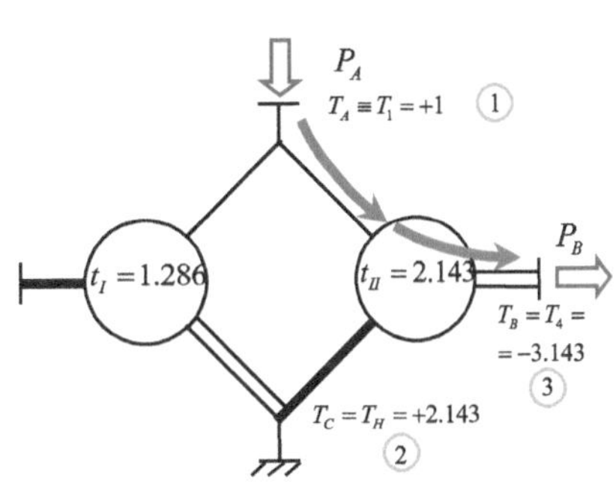

Second gear: Locked Cp.2 and Br.2

$$\omega_1 = \omega_A;\ \omega_4 = \omega_B;\ \omega_5 = 0$$

$$\sum T_i = T_A + T_B + T_C = +1.936 - 3.143 + 1.206 = 0$$

$$i_2 = \frac{\omega_A}{\omega_B} = -\frac{T_B}{T_A} = -\frac{-3.143}{+1.936} = +1.622$$

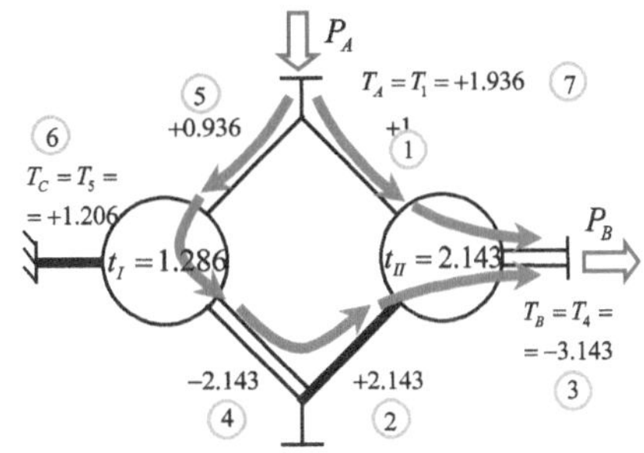

Third gear: Locked Cp.1 and Cp.2

$\omega_1 = \omega_H = \omega_A = \omega_4 = \omega_B$. PGT rotates as coupling.

$i_{III} = +1$

Fourth gear: Locked Cp.1 and Br.1

$$\omega_H = \omega_A;\ \omega_4 = \omega_B;\ \omega_5 = 0$$

$$\sum T_i = T_A + T_B + T_C = -4.429 + 3.143 + 1.286 = 0$$

$$i_{IV} = \frac{\omega_A}{\omega_B} = -\frac{T_B}{T_A} = -\frac{+3.143}{-4.429} = +0.7096$$

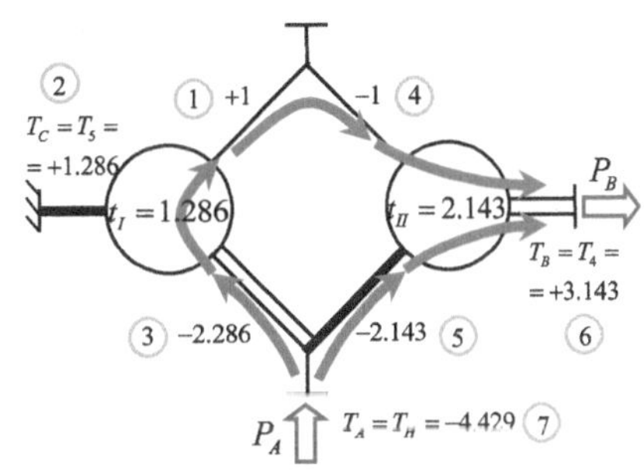

Reverse gear: Locked Cp.3 and Br.2

$$\omega_5 = \omega_A;\ \omega_4 = \omega_B;\ \omega_H = 0$$

$$\sum T_i = T_A + T_B + T_C = +1.286 + 3.143 - 4.429 = 0$$

$$i_R = \frac{\omega_A}{\omega_B} = -\frac{T_B}{T_A} = -\frac{+3.143}{+1.286} = -2.444$$

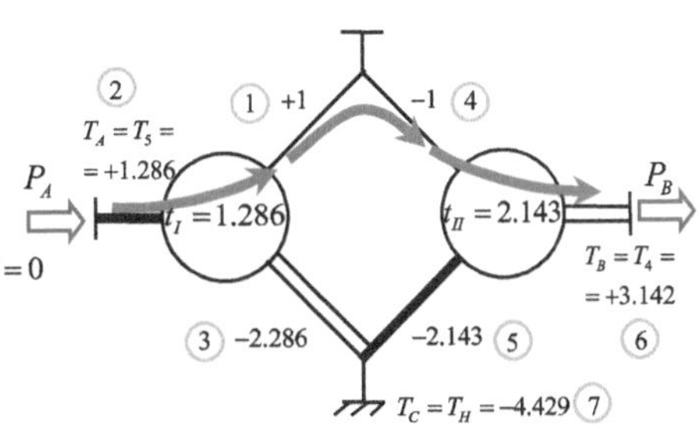

FIGURE 33.14
Determination of speed ratios of planetary change-gear from Figure 33.13.

34

Multi-Carrier Compound Planetary Gear Trains

As multi-carrier planetary gear trains (PGTs) are designated those composed of more than two component PGTs—three-carrier, four-carrier, etc. According to [166, 167], multi-carrier PGTs are designated as *higher* compound PGTs (see Chapter 28).

Only three- and four-carrier PGTs will be considered here. Figure 28.1 shows some structural schemes of these gear trains, which are used in this chapter.

34.1 Three-Carrier Compound PGTs

Some cases of different ways of combining the component PGTs and the obtained structural schemes are shown in the following few figures. Here too, the direction of transmitted power is indicated by solid arrows and of the relative (rolling) power—with dotted arrows.

The gear train shown in Figure 34.1 represents the simplest and most commonly used case of compound three-carrier PGT. Usually, the compound shaft formed by the three ring gears 3, 6, and 9 is fixed, unlike the same PGT (Figure 34.2) whose last carrier HIII is fixed. PGT in Figure 34.1 is obtained by adding another simple PGT to the sequentially connected two-carrier PGT (Figure 29.4). The power is transmitted sequentially, without internal division (as in Figure 34.2) or internal circulation. At fixed ring gears, the direction of relative (rolling) powers $P_{rel\,I}$, $P_{rel\,II}$, and $P_{rel\,III}$ is easily determined. In all three component PGTs, it is transmitted from the sun gears 1, 4, and 7 to the respective ring gears 3, 6, and 9. Since the input ideal and real torques are equal ($T'_A = T_A = +1$), the alternative way of determining efficiency can be applied (see Section 12.3).

In Figure 34.1 and in the following figures, calculations are given more succinctly than those given in Section 29.3. The same structural scheme shows not only ideal but also real torques, the latter being framed. The check of the correctness of torque calculations, and determination of speed (kinematic) ratio i_k, torque transmit ratio (torque transformation) i_T, and efficiency η are also presented. For the determination of real torques, and through them of efficiency, the correct determination of direction of the relative (rolling) powers $P_{rel\,I}$, $P_{rel\,II}$, and $P_{rel\,III}$ in the three component PGTs is very important. Here, as in Figures 29.4 and 29.6, the alternative way of efficiency determination is also given—only by the output torques (ideal T_B and real T'_B), since the ideal T_A and real T'_A input torques are equal $T'_A = T_A = +1$ (also see Section 12.3).

In the PGT shown in Figure 34.2 [30, 31], the compound shaft, connecting the three ring gears 3, 6, and 9, is an output shaft (in this case, ring gears are attached to a rope drum). It is not difficult to understand that this gear train is the most common sequentially connected three-carrier PGT from Figure 34.1, however, not the three ring gears, but the carrier of the third stage HIII is fixed. Determination of directions of the internal power, absolute and relative ($P_{rel\,I}$, $P_{rel\,II}$, and $P_{rel\,III}$), as in Figure 34.1, is not a problem. In all three component

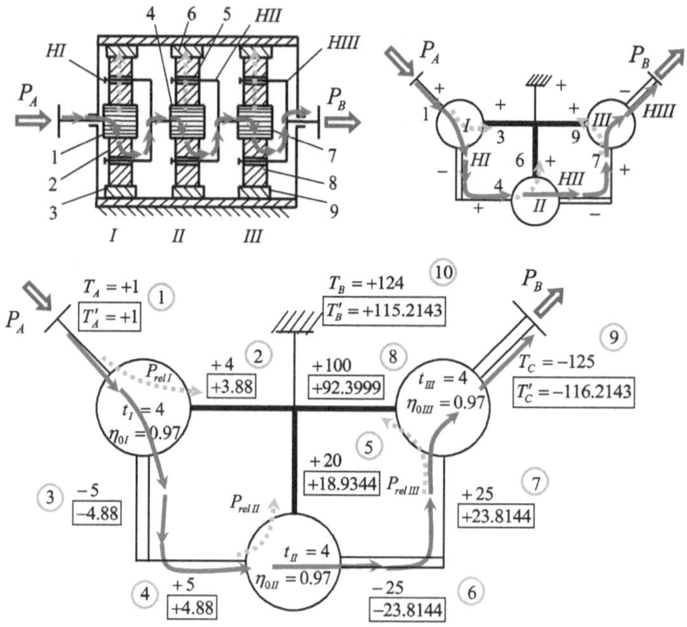

$$\sum T_i = T_A + T_B + T_C = +1 - 125 + 124 = 0$$

$$\sum T_i' = T_A' + T_B' + T_C' = +1 - 116.2143 + 115.2143 = 0$$

$$i_k = -\frac{T_B}{T_A} = -\frac{-125}{+1} = -+125$$

$$i_T = +\frac{T_B'}{T_A'} = +\frac{-116.2143}{+1} = -116.2143$$

$$\eta = \begin{Bmatrix} -\dfrac{i_T}{i_k} = -\dfrac{-116.2143}{+125} \\ \dfrac{T_B'}{T_B} = \dfrac{-116.2143}{-125} \end{Bmatrix} = 0.9297 \approx 93\,\%$$

FIGURE 34.1
Three-carrier PGT with three external shafts—one compound and two single shafts (with fixed three ring gears—in series connected compound PGTs).

PGTs, these powers are transmitted from sun gears 1, 4, and 7 to the respective ring gears 3, 6, and 9. Since the input ideal and real torques are equal $(T_A' = T_A = +1)$, the alternative way of determining efficiency can be applied, too (see Section 12.3).

It is useful for the reader to compare the above two PGTs by identifying and explaining the differences in speed ratios (the difference from one), the differences in efficiency, as well as the different directions of rotation of the output shaft.

The PGT, shown in Figure 34.3 [30, 31], can be considered to be derived from the sequentially connected two-carrier PGT, discussed in Section 29.3.1 (Figure 29.4), and these are the second (II) and third (III) component PGTs, with a further placed another PGT in front of them. This first (I) component PGT is with fixed carrier HI, which, together with the ring gears 6 and 9 of the second and third component PGTs, form a compound fixed shaft. This carrier (the external compound shaft, respectively) is fixed because the first (I) PGT undoubtedly possesses the highest speed of the three component PGTs. In such cases, to avoid the large centrifugal forces on the planets and on their bearings, the carriers are made

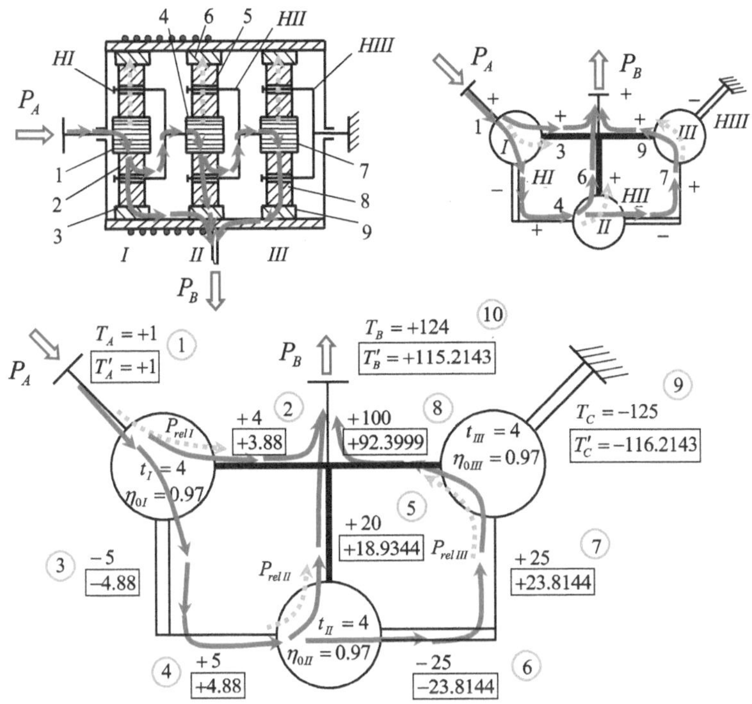

$$\sum T_i = T_A + T_B + T_C = +1 + 124 - 125 = 0$$

$$\sum T'_i = T'_A + T'_B + T'_C = +1 + 115.2143 - 116.2143 = 0$$

$$i_k = -\frac{T_B}{T_A} = -\frac{+124}{+1} = -124$$

$$i_T = +\frac{T'_B}{T'_A} = +\frac{+115.2143}{+1} = +115.2143$$

$$\eta = \left\{ \begin{array}{l} -\dfrac{i_T}{i_k} = -\dfrac{+115.2143}{-124} \\ \dfrac{T'_B}{T_B} = \dfrac{+115.2143}{+124} \end{array} \right\} = 0.9291 \approx 92.9\,\%$$

FIGURE 34.2
Three-carrier PGT with three external shafts—one compound and two single shafts (the output shaft is connected with the three ring gears).

stationary. As can be seen from the figure, the internal power flow is without division or circulation. At fixed carrier HI and ring gears 6 and 9, it is not difficult to understand that all three relative powers ($P_{rel\,I}$, $P_{rel\,II}$, and $P_{rel\,III}$) in the component PGTs are transmitted from the sun gears 1, 4, and 7 to the corresponding ring gears 3, 6, and 9.

In contrast to Figure 34.3, the external compound shaft is an input shaft in the gear train shown in Figure 34.4 [30, 31]. It consists of the three sun gears 1, 4, and 7 of the component PGTs. For this reason, the sun gears have the same angular velocity (the highest in the PGT), which means that the three relative powers ($P_{rel\,I}$, $P_{rel\,II}$, and $P_{rel\,III}$) here are transmitted from the sun gears 1, 4, and 7 to the corresponding ring gears 3, 6, and 9, which have lower angular speeds. The application of the alternative way for efficiency determination is not

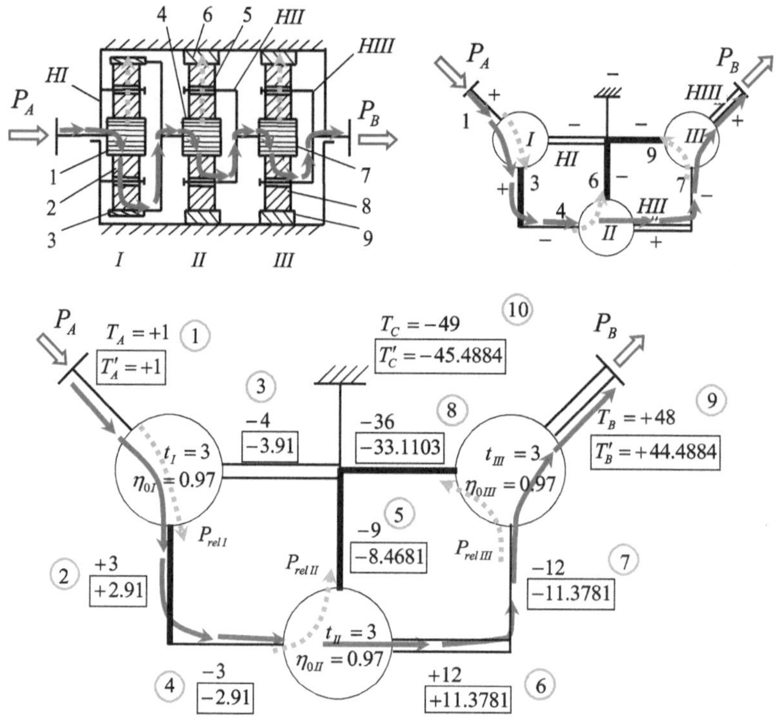

$$\sum T_i = T_A + T_B + T_C = +1 + 48 - 49 = 0$$

$$\sum T'_i = T'_A + T'_B + T'_C = +1 + 44.4884 - 45.4884 = 0$$

$$i_k = -\frac{T_B}{T_A} = -\frac{+48}{+1} = -48$$

$$i_T = +\frac{T'_B}{T'_A} = +\frac{+44.4884}{+1} = +44.4884$$

$$\eta = \begin{Bmatrix} -\dfrac{i_T}{i_k} = -\dfrac{+44.4884}{-48} \\ \dfrac{T'_B}{T_B} = \dfrac{+44.4884}{+48} \end{Bmatrix} = 0.9268 \approx 92.7\,\%$$

FIGURE 34.3
Three-carrier PGT with three external shafts—one compound and two single shafts (the fixed shaft is connected to all three component PGTs).

possible here, since the ideal input and output torques are not equal to the corresponding real ones, i.e., $T'_A \neq T_A$ and $T'_B \neq T_B$.

The gear train shown in Figure 34.5 is the tank "Leopard 2" gearbox. The PGT included has a similar structure as the PGT in Figure 34.4. By locking the three brakes *Br.1*, *Br.2*, and *Br.3*, three speed ratios (gears) are obtained. The fourth, direct gear is obtained by locking the friction clutch *Cp* and blocking the PGT.

Reversing the gearbox (front and rear) is achieved by the bevel gear that changes the direction of rotation of the output shaft when triggering the front (*Br.F*) or rear (*Br.R*) brakes, respectively. In this way, a total of eight speeds are obtained—four to the front and four to the rear.

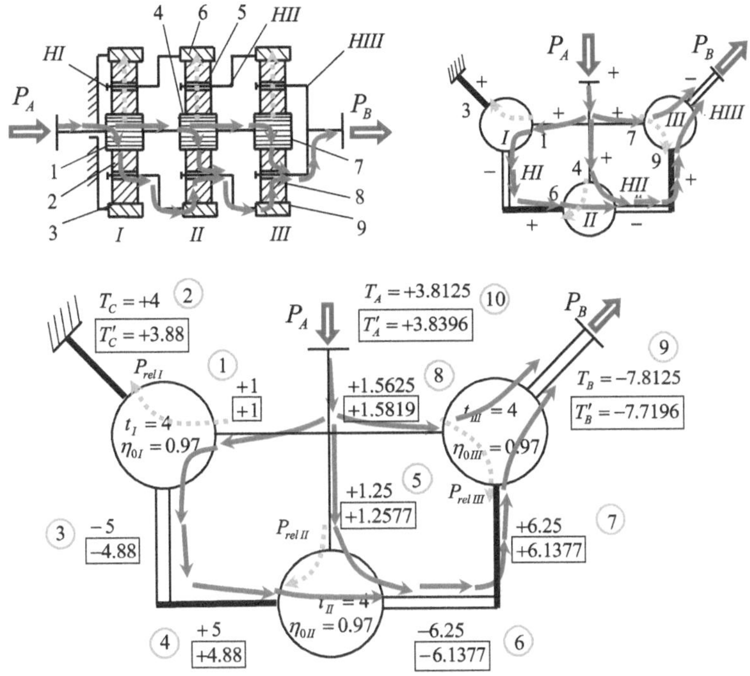

$$\sum T_i = T_A + T_B + T_C = +3,8125 - 7,8125 + 4 = 0$$

$$\sum T'_i = T'_A + T'_B + T'_C = +3,8396 - 7,7196 + 3,88 = 0$$

$$i_k = -\frac{T_B}{T_A} = -\frac{-7,8125}{+3,8125} = +2,049$$

$$i_T = +\frac{T'_B}{T'_A} = +\frac{-7,7196}{+3,8396} = -2,0105$$

$$\eta = \frac{i_T}{i_k} = \frac{-2,0105}{+2,049} = 0,9812 \approx 98,1\,\%$$

FIGURE 34.4
Three-carrier PGT with three external shafts—one compound and two single shafts (the input shaft is connected to all three component PGTs).

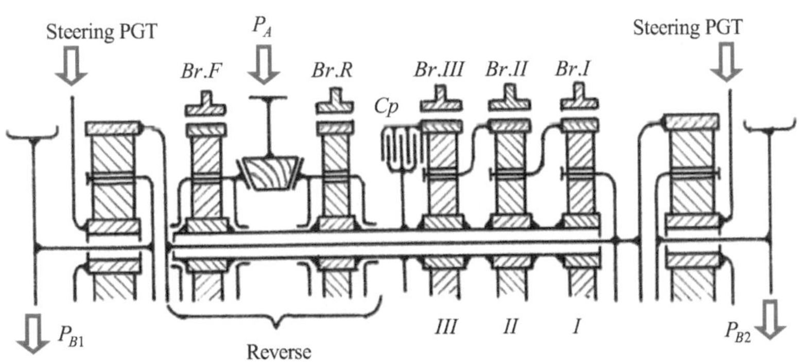

FIGURE 34.5
Gearbox (change-gear and steering system) of the tank "Leopard 2."

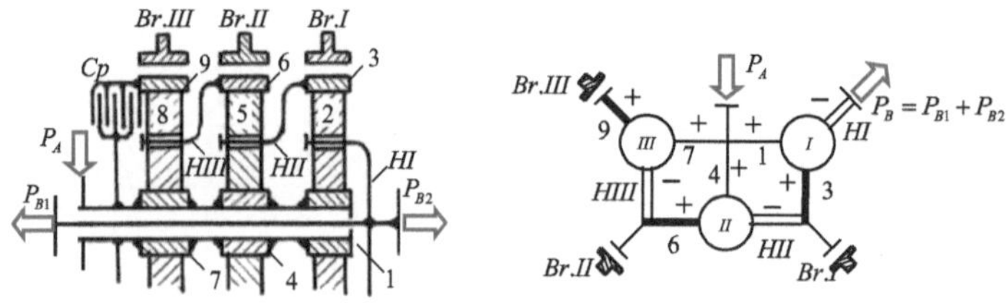

First gear:

$$\omega_1 = \omega_A;\ \omega_{HI} = \omega_B;\ \omega_3 = 0$$

$$\sum T_i = T_A + T_B + T_C = +1 - 4 + 3 = 0$$

$$i_1 = \frac{\omega_A}{\omega_B} = -\frac{T_B}{T_A} = -\frac{-4}{+1} = +4$$

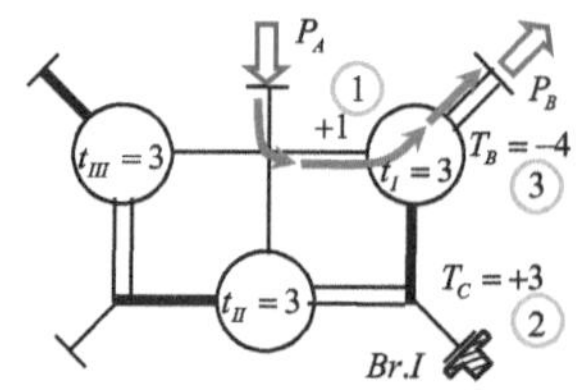

Second gear:

$$\omega_1 = \omega_4 = \omega_A;\ \omega_{HI} = \omega_B;\ \omega_6 = 0$$

$$\sum T_i = T_A + T_B + T_C = +1.75 - 4 + 2.25 = 0$$

$$i_2 = \frac{\omega_A}{\omega_B} = -\frac{T_B}{T_A} = -\frac{-4}{+1.75} = +2.26$$

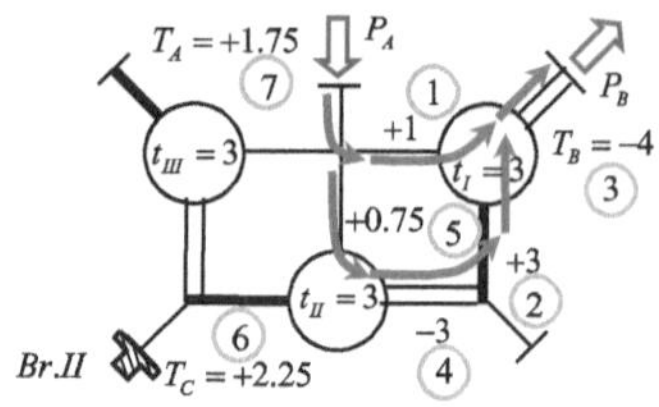

Third gear:

$$\omega_1 = \omega_4 = \omega_7 = \omega_A;\ \omega_{HI} = \omega_B;\ \omega_9 = 0$$

$$\sum T_i = T_A + T_B + T_C = +4.11 - 7.11 + 3 = 0$$

$$i_3 = \frac{\omega_A}{\omega_B} = -\frac{T_B}{T_A} = -\frac{-7.11}{+4.11} = +1.73$$

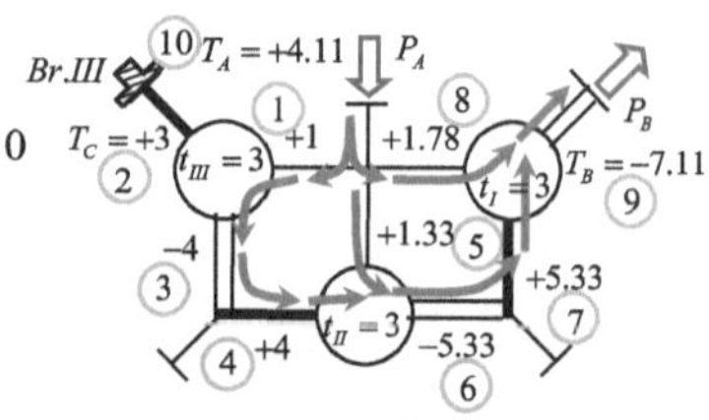

Fourth gear:

$$\omega_1 = \omega_4 = \omega_7 = \omega_A = \omega_{HI} = \omega_B.$$

PGT rotates as coupling.

$$i_4 = +1$$

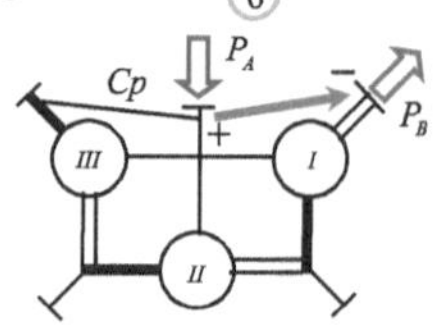

FIGURE 34.6
Determination of speed ratios of the three-carrier change-gear from the tank gearbox shown in Figure 34.5.

Through the two onboard simple PGTs, which sun gears receive movement from differential or hydraulic motion (not shown in the drawing), the tank steering is carried out.

Figure 34.6 shows kinematic scheme of the three-carrier PGT from Figure 34.5 only, with indicated elements and structural scheme with power flows for all three gear cases. The respective speed ratios are determined through the torque method. Unlike the previous example, the calculations, for convenience, are made for concrete values of the torque ratios of the component PGTs. For analysis purposes, they were selected $t_I = t_{II} = t_{III} = 3$.

As noted in previous discussions on change-gears Ravigneaux (Section 33.2 and Figure 33.13) and Mercedes-Benz® (Section 30.2 and Figure 30.4), the torque method allows for direct determination of torques on the individual gears. For the compound PGT in question, the percentage of load distribution between the three component PGTs at various gears (speeds) is determined in the following way by the magnitude of the specific external torques of Figure 34.6:

First gear:

$$\frac{T_1}{T_A} = 1$$

$$\frac{T_4}{T_A} = \frac{T_7}{T_A} = 0$$

Second gear:

$$\frac{T_1}{T_A} = \frac{T_1}{T_1 + T_4} = \frac{1}{1 + 0.75} = \frac{1}{1.75} = 0.57$$

$$\frac{T_4}{T_A} = \frac{T_4}{T_1 + T_4} = \frac{0.75}{1 + 0.75} = \frac{0.75}{1.75} = 0.43$$

$$\frac{T_7}{T_A} = 0$$

Third gear:

$$\frac{T_1}{T_A} = \frac{T_1}{T_1 + T_4 + T_7} = \frac{1.78}{1.78 + 1.33 + 1} = \frac{1.78}{4.11} = 0.43$$

$$\frac{T_4}{T_A} = \frac{T_4}{T_1 + T_4 + T_7} = \frac{1.33}{1.78 + 1.33 + 1} = \frac{1.33}{4.11} = 0.32$$

$$\frac{T_7}{T_A} = \frac{T_7}{T_1 + T_4 + T_7} = \frac{1}{1.78 + 1.33 + 1} = \frac{1.78}{4.11} = 0.24$$

Knowing the above distribution of torques on sun gears 1, 4, and 7 of the component PGTs, the relative load on planets 2, 5, and 8 as well as on ring gears 3, 6, and 9 can be determined. If from the operating experience, apart from the load, the relative duration of operation with the different speed ratios (at different gears) is also known, the load spectra of each gear can be determined, hence its equivalent load, equivalent stress, and ultimately the safety factors S_F and S_H (see Figure 34.18).

The structural scheme of the PGT shown in Figure 34.7 is different from those of the previous PGTs. According to the algebraic signs of the compound shaft, the directions of the branched power flows are determined. And the determination of relative (rolling) power ($P_{rel\,I}$, $P_{rel\,II}$, and $P_{rel\,III}$) direction is not a problem, as the third carrier HIII is fixed and the speeds of the other two carriers HI and HII are lower than those of the respective sun gears 1 and 4. In this case, as it is known, relative (rolling) powers are transmitted from the sun gears 1, 4, and 7 to the respective ring gears 3, 6, and 9. Determination of ideal torque T_{HIII} in the third (III) component PGT is a little more special. In this gear train, it is determined by the equilibrium of the three torques on the compound shaft, two of which are known—$T_{HI} = -5$ on the carrier HI and $T_6 = -16$ on the ring gear 6. Thus, the torque $T_7 = +21$ is determined, which together with $T_9 = +84$ determines the torque $T_{HIII} = T_C = -105$.

In the following figures, three-carrier PGTs are shown, but with a structure different from that of the previous PGTs (there is no triple compound shaft).

The gear shown in Figure 34.8 [30, 31] according to [167] is denoted as a *symmetrical compound PGT for precise speed ratios*. Symmetrical compound PGTs can realize arbitrary

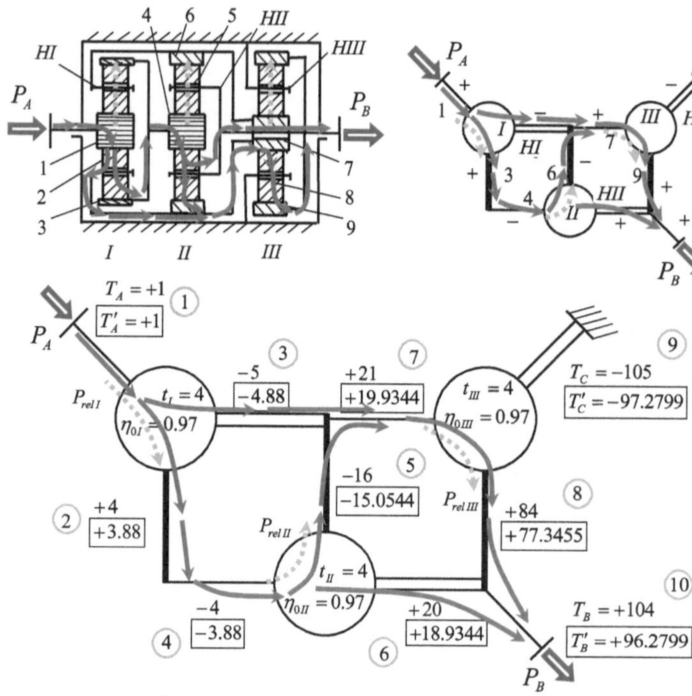

$$\sum T_i = T_A + T_B + T_C = +1 + 104 - 105 = 0$$

$$\sum T'_i = T'_A + T'_B + T'_C = +1 + 96.2799 - 97.2799 = 0$$

$$i_k = -\frac{T_B}{T_A} = -\frac{+104}{+1} = -104$$

$$i_T = +\frac{T'_B}{T'_A} = +\frac{+96.2799}{+1} = +96.2799$$

$$\eta = \left\{ \begin{array}{l} -\dfrac{i_T}{i_k} = -\dfrac{+96.2799}{-104} \\ \dfrac{T'_B}{T_B} = \dfrac{+96.2799}{+104} \end{array} \right\} = 0.9258 \approx 92.6\,\%$$

FIGURE 34.7
Three-carrier PGT with three external shafts—one compound and two single shafts (the output shaft is connected to two of the component PGTs).

speed ratios. The figure shows how to determine ideal and real torques, speed ratio i_k, torque transmit ratio i_T, and efficiency in this gear train. This gear, except as a reducer, is also used as a multiplier for wind turbines [87–89].

Determination of relative (rolling) power ($P_{rel\,I}$, $P_{rel\,II}$, and $P_{rel\,III}$) directions of all three component gears needs some clarification. In the third (III) component gear, ring gear 9 is fixed. As it is known from Chapter 10, in a simple $\overline{\text{AI}}$-PGT with a fixed ring gear, the relative (rolling) power ($P_{rel\,III}$ in this case) is transmitted from sun gear 7 to ring gear 9. In the second (II) component PGT, the situation is even simpler, as the carrier HII is fixed and gear train works as a non-planetary one. Naturally, relative (rolling) power $P_{rel\,II}$ is transmitted from sun gear 4 to ring gear 6. The speed of sun gear 1 is the highest in the first (I) component PGT (as well as in all the train). Thus, relative (rolling) power $P_{rel\,II}$ is transmitted from sun gear 1 to ring gear 3.

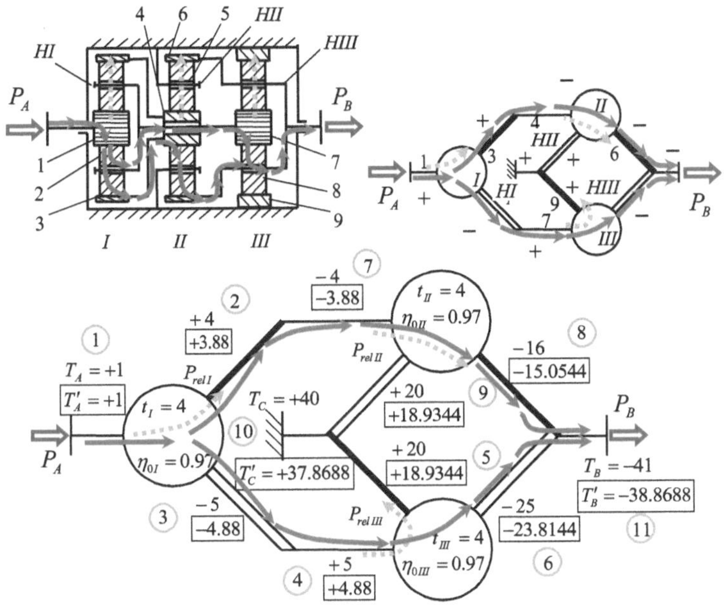

$$\sum T_i = T_A + T_B + T_C = +1 - 41 + 40 = 0$$

$$\sum T_i' = T_A' + T_B' + T_C' = +1 - 38.8688 + 37.8688 = 0$$

$$i_k = -\frac{T_B}{T_A} = -\frac{-41}{+1} = +41$$

$$i_T = +\frac{T_B'}{T_A'} = +\frac{-38.8688}{+1} = -38.8688$$

$$\eta = \left\{ \begin{array}{l} -\dfrac{i_T}{i_k} = -\dfrac{-38.8688}{+41} \\ \dfrac{T_B'}{T_B} = \dfrac{-38.8688}{-41} \end{array} \right\} = 0.9480 = 94.8\,\%$$

FIGURE 34.8
Three-carrier PGT with three external shafts—one single (input) and two compound shafts (both connected to two of the component PGTs).

Figure 34.9 shows a structural scheme of the PGT from Figure 34.8, working as a multiplier. Assuming that in both cases, the basic efficiency of the component PGTs is $\eta_0 = 0.97$, the multiplier efficiency is a little lower.

Compared to the previous symmetrical gear in Figure 34.8 (both external compound shafts are connected to the same couple of the component PGTs), the PGT shown in Figure 34.10 [30, 31] has an output shaft located elsewhere in the structural scheme—the external compound shafts are connected to different couples of the component PGTs.

Determining the direction of relative (rolling) powers ($P_{rel\,I}$, $P_{rel\,II}$, and $P_{rel\,III}$) in this gear train is even easier since the second (II) component PGT is with a fixed ring gear 6 and the third (III) PGT is with fixed carrier HIII. This determination is similar to the previous case (Figure 34.8). In the first (I) component PGT, things are even more obvious. Carrier HI, which is connected to the output shaft B, rotates more slowly than the input shaft A, and the sun gear 1, respectively, rotates most rapidly. In this situation, the relative (rolling) power $P_{rel\,I}$ is transmitted from sun gear 1 to ring gear 3.

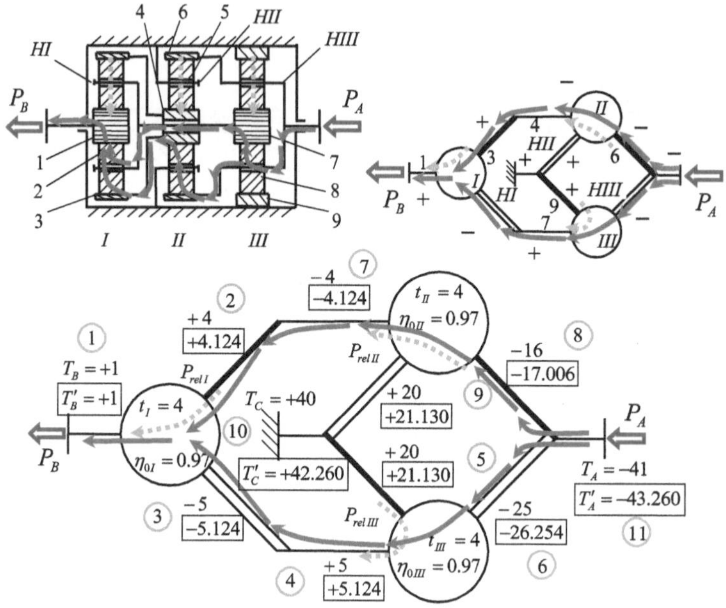

$$\sum T_i = T_A + T_B + T_C = +1 - 41 + 40 = 0$$

$$\sum T'_i = T'_A + T'_B + T'_C = -43.260 + 1 + 42.260 = 0$$

$$i_k = -\frac{T_B}{T_A} = -\frac{+1}{-41} = +0.02439$$

$$i_T = +\frac{T'_B}{T'_A} = +\frac{+1}{-43.260} = -0.023116$$

$$\eta = \left\{ \begin{array}{l} -\dfrac{i_T}{i_k} = -\dfrac{-0.023116}{+0.02439} \\ \dfrac{T'_A}{T_A} = \dfrac{-41}{-43.260} \end{array} \right\} = 0.9477 = 94.77\%$$

FIGURE 34.9
The three-carrier PGT with three external shafts—one single and two compound shafts from Figure 34.8, working as a multiplier.

In the previous two examples of Figures 34.8 and 34.10, there is no internal division of power. However, such a division exists in the PGT depicted in Figure 34.11. Therefore, the determination of relative (rolling) power ($P_{rel\,I}$, $P_{rel\,II}$, and $P_{rel\,III}$) directions is more specific, which will be done here by the *method of samples* [209] and which is shown in Figures 34.12 and 34.13 (see Section 29.4). Since in this case, it is necessary to determine the direction of relative power $P_{rel\,II}$ in the second (II) component PGT only, for its basic efficiency undervalued, e.g., $\eta_{0II} = 0.5$ is assumed and for the basic efficiency of the other two component PGTs, $\eta_{0I} = \eta_{0III} = 1$ is accepted. The direction of $P_{rel\,II}$ is randomly chosen, the real torques are determined, and efficiency η of the compound PGT is calculated. If an absurd $\eta > 1$ is obtained (Figure 34.12), the chosen direction is not correct, and vice versa, if $\eta < 1$ is obtained (Figure 34.13)—the direction is correct. With this correct direction of the relative power $P_{rel\,II}$ in Figure 34.11, efficiency is calculated.

The method of samples is very simple, understandable, and much more convenient for the designer's practical work than the method of partial derivatives [112].

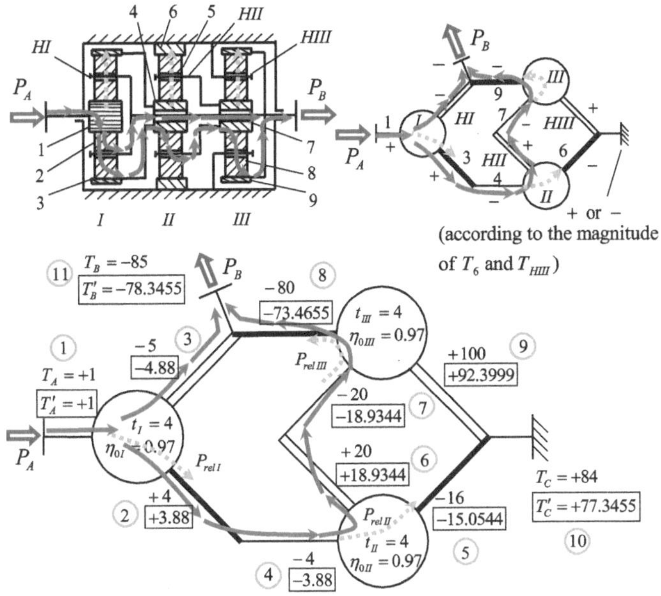

$$\sum T_i = T_A + T_B + T_C = +1 - 85 + 84 = 0$$

$$\sum T'_i = T'_A + T'_B + T'_C = +1 - 78.3455 + 77.3455 = 0$$

$$i_k = -\frac{T_B}{T_A} = -\frac{-85}{+1} = +85$$

$$i_T = +\frac{T'_B}{T'_A} = +\frac{-78.3455}{+1} = -78.3455$$

$$\text{К. п. д.: } \eta = \left\{ \begin{array}{l} -\dfrac{i_T}{i_k} = -\dfrac{-78.3455}{+85} \\ \dfrac{T'_B}{T_B} = \dfrac{-78.3455}{-85} \end{array} \right\} = 0.9217 \approx 92.2\,\%$$

FIGURE 34.10
Three-carrier PGT with three external shafts—one single (input) and two compound shafts (both are connected to two of the component PGTs).

Figure 34.14 shows a three-carrier PGT [30, 31, 88, 89] used in wind turbines as a multiplier, unlike the previous ones that work as reducers. All torques (ideal and real), speed ratio i_k, and efficiency η are determined. Since the output ideal and real torques are equal ($T_B = T'_B = +1$), efficiency can be defined as a ratio of the input ideal $T_A = -35.006$ and real $T'_A = -36.073$ torques.

34.2 Four-Carrier Compound PGTs

While three-carrier PGTs are mainly used as reducers, multipliers, and change-gears, four-carrier PGTs are used only as change-gears.

The following figures are examples of change-gears [30, 21]. Figure 34.15 (see Figure 2.11) presents the well-known Wilson transmission used in the motor vehicle [161].

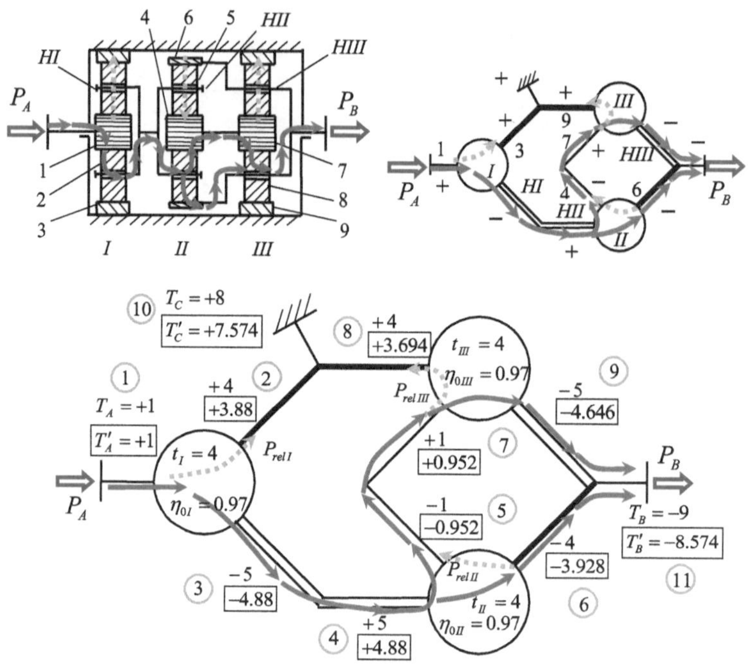

$$\sum T_i = T_A + T_B + T_C = +1 - 9 + 8 = 0$$

$$\sum T'_i = T'_A + T'_B + T'_C = +1 - 8.574 + 7.574 = 0$$

$$i_k = -\frac{T_B}{T_A} = -\frac{-9}{+1} = +9$$

$$i_T = +\frac{T'_B}{T'_A} = +\frac{-8.574}{+1} = -8.574$$

$$\eta = \left\{ \begin{array}{l} -\dfrac{i_T}{i_k} = -\dfrac{-8.574}{+9} \\ \dfrac{T'_B}{T_B} = \dfrac{-8.574}{-9} \end{array} \right\} = 0.9527 \approx 95.3\,\%$$

FIGURE 34.11
Three-carrier PGT with three external shafts—one single (input) and two compound shafts (both connected to two of the component PGTs).

Figure 34.16 shows a tractor change-gear and Figure 34.17—a heavy vehicle change-gear realizing four speed ratios (gears) in one direction. How to determine the load spectrum of a particular element of this PGT is shown in Figure 34.18. In this case, it is shown how different loads and load spectrum of the sun gear of the fourth (IV) component PGT are determined and how to determine its safety factor S by the equivalent load at certain duration φ of the individual loads. By the load on this sun gear, load spectra of the other elements of the fourth (IV) component PGT can be determined. Determination of the efficiency for all gears of this change-gear is discussed in [137]. In general, the use of the torque method analysis of PGT allows for direct determination of load on all gear train elements without passing through the kinematic analysis of Willis's or Kutschbach's methods. This is undoubtedly a valuable asset of the torque method that the other analysis methods lack.

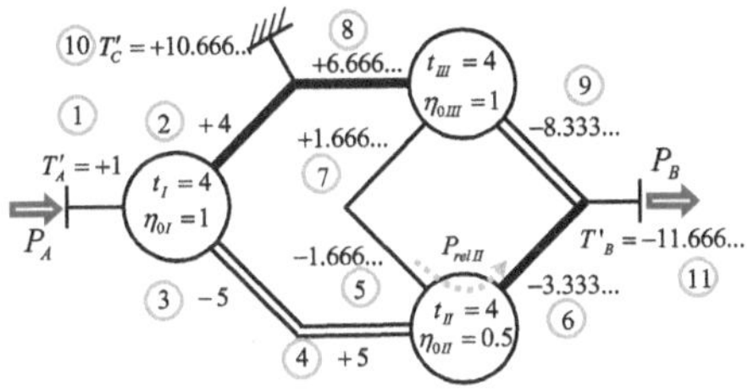

Equilibrium of real external torques on II-nd component PGT only:

$$\sum T'_i = T'_4 + T'_6 + T'_{HII} = 0$$

$$T'_4 + \eta_{0II} \cdot t_{II} \cdot T'_4 + 5 = 0$$

$$T'_4 = \frac{-5}{1 + \eta_{0II} \cdot t_{II}} = \frac{-5}{1 + 0.5 \cdot 4} = -1.666...$$

$$T'_6 = \left\{ \begin{matrix} -T'_{HII} - T'_4 = -5 - (-1.666) \\ \eta_0 \cdot t_{II} \cdot T'_4 = 0.5 \cdot 4(-1.666) \end{matrix} \right\} = -3.333$$

Equilibrium of real external torques on compound PGT:

$$\sum T'_i = T'_A + T'_B + T'_C = +1 - 11.666 + 10.666 = 0$$

Torque transmit ratio:

$$i_T = \frac{T'_B}{T'_A} = -\frac{-11.666...}{+1} = -11.666...$$

Efficiency:

$$\eta = \left\{ \begin{matrix} -\frac{i_T}{i_k} = -\frac{-11.666...}{+9} \\ \frac{T'_B}{T_B} = \frac{-11.666}{-9} \end{matrix} \right\} = 1.296 > 1 \quad \text{– An absurd value!!!}$$

FIGURE 34.12
Check the direction of relative (rolling) power $P_{rel\,II}$ in the second component PGT at wrongly chosen direction.

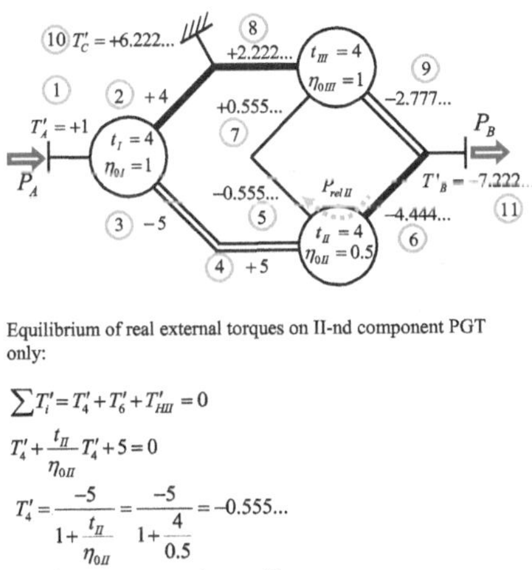

Equilibrium of real external torques on II-nd component PGT only:

$$\sum T'_i = T'_4 + T'_6 + T'_{HII} = 0$$

$$T'_4 + \frac{t_{II}}{\eta_{0II}} T'_4 + 5 = 0$$

$$T'_4 = \frac{-5}{1 + \frac{t_{II}}{\eta_{0II}}} = \frac{-5}{1 + \frac{4}{0.5}} = -0.555...$$

$$T'_6 = \left\{ \begin{matrix} -T'_{HII} - T'_4 = -5 - (-0.555) \\ \frac{t_{II}}{\eta_{0II}} T'_4 = \frac{4}{0.5}(-0.555) \end{matrix} \right\} = -4.444$$

Equilibrium of real external torques on compound PGT:

$$\sum T'_i = T'_A + T'_B + T'_C = +1 - 7.222 + 6.222 = 0$$

Torque transmit ratio:

$$i_T = \frac{T'_B}{T'_A} = -\frac{-7.222...}{+1} = -7.222...$$

Efficiency:

$$\eta = -\frac{i_T}{i_k} = -\frac{-7.222...}{+9} = 0.802 < 1 \quad \text{– A real value!!!}$$

FIGURE 34.13
Check the direction of relative (rolling) power $P_{rel\,II}$ in the second component PGT at correctly chosen direction.

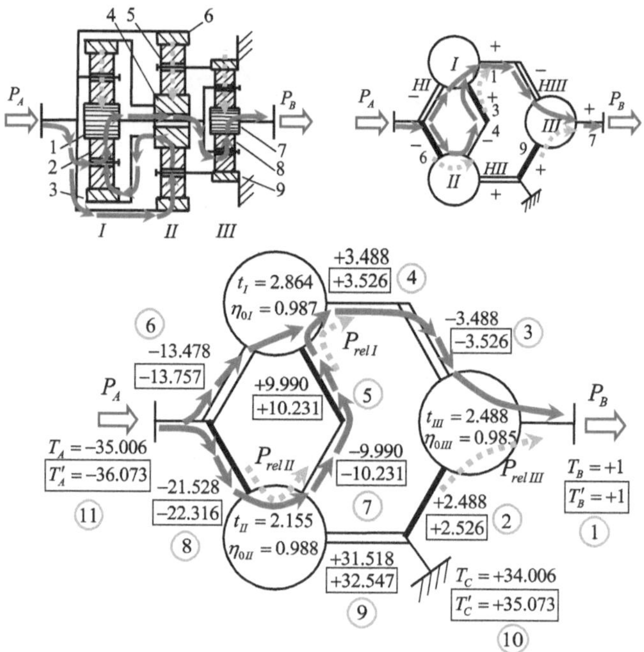

$$\sum T_i = T_A + T_B + T_C = -35.006 + 1 + 34.006 = 0$$

$$\sum T'_i = T'_A + T'_B + T'_C = +3.8396 - 7.7196 + 3.88 = 0$$

$$i_k = \frac{T_B}{T_A} = -\frac{+1}{-35.006} = +0.028566$$

$$i_T = +\frac{T'_B}{T'_A} = +\frac{+1}{-36.073} = -0.02772$$

$$\eta = \begin{cases} -\dfrac{i_T}{i_k} = -\dfrac{-0.02772}{+0.028566} \\ \dfrac{T_A}{T'_A} = \dfrac{-35.006}{-36.073} \end{cases} \approx 97.04\%$$

FIGURE 34.14
Three-carrier PGT with three external shafts—one compound and two single shafts (the fixed shaft is connected to two of the component PGTs), working as a multiplier of a wind turbine.

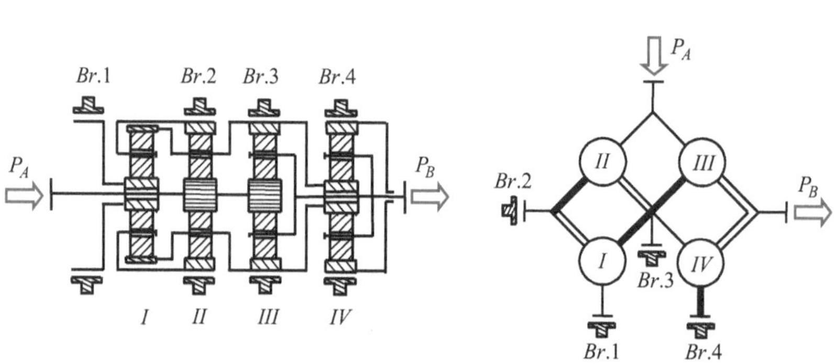

FIGURE 34.15
Change-gear Wilson for vehicles.

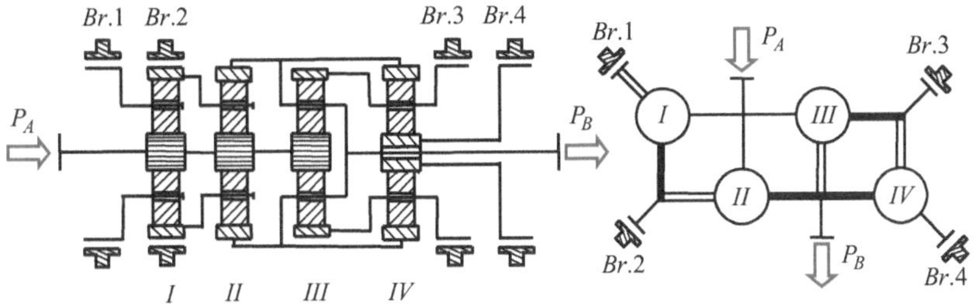

FIGURE 34.16
Change-gear of tractors.

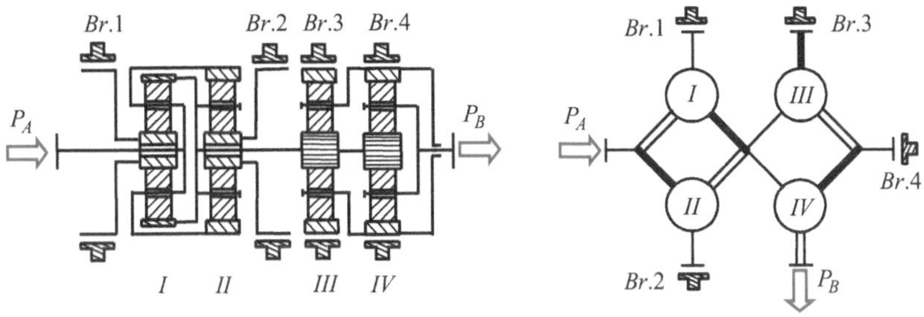

FIGURE 34.17
Change-gear of heavy vehicles.

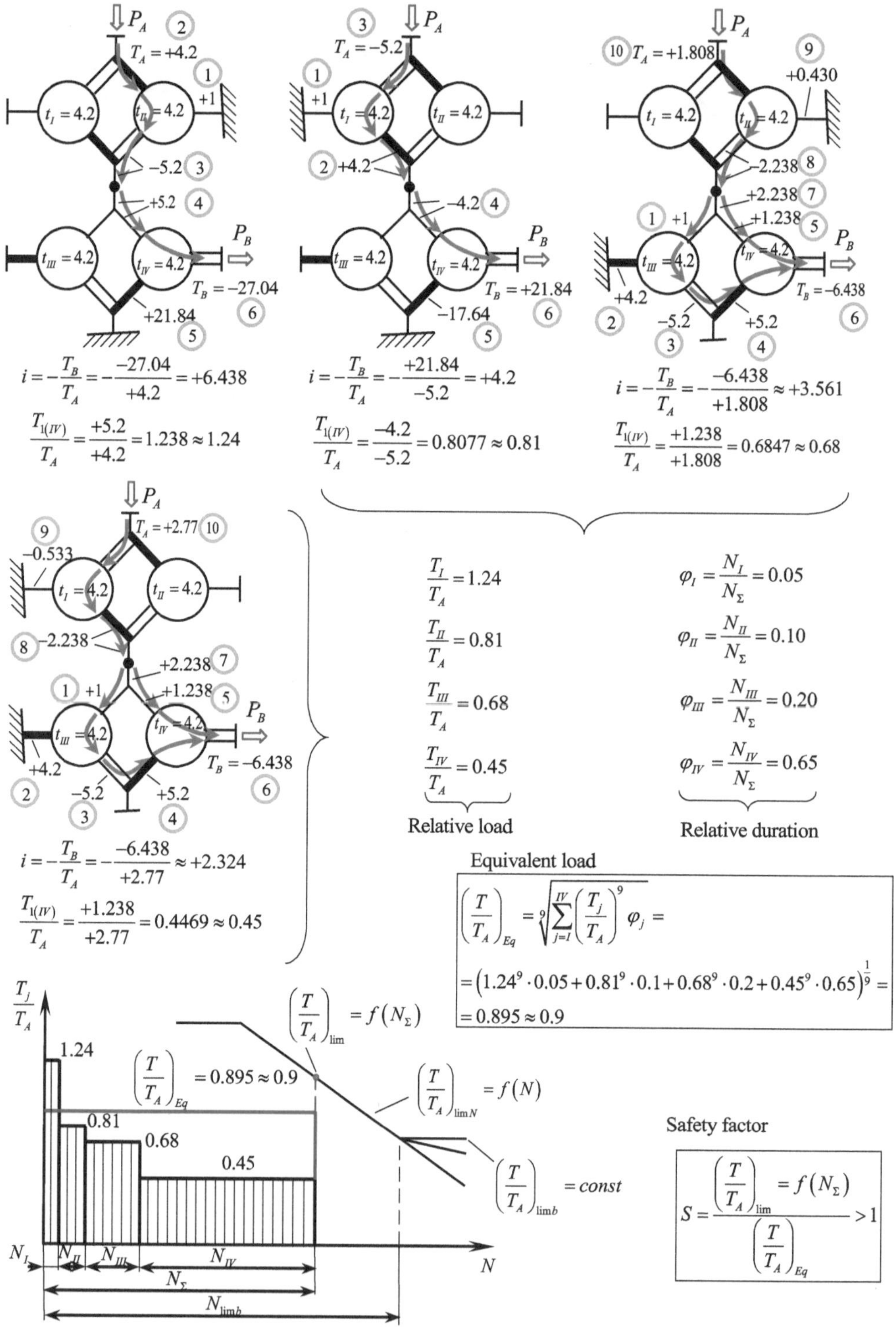

FIGURE 34.18
Determination of load spectrum of the sun gear of the fourth (IV) component PGT of the change-gear from Figure 34.17.

35

Examples for Application of Compound Planetary Gear Trains

This chapter contains examples of arrangements of two-carrier, three-carrier, and four-carrier compound planetary gear trains (PGTs) which are applicable in various areas of technique. Examples here are also accompanied by comments aiming at facilitating the designers and directing them to successful and time-tested technical solutions.

35.1 Two-Carrier Compound PGTs

Figure 35.1 shows a two-carrier PGT with the same module of all gears, which is favorable for its manufacturing. Since the load on the second stage is decisive, the first stage is not completely loaded despite the smaller width of gears. The first-stage carrier forms a one unit with the second-stage sun gear (internal compound shaft) and is not floating. The first-stage sun gear is not seated in the carrier but is cantilevered in the housing. The single-row ball bearing of the first-stage planets allows for their self-aligning, which reduces uneven load distribution on the gear face (considered by face load factor K_β). This is not applicable for second-stage planet bearing whose needle roller bearing does not allow self-aligning.

In contrast to the previous gear train, in the PGT of Figure 35.2, each stage is dimensioned according to its load. Here, too, the internal shaft consists of the first-stage carrier and second-stage sun gear, and the first-stage sun gear is cantilevered in the housing. The unfavorable thing is that needle roller bearings do not allow planets to self-align and the first-stage sun gear is cantilevered and not floating (face load factor K_β!).

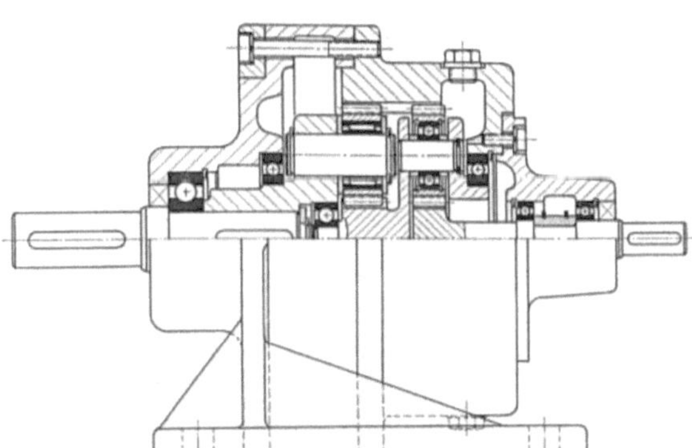

FIGURE 35.1
Two-carrier PGT with the same module of all gears.

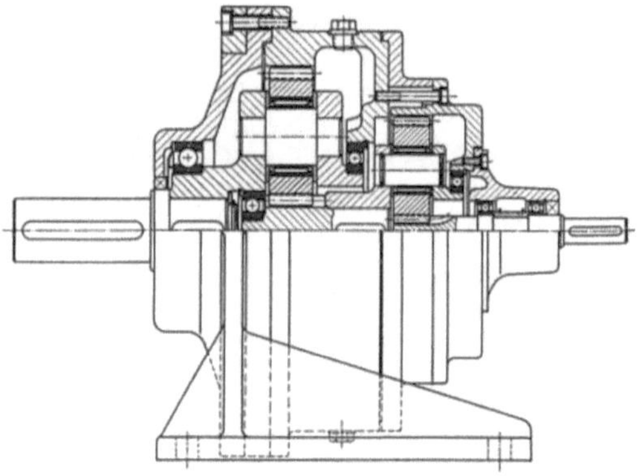

FIGURE 35.2
Two-carrier PGT with different modules of the two stages.

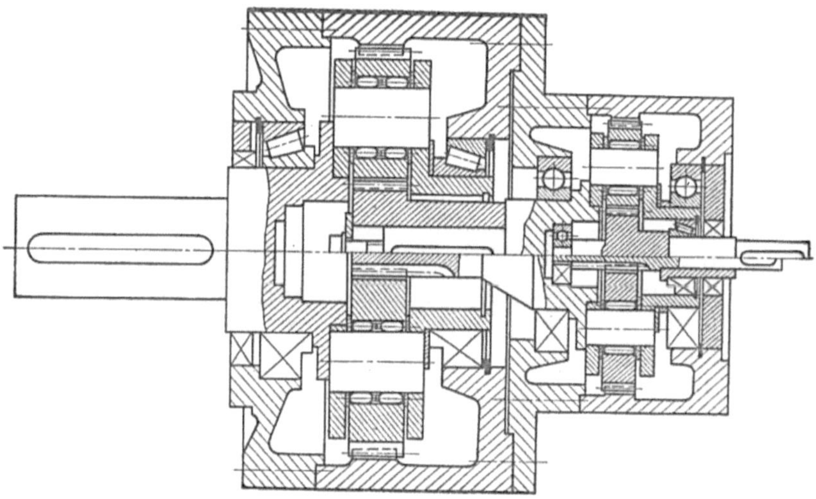

FIGURE 35.3
Two-carrier PGT with no floating element.

Figure 35.3 also shows a PGT with no floating element. This allows for bearing the sun gear on the first stage of its carrier. Apparently, for equalizing load distribution between planets, high accuracy is needed.

The PGT in Figure 35.4 also has no floating element. The proper adjustment of the clearance in tapered roller bearings may allow for certain aligning movements of elements and decrease in uneven load distribution on the planets (mesh load factor $K_{\gamma\,max}$).

Figure 35.5 shows a two-carrier PGT for extruder. In this gear train, the diameters of the respective gears of the two stages are the same, but the modules and their widths are different, depending on the load on the respective stages, so that the first stage is not completely loaded, as with the gear train in Figure 35.1. For technological reasons, the two ring gears are manufactured as separate details. This, however, requires strict parallelism of their front surfaces. The first-stage carrier and the second-stage sun gear form a common element (assembly) which, however, is not floating. Therefore, in this gear train, accuracy is crucial to obtain satisfactory load distribution between planets.

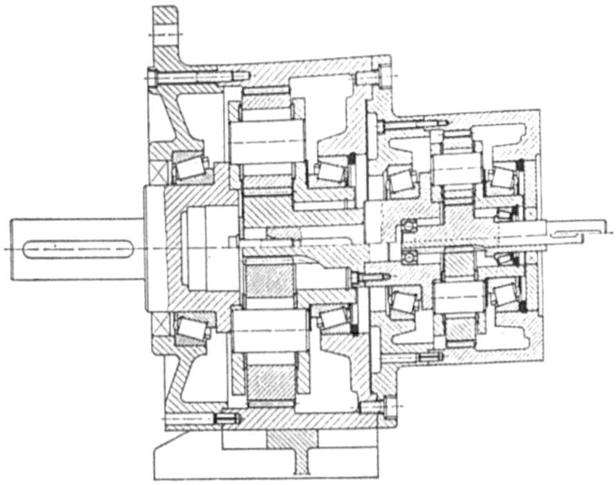

FIGURE 35.4
Two-carrier PGT with tapered roller bearings.

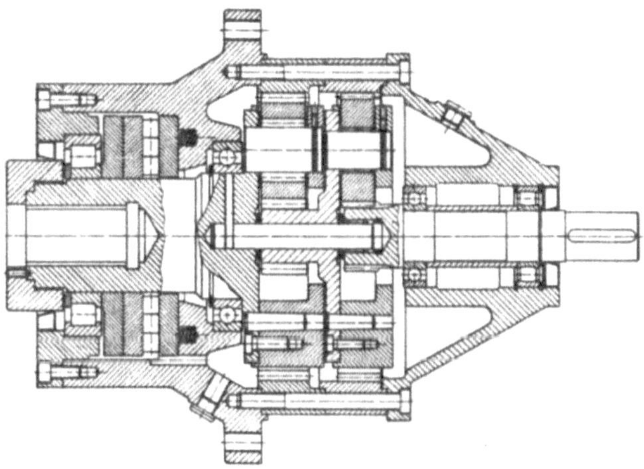

FIGURE 35.5
Two-carrier PGT for extruder with the same diameters of the respective gears of the two stages.

The two-carrier PGT shown in Figure 35.6 is a part of a vertically working geared motor which drives a mixer. What is typical for it is that the first-stage carrier is fixed and gear couplings are used to provide the possibility of aligning movements of sun gears and ring gears of the two stages.

Figure 35.7 shows a three-stage reducer from the drive of a ball mill drum, whose second and third stages are planetary ones. The arrangement of planetary stages is a good example of how thanks to double-articulated gear couplings and single-row bearings of planets there are prerequisites for the best possible aligning both of load between planets (mesh load factor $K_{\gamma\,max}$) and of load distribution over the gear face (face load factor K_{β}). The solution of the third stage is interesting where a possibly longer sleeve of the double-articulated gear coupling is obtained, which is favorable for the higher sensitivity of the distribution device (see Section 8.6.5 and Figure 8.17). The planets of both stages have only one bearing each, which is favorable.

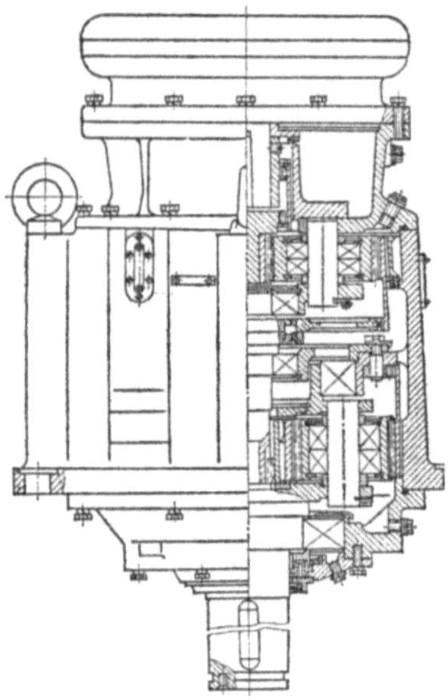

FIGURE 35.6
Two-carrier PGT with floating sun gears and ring gears.

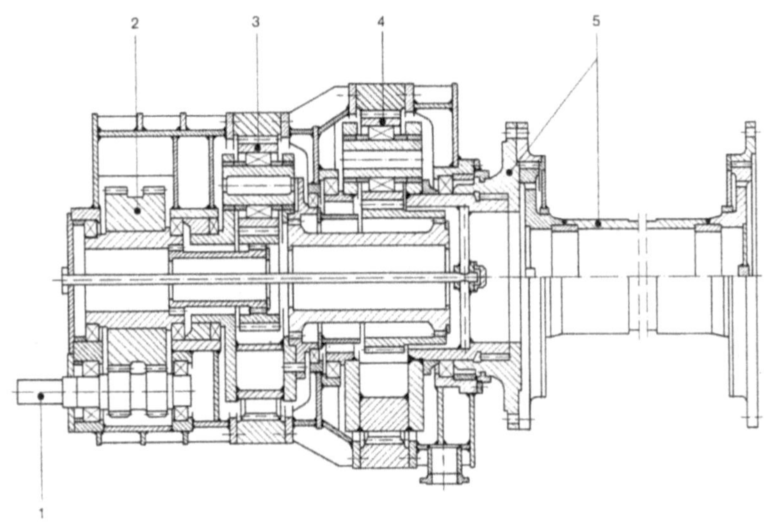

FIGURE 35.7
Three-stage reducer with two planetary stages from the driving of a ball mill drum.

Figure 35.8 shows a helicopter PGT—the main reducer of the carrying screw (propeller), performed in compliance with variant 26 ≡ 62 from Table 29.1; i.e., it is sequentially connected compound two-carrier PGT (Figure 29.4). In this gear train, the planets of both stages are seated in the respective carrier. Equalizing of load distribution between planets is achieved by the very high precision of manufacturing. The fact that the gear train works with vertical main geometric axis eliminates the impact of gravity of individual elements on load distribution.

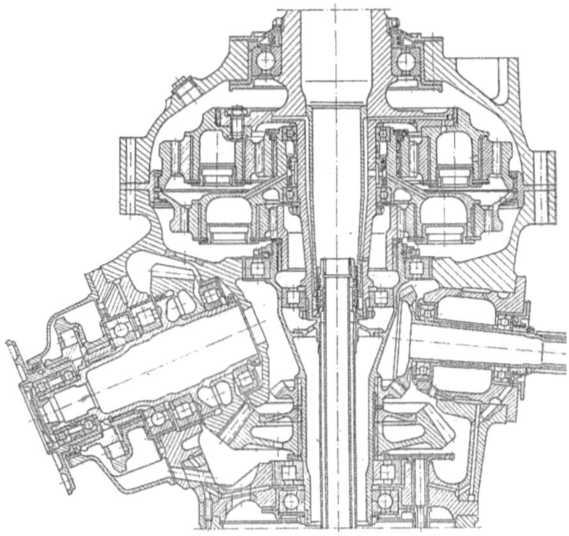

FIGURE 35.8
Two-carrier PGT—main reducer of a helicopter (variant 26 ≡ 62 from Table 29.1).

The PGT in Figure 35.9 is performed according to variant 16 ≡ 61 from Table 29.1 and is similar to the gear train in Figure 29.6, as well as to the second left kinematic scheme in Figure 29.7. In it, the planets themselves play the role of outer races of bearings.

The airplane PGT shown in Figure 35.10 (gearbox of Il-18) is a special "lightweight" arrangement, where all elements are made as light as possible [236]. The planet bearings are really two, but they are certainly chosen for each planet with the same radial clearance.

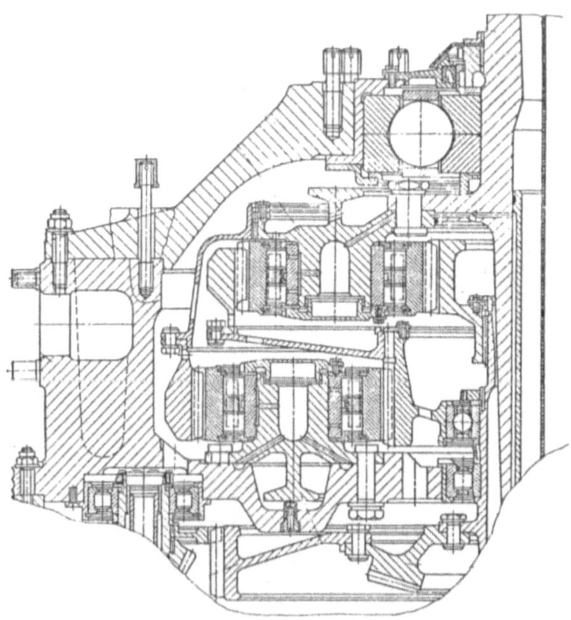

FIGURE 35.9
Two-carrier PGT—main reducer of a helicopter (variant 16 ≡ 61 from Table 29.1).

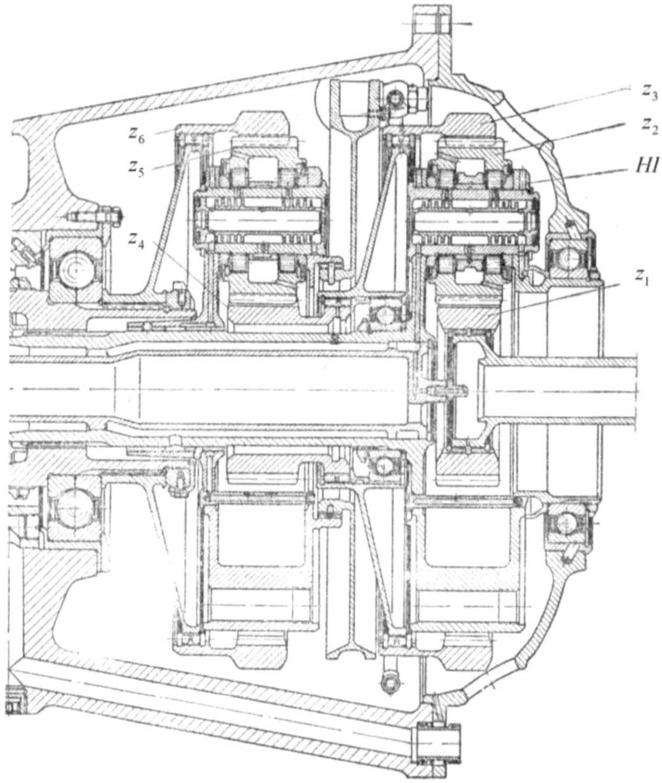

FIGURE 35.10
Two-carrier PGT—gearbox of airplane Ilyushin Il-18 (variant 16 ≡ 61 from Table 29.1).

For equalizing the load distribution between planets, the sun gear of the first stage is floating, and the two ring gears are pliable. Accuracy of manufacturing of this responsible gear train is technically the highest. The gear train in question has internal power division (variant 16 ≡ 61 from Table 29.1) and is theoretically described in Section 29.3.3 (the leftmost kinematic scheme in Figure 29.7).

35.2 Three-Carrier Compound PGTs

The three-carrier PGTs shown in Figure 35.11 have no floating element and have much in common with the two-carrier gear train of Figure 35.3, so what is said about it also applies to these gear trains.

Figure 35.12 shows a three-carrier PGT with a common ring gear, which is very convenient from a technological point of view. The three component PGTs are sequentially connected, and the common ring gear is a fixed element. The same structural scheme is presented in Figure 34.2, but there the common ring gear is the output element and the third-stage carrier—the fixed one. The arrangement shown here is suitable for low powers, since at the same geometry of gears, the ones of the first two stages are not sufficiently loaded.

Figure 35.13 shows a three-carrier PGT (reducer) of the hoisting mechanism of a 25-tonne hoist ("Podemcrane", Gabrovo, Bulgaria). The component PGTs (with fixed ring gears) are

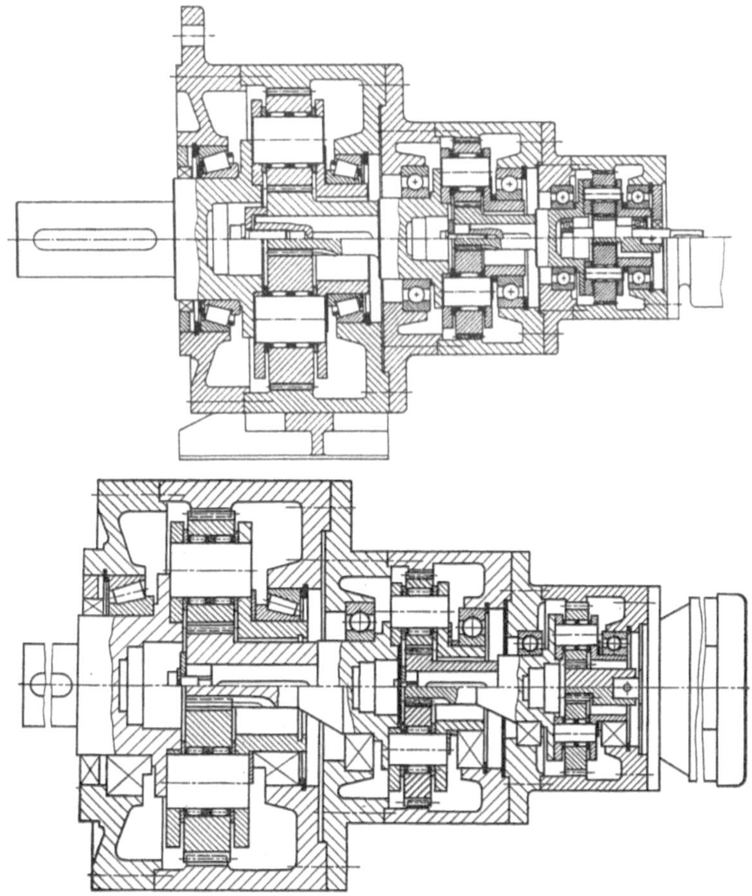

FIGURE 35.11
Three-carrier PGTs with no floating element.

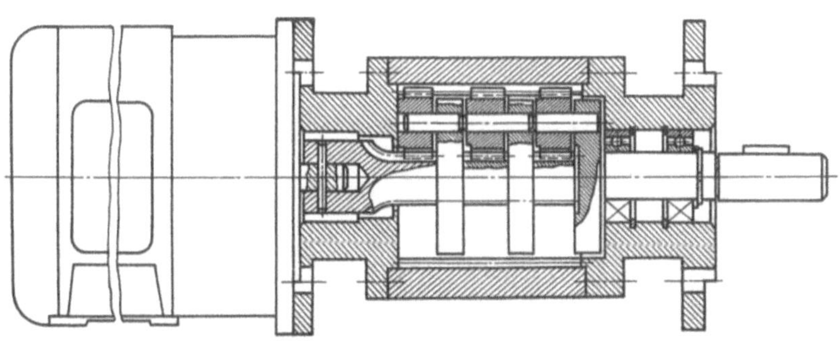

FIGURE 35.12
Three-carrier PGTs with a common ring gear.

sequentially connected. By changing the number of teeth of only the gears of the first stage, a series (family) of reducers are achieved with the speed ratios of 78, 96, 125, and 158 for a motor power of 10 to 30 kW and input rotation speeds of $n_A = 1{,}440$ and $n_A = 950\,\text{min}^{-1}$. Torque ratios of the component PGTs are from $t = 2.85$ to $t = 4.82$. The first stage has $k_I = 3$

planets, the second one $k_{II} = 4$, and the third one $k_{III} = 5$. The attempt to reduce the axial gauge has led to the inability to use a floating equalizing (distribution) device with high sensitivity. Floating elements are the sun gears of the three stages, the ones of the second and third stages being together with the respective carrier of the first and second stages. With the exception of the long sleeve of the single articulated gear coupling of the first-stage sun gear, the other gear couplings between the carrier and the respective sun gear are simply splice compounds. The floating assemblies of carrier and sun gear not only contribute for certain equalizing of load distribution between planets but also simplify the arrangement. Thus, this reducer features a great compactness and specific load carrying capacity.

Three-carrier and four-carrier PGTs are often used as automatic transmissions (change-gears) in vehicles [161]. Figure 35.14 shows the kinematic scheme of a three-carrier

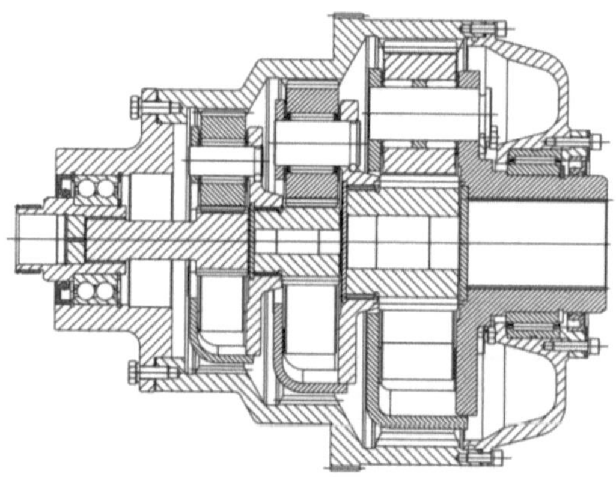

FIGURE 35.13
Three-carrier PGT (reducer) of hoisting mechanism of electric hoist.

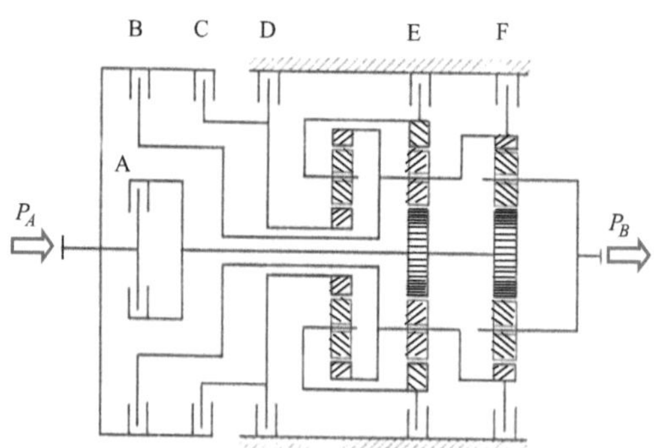

Gear	A	B	C	D	E	F
1	•					•
2	•				•	
3	•			•		
4	•	•				
5		•		•		
R			•			•

FIGURE 35.14
Kinematic scheme of a three-carrier planetary change-gear for buses, heavy and special vehicles.

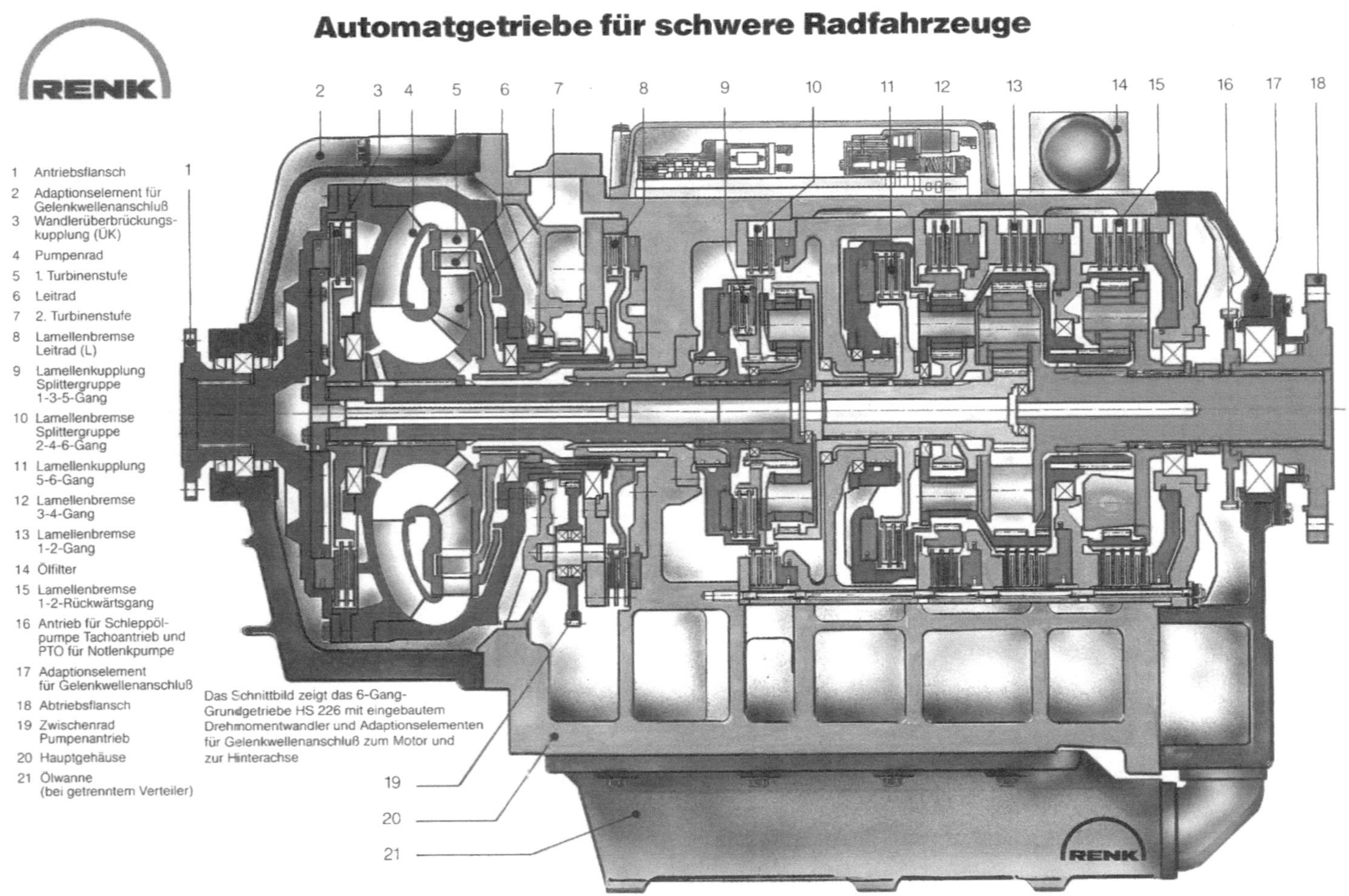

FIGURE 35.15
Arrangement of change-gear Renk® for heavy vehicles with a four-carrier PGT. (Courtesy of RENK Aktiengesellschaft, Augsburg, Germany.)

planetary change-gear ZF Ecomat®, which is used in buses, and heavy and special vehicles. From the table, it can be seen which brakes lock to realize various speed ratios (gears). Making structural schemes and determining the power flows would be an appropriate task for the reader who is aware of the contents of this book. Kinematic and power analysis of this PGT is made in [137].

35.3 Four-Carrier Compound PGTs

Figure 35.15 shows the arrangement of an automatic four-carrier planetary change-gear Renk® for heavy vehicles.

Part V

Supplements

36

Bevel, Worm, and Crossed-Helical Simple (Single-Carrier) Planetary Gear Trains

Various types of bevel planetary gear trains (PGTs) exist, but one of them is most commonly used. This is the gear train shown in Figure 36.1. From kinematic point of view, it is equivalent to those, given on the same figure $\overline{AAA}$—and IAI-PGTs (see Chapter 25). What is common between the gear trains depicted in Figure 36.1 is their specific application as a symmetrical inter-wheel or inter-axle differential in vehicles due to the fact that their basic speed ratio is $i_0 = -1$. They can also be used as an asymmetric inter-axle differential in $i_0 \neq -1$. For this reason, they are probably the most mass-produced PGTs, as millions and millions of cars are produced annually in the world and, in the majority of cases, they have conical differentials.

Although very rarely, apart from differential, bevel PGTs are also used as embedded in wheels gearboxes of heavy vehicles (Figure 36.2). It is a good example that PGTs in general,

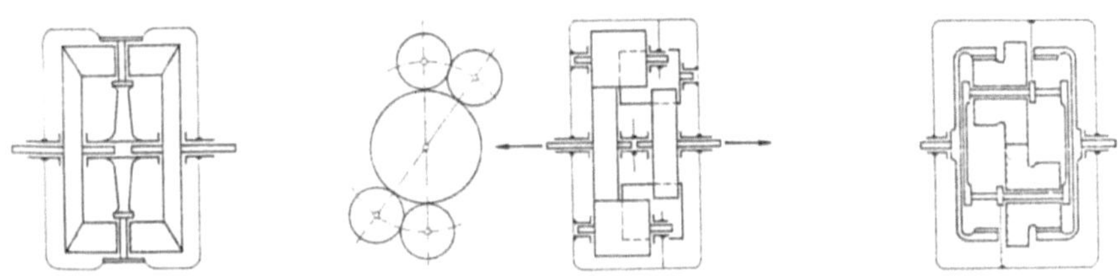

FIGURE 36.1
Symmetrical bevel and cylindrical inter-wheel differential for vehicles.

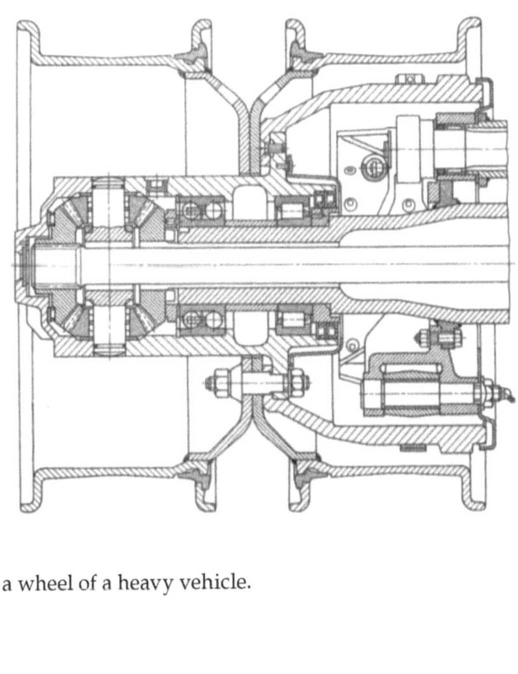

FIGURE 36.2
Bevel planetary reducer in a wheel of a heavy vehicle.

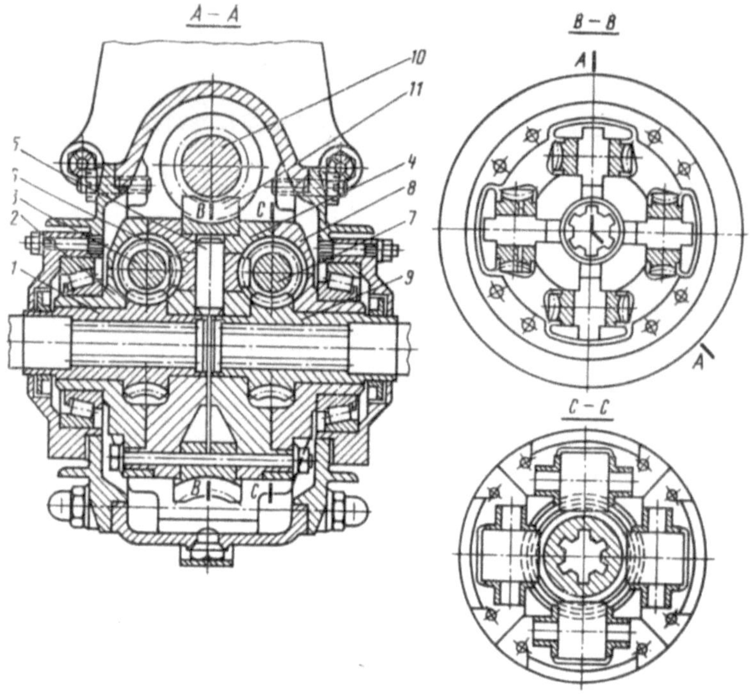

FIGURE 36.3
Symmetrical worm differential for vehicles.

including bevel PGTs, can perform various functions such as reducers, multipliers, and summation or division differentials.

Still less often worm gear differentials are used in the automotive industry (Figure 36.3). These differentials have increased internal friction, which is favorable for the passability of the vehicle.

More information about bevel PGTs one can find in [161, 174].

37

Special Planetary Gear Trains

The planetary gear trains (PGTs) shown in Figure 37.1 represents a particularly complex variation of the typical PGTs. Characteristic of this so-called *bi-planetary* (double planetary) gear trains is the use of two or more carriers that are connected so that the planets rotate not in respect to two axes but in respect to three or more axes [75, 235]. In this way, they perform complex movements and describe complex trajectories that prove to be useful for some machines, e.g., for coal cutting (Figure 37.2) [235] and tunnel boring machines.

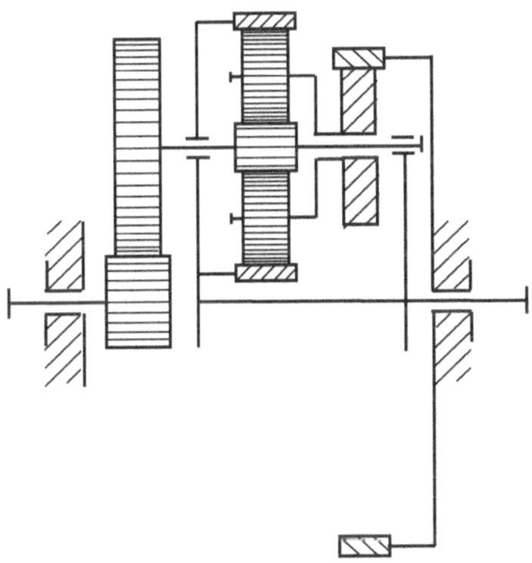

FIGURE 37.1
Bi-planetary gear train.

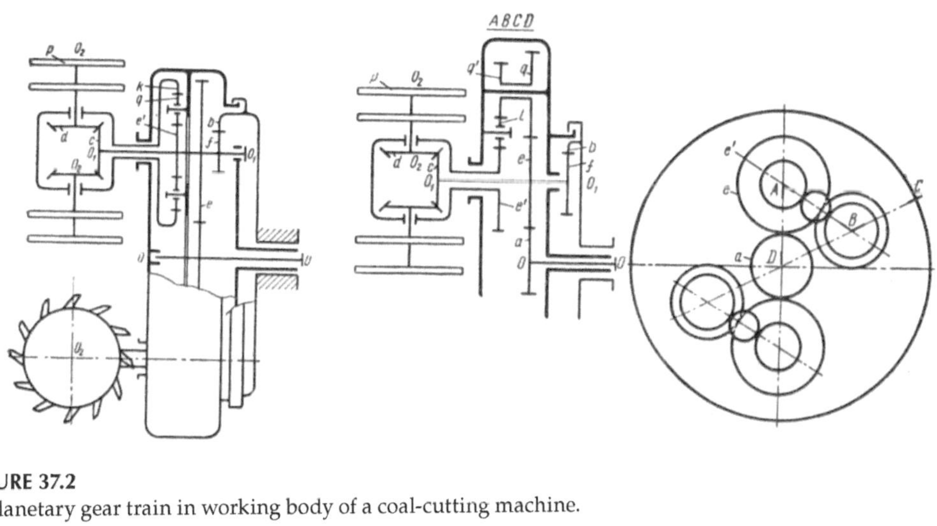

FIGURE 37.2
Bi-planetary gear train in working body of a coal-cutting machine.

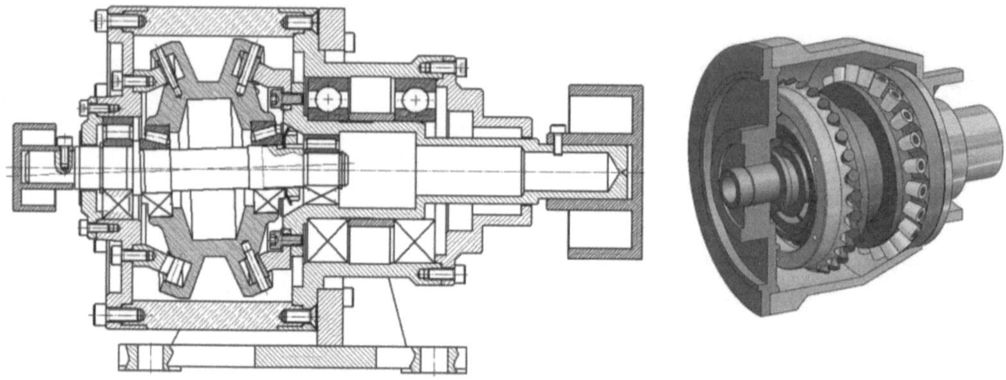

FIGURE 37.3
Precessional PGT. (Courtesy of Valeriu Dulgheru.)

Figure 37.3 shows the so-called precessional PGT. The name comes from the planets precession movement. This particular type of PGTs is very seriously investigated in the Academy of Sciences of Moldova [50].

The complicated structure and complex kinematics of bi-planetary gear trains make their analysis more complex than that of the usual PGT [75]. The torque method is very suitable for their analysis.

38

Involute Gears with Asymmetric Teeth

Gear wheels with asymmetric teeth are suitable for use when only one tooth flank is loaded [71–73, 146]. This is achieved both in unidirectional rotating gear trains (fans, pumps, airplanes, and helicopters) and in hoisting mechanisms (where both in lift-on and in lift-down the same side of the tooth is loaded).

For gears with asymmetric teeth, a different pressure angle is used for the two tooth profiles (Figure 38.1). In this way, a high transverse contact ratio ε_α and a high pressure angle α of the drive (main, working) flank are simultaneously obtained, and a pointed tooth tip is avoided. High pressure angle reduces the contact stresses in the drive flank, and pressure angle of the coast (adjacent, nonworking) flank α_N is chosen to provide a sufficiently wide tooth root, which also leads to a reduction in the bending stress in the critic section.

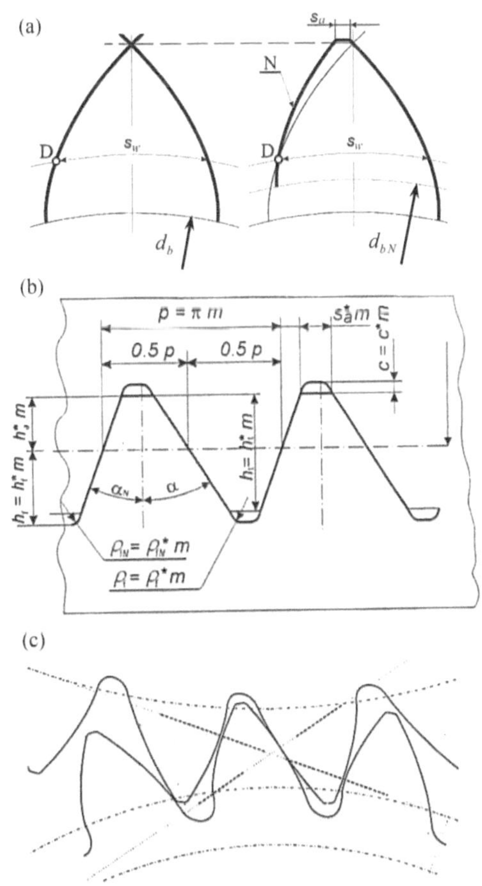

FIGURE 38.1
Involute asymmetric teeth: (a) Symmetric and asymmetric teeth; (b) basic rack; (c) meshing.

TABLE 38.1

Parameters of Basic Rack of Symmetric (No. 0) and Asymmetric (No. 1, 2, and 3) Teeth

No.	α_n	α_{nN}	h_a^*	s_a^*	ρ_f^*	ρ_{fN}^*	c^*
0	20°	20°	1	var	0.38	0.38	0.25
1	33°	20°	1	var	0.20	0.20	0.25
2	25°	20°	var	0.35	0.30	0.30	0.25
3	33°	25°	1	var	0.30	0.30	0.20

Both flank profiles are formed by two different involutes with different basic circles (Figure 38.1a), like that of the coast profile N having a larger diameter $(d_{bN} > d_b)$. The thickness of the asymmetric tooth on the operating circle (which is usually a reference one) s_w is the same as the symmetric tooth, and in the case of meshing with zero backlash, it is equal to half of the pitch.

Table 38.1 shows recommended values for asymmetric tooth basic rack compared to symmetric ones (Variant No. 0). Variant No. 1 is appropriate for heavily loaded gears, such as for hoisting machines. Variant No. 2 is appropriate for high-speed gears, e.g., for aircraft (Figure 38.2). Variant No. 3 is generally recommended for AI-planetary gear train (AI-PGT). What is special about the latter is that the both flanks of planet teeth are loaded. Drive flank of the planet tooth $(\alpha = 33°)$ meshes with a sun gear tooth and coast flank $(\alpha_N = 25°)$—with a ring gear tooth. Thus, the pressure angle of external meshing is higher than that of the internal meshing, which is favorable [71].

ISO 6336 [112] or [94, 215] can be used for load capacity calculation of cylindrical gears with asymmetric teeth, and most factors can be used without hindrance. There are only three factors to change:

Y_{FS}—form factor, taking into account the influence of tooth form as well as the stress concentration at root fillet.

FIGURE 38.2
Ilyushin Il-114 commuter airplane (a), whose gearbox is a two-carrier PGT with asymmetric teeth; sun gear (b) and carrier with planets assembly (c) of the second stage of the gearbox.

Z_H—zone factor, considering tooth flank curvature at the pitch point.

c—mesh stiffness per unit face width.

For these factors, separate diagrams have been developed that can be used by designer [71–73].

Application of the asymmetric tooth gears allows to significantly increase bending strength and pitting durability compared to the similar symmetric tooth gears as follows [72]:

- Bending strength about 28% (in optimized radii of root filet ρ_f and ρ_{fN})
- Pitting durability about 18%

Increasing load capacity reduces the size (and mass) of the PGT, which is of great importance, especially in aviation.

The unfavorable aspect of the asymmetric teeth is that a special tool is required, making it difficult for them to be widely used in general mechanical engineering.

More details about asymmetric teeth apart from [94, 215] can be found in the following sources: [2, 3, 51, 123, 127, 128, 176, 243].

References

1. Alexandrov, I. On loses determination in mechanical transmissions. *Vestnik Mashinostroeniya.* 1998, No. 6, pp. 12–14 (in Russian).
2. Alipiev, O. Asymmetric involute-lantern meshing formed by identical spur gears with a small number of teeth. *VDI-Berichte.* 2010, Vol. 2108, pp. 925–940. ISSN 0083-5560.
3. Alipiev, O. and S. Antonov. Main thesis in the geometrical theory of the involute meshing, shaped by a generalized basic rack. *Proceedings of the 3rd International Conference Power Transmissions'09*, edited by A. Mihailidis, Kallithea [Greece], Oct. 1–2, 2009, pp. 43–50. ISBN 978-960-243-662-2.
4. *ALLIANZ-Handbuch der Schadenverhütung*, 3. Auflage. Düsseldorf: VDI-Verlag, 1984.
5. ANSI/AGMA 6123-CXX. Design manual for enclosed epicyclic gear drives, 2014.
6. Arnaudow, K. Über einige Grundfragen des Lastausgleiches bei Planetengetrieben. *Maschinenbautechnik.* 1963, No. 3, pp. 433–437.
7. Arnaudov, K. On some principles in planetary gear trains loading determination. *Mashinostroene.* 1963, No. 2, pp. 17–21 (in Bulgarian).
8. Arnaudow, K. Kinematics and dynamics of the load distribution in planetary gear drives with supposedly elastic members. *Proceedings of the International Scientific Symposium on Planetary Gear Drives,* Miskolc [Hungary], 1966, pp. 229–234.
9. Arnaudow, K. Untersuchung des Lastausgleiches in Planetengetrieben. *Dissertation.* Dresden: Technische Universität, 1968.
10. Arnaudow, K. Untersuchung des Lastausgleiches in Planetengetrieben. *Maschinenbautechnik.* 1973, No. 6, pp. 275–280; No. 8, pp. 361–366.
11. Arnaudov, K. Investigation of load sharing in planetary gear trains. *Express-Information "Machines Elements".* 1973, No. 42, pp. 23–40 (Translation from *Maschinenbautechnik.* 1973, No. 6, pp. 275–280; No. 8, pp. 361–366. Paper-extract from author's Dissertation) (in Russian).
12. Arnaudov, K. Method for planet pins calibration in planetary gear trains. BG Patent No. 22279, F 16H 1/48.
13. Arnaudov, K. Engineering analysis of compound (two carrier) planetary gear trains. *Symposium "Heavy Machinery",* Varna [Bulgaria], 1984 [microfiche (48x)] (in Bulgarian).
14. Arnaudov, K. Experimental study of spectra of internal loading in planetary gear trains. *Proceedings of the 6th Congress on Mechanics,* Varna [Bulgaria], 1989, pp. 65–70 (in Bulgarian).
15. Arnaudov, K. Theoretical-experimental investigation of internal loads distribution in planetary gear trains. *Habilitation Thesis.* Sofia: Bulgarian Academy of Sciences, 1990 (in Bulgarian).
16. Arnaudow, K. Die Empfindlichkeit der Ausgleichvorrichungen der Planetengetriebe und ihre Auswirkung auf die Lastverteilung. 3. Congres Mondial des Engrenages et des Transmissions, Paris, 1992, pp. 1293–1302.
17. Arnaudov, K. Experimental determination of the efficiency of planetary gears. *Proceedings of the 4th International Tribology Conference,* Perth [Australia], 1994.
18. Arnaudow, K. Einfaches Verfahren zur Ermittlung des Übersetzungsverhältnisses zusammengesetzter Planetengetriebe. *VDI-Berichte.* 1996, Vol. 1230, pp. 313–328. ISSN 0083-5560.
19. Arnaudov, K., A. Binev and P. Tsenov. Experimental investigation of load sharing in planetary gear trains. *Proceedings of the 3rd Scientific-Technical Conference "Gear Trains'85",* Varna [Bulgaria], 1985 [microfiche (48x)] (in Bulgarian).
20. Arnaudov, K., A. Binev and P. Tsenov. Quantification of the uneven distribution of internal loads in the load capacity calculation of planetary gear trains. *Proceedings of the 6th Congress on Mechanics,* Varna [Bulgaria], 1989, pp. 59–64 (in Bulgarian).
21. Arnaudow, K., P. Genova and L. Dimitrov. For a unified and correct IFToMM terminology in the area of gearings. *Mechanisms and Machine Theory.* 2005, Vol. 40, pp. 993–1001.

22. Arnaudov, K. and D. Karaivanov. Engineering analysis of the coupled two-carrier planetary gearing through the lever analogy. *Proceedings of the International Conference on Mechanical Transmissions*, China Machine Press, Chongqing [China], Apr. 5–9, 2001, pp. 44–49.
23. Arnaudow, K. and D. Karaivanov. Die Blindleistung in Planetengetrieben. *Proceedings of the Conference on Research and Development of Machine Elements and Systems IRMES'04*, Kragujevac [Serbia & Montenegro], Sep. 16–17, 2004, pp. 587–594.
24. Arnaudow, K. and D. Karaivanov. Raum- und Massesparende Zahnradgetriebe. *Proceedings of the 3rd Conference about Construction, Shaping & Design*, Novi Sad [Serbia & Montenegro], May 19, 2004, pp. 73–78.
25. Arnaudow, K. and D. Karaivanov. Systematik, Eigenschaften und Möglichkeiten von zusammengesetzten Mehrsteg-Planetengetrieben. *Antriebstechnik*. 2005, No. 5, pp. 58–65.
26. Arnaudov, K. and D. Karaivanov. Higher compound planetary gear trains. *VDI-Berichte*. 2005, Vol. 1904, No. 1, pp. 327–344. ISSN 0083-5560.
27. Arnaudov, K. and D. Karaivanov. The Wolfrom gear train – a case of highest-complexity related modifications of the tooth meshing. *Proceedings of the ASME International Power Transmissions and Gearing Conference (PTG)*, San Diego, CA [USA], Aug. 30–Sep. 2, 2009, ISBN (CD-ROM) 978-0-7918-3856-3.
28. Arnaudov, K. and D. Karaivanov. The complex compound multi-carrier planetary gear trains – a simple study. *VDI–Berichte 2108-2*, 2010, pp. 673–684. ISSN 0083-5560.
29. Arnaudov, K., D. Karaivanov and L. Dimitrov. Some practical problems of distribution and equalization of internal loads in planetary gear trains. *Mechanisms and Machine Science*. 2013, 13; Power Transmissions, *Proceedings of the 4th International Conference*, edited by G. Dobre, Sinaia [Romania], June 20–23, 2012, pp. 585–596. ISSN 2211-0984.
30. Arnaudov, K. and D. Karaivanov. *Planetary Gear Trains*. Sofia: Bulgarian Academy of Sciences Publ. "Prof. Marin Drinov", 2017, 368 p. ISBN 978-954-322-885-0 (in Bulgarian).
31. Arnaudov, K. and D. Karaivanov. *Torque Method for Analysis of Compound Planetary Gear Trains*. Beau Bassin: LAP LAMBERT Academic Publishing, 2017, 92 p. ISBN 978-3659806841.
32. Arnaudov, K., E. Kolev and L. Lyubenov. Design of special reducer with two speed ratios for NPP "Kozloduy". *Proceedings of the 4th Conference "Machinery and Machine Elements"*, Sofia [Bulgaria], 2008, pp. 47–57 (in Bulgarian).
33. Arnaudow, K. and P. Zenow. Kinematic and dynamic conditions of the load compensation in planetary gear drive. *Proceedings of the International Scientific Symposium on Planetary Gear Drives*, Miskolc [Hungary], 1966, pp. 115–139.
34. Arnaudov, K. and others. Experimental determination of planets loading in a planetary gear train. *Proceedings of the Scientific Conference "Gear Trains"*, Gabrovo [Bulgaria], 1974, pp. 132–139 (in Bulgarian).
35. Arnaudov, K. and others. Experimental investigation of two-stage planetary gear train. *Vestnik Mashinostroeniya*. 1984, No. 12, pp. 11–12 (in Russian).
36. Arnaudov, K. and others. On the question of uniform international terminology in the field of gear trains. *Vestnik Mashinostroeniya*. 1990, No. 12, pp. 14–17 (in Russian).
37. Ayrapetov, E. and M.D. Genkin. Determination of planets loading unevenness. *Vestnik Mashinostroeniya*. 1966, No. 8, pp. 22–24 (in Russian).
38. Ayrapetov, E. and M.D. Genkin. Increasing of efficiency of the equalizing devices of planetary gear trains. *Vestnik Mashinostroeniya*. 1967, No. 10, pp. 41–43 (in Russian).
39. Ayrapetov, E. and M.D. Genkin. On an experimental determination of planet loading in planetary mechanisms. *Vestnik Mashinostroeniya*. 1971, No. 11, pp. 22–24 (in Russian).
40. Bahk, C.-J. and R.G. Parker. A study on planetary gear dynamics with tooth profile modification. *Proceedings of the ASME International Design Engineering Technical Conferences & Computers and Information in Engineering Conference IDETC/CIE 2011*, Washington, DC [USA], Aug. 28–31, 2011, DETC2011-47346.
41. Balbayev G. *New Planetary Gearbox: Design and Testing*. Beau Bassin: LAP LAMBERT Academic Publishing, 2015, 124 p. ISBN 978-3659806841.

42. Baumann, F. Theoretische Untersuchungen zur Lastaufteilung und Lastverteilung in Planetengetrieben. *Dissertation*. Dresden: Technische Universität, 2012.
43. Baumann, F., H. Graf and H. Linke. Extended load capacity calculation of planetary gears. *VDI-Berichte*. 2013, Vol. 2199, No. 1, pp. 1139–1153. ISSN 0083-5560.
44. Baumann, F. and J. Woller. Calculation approach for load capacity calculation of the tooth root of thin walled planet gears for planetary gears with high peripheral speeds. *VDI-Berichte*. 2015, Vol. 2251, No. 1, pp. 89–103. ISSN 0083-5560.
45. Boguski, B., A. Kahraman and T. Nishimo. A new method to measure planet load sharing and sun gear radial orbit of planetary gear sets. *Proceedings of the ASME International Design Engineering Technical Conferences & Computers and Information in Engineering Conference IDETC/CIE 2011*, Washington, DC [USA], Aug. 28–31, 2011. DETC2011-47196.
46. Bolotovskiy, I. and others. *Cylindrical Involute Gear Sets with Internal Meshing*. Moscow: Mashinostroenie, 1977 (in Russian).
47. Bondarenko, S. and others. Influence of the central wheels design of the planetary gear trains on the uniformity of load distribution in the meshing. *Vestnik Mashinostroeniya*. 1989, No. 3, pp. 24–26 (in Russian).
48. Bondarenko, S. and others. The effect of the accuracy of planetary gears manufacturing on the load distribution between planets. *Vestnik Mashinostroeniya*. 1989, No. 2, pp. 15–16 (in Russian).
49. Bondarenko, S. and A. Krylov. Experimental study of two-stage planetary gear train. *Vestnik Mashinostroeniya*. 1989, No. 7, pp. 17–20 (in Russian).
50. Bostan, I. and V. Dulgheru. Development of planetary precessional transmissions with multi-couple gear. *Mechanisms and Machine Science*. 2013, 13, Power Transmissions, *Proceedings of the 4th International Conference*, edited by G. Dobre, Sinaia [Romania], June 20–23, 2012, pp. 597–607. ISSN 2211-0984.
51. Brown, F.W., S.R. Davidson, D.B. Hanes, D.J. Weirles and A. Kapelevic. Analysis and testing of gears with asymmetric involute tooth form and optimized fillet form for potential application in helicopter main drives. AGMA Technical Paper, 2010. ISBN 978-1-55589-989-9.
52. Bruser, P. and G. Grüschow. Optimierung von Planetengetrieben. *Antriebstechnik*. 1989, No. 2, pp. 64–67.
53. Buiga, O. and S. Haragas. Single-row planetary gearbox gearings optimization using genetic algorithms. *Mechanisms and Machine Science*. 2013, 13, Power Transmissions, *Proceedings of the 4th International Conference*, edited by G. Dobre, Sinaia [Romania], June 20–23, 2012, pp. 447–456. ISSN 2211-0984.
54. Busch, J. Two-speed epicyclic final drives for 14-ton amphibious vehicle. PN, 1985. ASIN: B00CC2FACM.
55. Carlsson, B., H. Larsson and B. Jakobsson. Wirkungsgrad und Selbsthemmung einfacher Umlaufgetriebe. *Transactions of Chalmers University of Technology*. Gothenburg. 1957, No. 186, pp. 1–48.
56. Chapron, M. and others. On optimum tooth profile modifications to minimize dynamic mesh forces in planetary gears. *VDI-Berichte*. 2015, Vol. 2255, No. 1, pp. 67–78. ISSN 0083-5560.
57. Clarenbach, J. *Lorenz-Verzahnwerkzeuge*. 3. Auflage. Ettlingen: Firma "Lorenz" GmbH, 1977.
58. Collins, J.A, H. Busby and G. Staab. *Mechanical Design of Machine Elements and Machines*. 2nd ed. Hoboken, NJ: John Wiley & Sons, 2010, 889 p. ISBN 978-0-470-41303-6.
59. Concli, F. and C. Gorla. A new methodology for the prediction of the no-load losses of gears: CFD an experimental investigation of the efficiency of a planetary gearbox. *VDI-Berichte*. 2013, Vol. 2199, No. 2, pp. 1125–1137. ISSN 0083-5560.
60. Concli, F. and J. Coenen. Low-lost gears for precision planetary gearboxes: Influence of the gear design on the meshing and the churning power losses. *VDI-Berichte*. 2015, Vol. 2255, No. 1, pp. 53–65. ISSN 0083-5560.
61. Cottin, D. and E. Puls. *Angewandte Betriebsfestigkeit*. Leipzig: VEB Deutscher Verlag für Grundstoffindustrie, 1985.
62. DIN 3979. Zahnschäden an Zahnradgetrieben, Bezeichnung, Merkmale, Ursahen, 1979.

63. DIN 3990. Teil 11 Tragfähigkeitsberechnung von Stirnrädern – Anwendungsnorm für Industriegetriebe, Detail-Methode, 1989.
64. DIN 3992. Profilverschiebung bei Stirnrädern mit Außenverzahnung.
65. DIN 3993. Teil 1 bis 4 Geometrische Auslegung von zylindrischen Innenradpaaren mit Evolventenverzahnung.
66. DIN 51519. ISO-Viskositätsklassifikation für flüssige Industrie-Schmierstoffe, 1976.
67. Dobreva, A. Dimensionierung von Exzenter-Umlaufrädergetrieben mit geringer Zähnezahldifferenz. *Dissertation.* Magdeburg: Technische Universität "Otto von Guericke", 1989.
68. Dobreva, A. and S. Stoyanov. *Optimization Research of Gear Trains with Internal Meshing.* Ruse: University of Ruse Publ., 2012.
69. Dolchinkov, R. Profiling of gear-cutting wheel for profile wheel processing of cyclo-transmission with two meshing lines. *Proceedings of the Conference on Research and Development of Machine Elements and Systems IRMES'95,* Niš [Yugoslavia], 1995.
70. Dolchinkov, P. *Planetary Mechanisms with Gear and Friction Wheels.* Burgas: Feniks Print, 2015 (in Bulgarian).
71. Dorofeev, V., K. Arnaudov and B. Dorofeev. Increasing the strength of gears by using asymmetric teeth. *Proceedings of VIII International Congress "Machines, Technologies, Materials" 2011,* V. 3, Varna [Bulgaria], 2011, pp. 83–87 (in Russian).
72. Dorofeev, V., D. Dorofeev and K. Arnaudov. Selection of optimal geometry and stress calculation on wheel teeth with asymmetrical tooth profile. *VDI-Berichte.* 2013, Vol. 2199, No. 2, pp. 1341–1353. ISSN 0083-5560.
73. Dorofeev, V. and others. Theoretical and experimental dynamic investigation of gear with the advanced asymmetric profile with presence of profile errors and runout in gears. *VDI-Berichte.* 2015, Vol. 2255, No. 1, pp. 251–259. ISSN 0083-5560.
74. Dresig, H. and F. Holzweißig. *Dynamics of Machinery. Theory and Applications.* Berlin, Heidelberg: Springer-Verlag, 2010. ISBN 978-3-540-89939-6.
75. Drewniak, J., P. Garlicka and A. Kolber. Design for a bi-planetary gear train. *Scientific Journal of Silesian University of Technology. Series Transport.* 2016, Vol. 91, pp. 5–17. ISSN 0209-3324.
76. Duda, M. Der geometrische Verlustbeiwert und die Verlustunsymmetrie bei geradverzahnten Stirnradgetrieben. *Forschung im Ingenieurwesen.* 1971, Vol. 37, No. 1, pp. 7–14. VDI-Verlag.
77. Ebner, M., T. Lohner, K. Michaelis, B.-R. Höhn and K. Stahl. Self-lubricating gears with oil-impregnated sintered materials. *VDI-Berichte.* 2017, Vol. 2294, No. 1, pp. 17–28. ISSN 0083-5580.
78. Eckhardt, F. *Stationäre Zahnradgetriebe. Schmierung und Wartung.* Hamburg: Mobil Oil AG.
79. Edwards, P. Tank steering systems (differentials, the theory and practice). *Constructor Quarterly.* Sept. 1988, No.1, pp. 47–48.
80. Ehrlenspiel, K. Die konstruktive Lösung des Lastausgleiches. *Konstruktion.* 1979, Vol. 31, No. 1, pp. 12–18.
81. Eritenel, T. and R.G. Parker. Vibration models of helical planetary gears. *Proceedings of the ASME International Power Transmissions and Gearing Conference (PTG),* San Diego, CA [USA], Aug. 30–Sep. 2, 2009. ISBN (CD-ROM) 978-0-7918-3856-3.
82. Erixon, T.M. and R.G. Parker. Planetary gear modal properties and dynamic response: Experiments and analytical simulation. *Proceedings of the ASME International Design Engineering Technical Conferences & Computers and Information in Engineering Conference IDETC/CIE 2011,* Washington, DC [USA], Aug. 28–31, 2011, DETC2011-47142.
83. Eschmann, P., L. Hasbargen and K. Weigand. *Die Wälzlagerpraxis. Handbuch für die Berechnung und Gesaltung von Lagerungen.* München: R. Oldenburg Verlag, 1978.
84. EUROTRANS. *Wörterbuch der Kraflübertragungselemente.* Band 1 – Zahnräder. Berlin: Springer-Verlag, 1982.
85. Förster, H.J. Zur Berechnung des Wirkungsgrades von Planetengetrieben. *Konstruktion.* 1969, Vol. 21, No. 5, pp. 165–178.
86. de Gevighey, J.D. and others. Experimental investigation on no-load dependent power losses in a planetary gear set. *VDI-Berichte.* 2013, Vol. 2199, No. 2, pp. 1101–1112. ISSN 0083-5560.

87. Giger, U. Neuer Ansatz für sehr große Windkraftgetriebe. *Wind Kraft Journal*. 2007, No. 5, pp. 2–4.
88. Giger, U. and K. Arnaudov. Redesign of a gearbox for 5MW wind turbines. *Proceedings of the ASME International Design Engineering Technical Conferences & Computers and Information in Engineering Conference IDETC/CIE 2011*, Washington, DC [USA], Aug. 28–31, 2011, DETC2011-47492.
89. Giger, U. and K. Arnaudov. New drive train design for ultra large 15 MW wind turbines. *VDI-Berichte*. 2013, Vol. 2199, No. 1, pp. 101–112. ISSN 0083-5560.
90. Gorobets, I. On the load distribution in planetary gear trains of type 2K-H with a floating sun gear. *Collection "Machine parts" No. 15*. Kiev: Tekhnika, 1972, pp. 8–16 (in Russian).
91. Gorobets, I. On the load distribution in meshing of two-stage planetary gearboxes. *Collection "Machine parts" No. 17*. Kiev: Tekhnika, 1973, pp. 5–9 (in Russian).
92. Gorobets, I. A study of load distribution between planets in two-stage planetary gearboxes with floating central elements. *Collection "Machine parts" No. 18*. Kiev: Tekhnika, 1973, pp. 3–9 (in Russian).
93. Gorobets, I. and M. Osipenko. Experimental determination of the meshing loads of two-stage planetary gearboxes. *Collection "Machine parts" No. 29*. Kiev: Tekhnika, 1979, pp. 12–18 (in Russian).
94. GOST 21354-87. Cylindrical involute gear trains with external meshing. Load capacity calculations (in Russian).
95. Gröber, S.F. Leistungsverluste in Umlaufrädergetrieben. *Industrie-Anzeiger*. 1968, No. 48, pp. 28–32.
96. Guo, Y. and R.G. Paeker. Dynamic modeling and analysis of planetary gear involving tooth wedging and bearing clearance nonlinearity. *Proceedings of the ASME International Power Transmissions and Gearing Conference (PTG)*, San Diego, CA [USA], Aug. 30–Sep. 2, 2009. ISBN (CD-ROM) 978-0-7918-3856-3.
97. Guo, Y. and R.G. Parker. Back-side contact gear mesh stiffness. *Proceedings of the ASME International Design Engineering Technical Conferences & Computers and Information in Engineering Conference IDETC/CIE 2011*, Washington, DC [USA], Aug. 28–31, 2011, DETC2011-48055.
98. Guo, Y. and others. Theoretical and experimental study on gear-coupling contact and loads considering misalignment, torque, and friction influences. *Mechanism and Machine Theory*. 2016, Vol. 98, pp. 242–262. ISSN 0094-114X.
99. Handschuh, R., NASA Technical Reports Server and others. New Design and Improvement of Planetary Gear Trains. Washington, DC: BiblioGov, 2013, 36 p. ISBN 978-1287266594.
100. Haniszewski, T. Modeling the dynamics of cargo lifting process by overhead crane for dynamic overload factor estimation. *JVE International Ltd. Journal of Vibroengineering*. 2017, Vol. 19, No. 1, pp. 75–86. ISSN 1392-8716.
101. Hantschack, F. Zahnfußtragfähigkeit von Innenverzahnungen mit Kranzeinfluß bei Planetengetrieben. *Dissertation*. Dresden: Technische Universität, 2011.
102. Hardy, H.W. *Planetary Gearing: Design and Efficiency*. Andesite Press, 2017, 66 p. ISBN 978-13759190.
103. Heider, M. and others. Vibration excitation of a planetary gear stage. *VDI-Berichte*. 2013, Vol. 2199, No. 2, pp. 1091–1100. ISSN 0083-5560.
104. Henriot, G. *Gear and Planetary Gear Trains*. Publisher: Brevini [Italy], 1993.
105. Hidaka, T. and others. Dynamic behavior of planetary gear. *Bulletin of JSME*. 1976, Vol. 19, No. 132, pp. 690–698; 1976, Vol. 19, No. 138, pp. 1563–1570; 1977, Vol. 20, No. 150, pp. 1663–1672; 1979, Vol. 22, No. 167, pp. 877–884; 1979, Vol. 22, No. 169, pp. 1017–1025; 1979, Vol. 22, No. 169, pp. 1026–1033; 1979, Vol. 22, No. 170, pp. 1142–1149.
106. Hill, F. Wirkungsgrad von Stirnrad- und Umlaufgetrieben. Technisch-wissenschaftliche Veröffentlichungen. *Zahnradfabrik Friedrichshafen*. 1965, No. 6, pp. 23–28.
107. Hill, F. Einbaubedingungen bei Planetegetrieben. *Konstruktion*. 1967, Vol. 19, No. 10, pp. 393–394.
108. Höhn, B.-R., K. Stahl and P. Gwinner. Improved efficiency for high-ratio planetary gear transmission for wind turbines – low-loss Wolfrom transmission for wind turbines. *VDI-Berichte*. 2013, Vol. 2199, No. 2, pp. 1113–1124. ISSN 0083-5560.

109. Hortel, H. and A. Škuderova. Zur inneren Dynamik von hochtourigen hochbeanspruchten Planetengetrieben. *Proceedings of the 3rd International Conference Power Transmissions'09*, edited by A. Mihailidis, Kallithea [Greece], Oct. 1–2, 2009, pp. 369–378. ISBN978-960-243-662-2.
110. Huang, K.J. and S.R. Zhang. A finite element investigation to modal and dynamic behaviors of planetary gearings concerning the effect of bearing and carrier stiffnesses. *Proceedings of the ASME International Design Engineering Technical Conferences & Computers and Information in Engineering Conference IDETC/CIE 2011*, Washington, DC [USA], Aug. 28–31, 2011, DETC2011-47136.
111. Imwalle, D.E. Load equalization in planetary gear systems. ASME-Publication, 1972, 72-PTG-16.
112. ISO 6336. Calculation of load capacity of spur and helical gears, 2006.
113. ISO 9085. Calculation of load capacity of spur and helical gears – application for industrial gears, 2002.
114. ISO 10825. Gears – wear and damage of gear teeth – terminology, 1995
115. Ivanov, V. and I. Sidorenko. *Designing Machine Elements and Structural Components in the Mechanical Desktop Environment*. Odessa [Ukraine]: Nauka i Tehnika, 2005. ISBN 966-8335-27-9 (in Russian).
116. Jahn, C. Theoretische und experimentelle Untersuchungen zur Zahnfußtragfähigkeit von Innenverzahnungen. *Dissertation*. Dresden: Technische Universität, 1997.
117. Jahn, C., H. Linke and R. Kupfer. Zahnfuß – und Zahnkranzbeanspruchung bei Innenverzahnungen von Planetengetrieben. *Antriebstechnik*. 1998, No. 7, pp. 65–68; No. 10, pp. 70–74; No. 11, pp. 69–73.
118. Jakobsson, B. Torque distribution, power flow and zero output conditions of epicyclic gear train. *Transactions of Chalmers University of Technology*. Gothenburg. 1960, No. 224, pp. 1–55.
119. Jarchow, F. and H.T. Wagner. Lastverteilung in Planetengetrieben. Einflüße von Drehmoment, Drehzahl, Gestaltung und Toleranzen. Vergleich zwischen Rechnung und Versuch. *FVA-Forschungsreport*, 1980.
120. Jelaska, D. *Gears and Gear Drives*. Chichester: John Wiley & Sons, 2012, 444 p. ISBN 9781119941309.
121. Jensen, P.W. Raumbedarf und Wirkungsgrad zusammengesetzter Planetengetriebe mit einstufigem Planetenrad. *Konstruktion*. 1969, Vol. 21, No. 5, pp. 178–184.
122. Kahraman, A. and H. Ligata. Effects of external splines on deformations of ring gears of planetary gear sets. *VDI-Berichte*. 2017, Vol. 2294, No. 1, pp. 341–352. ISSN 0083-5580.
123. Kapelevich, A. *Direct Gear Design*. Boca Raton, FL: CRC Press, 2013, 299 p. ISBN 978-1-4398-7618-3.
124. Kapelevich, A. High gear ratio epicyclic drives analysis. *Gear Technology*. June 2014, pp. 62–67.
125. Kapelevich, A. Analysis and optimization of asymmetric epicyclic gears. *VDI-Berichte*. 2015, Vol. 2255, No. 1, pp. 239–249. ISSN 0083-5560.
126. Kapelevich, A. and V. Ananiev. Gear transmission density maximization. *Proceedings of the ASME International Design Engineering Technical Conferences & Computers and Information in Engineering Conference IDETC/CIE 2011*, Washington, DC [USA], Aug. 28–31, 2011, DETC2011-47021.
127. Kapelevich, A. and R. Kleiss. Direct gear design for spur and helical gears. Gear Technology. Sept./Oct. 2002, pp. 29–35. ISSN 0743-6858.
128. Kapelevich, A. and Y. Shekhtman. Tooth fillet profile optimization for gears with symmetric and asymmetric teeth. AGMA Fall Technical Meeting, San Antonio, TX, Oct. 12–14, 2008 (08FTM06).
129. Karademir, M.A. Zahnsteifigkeiten in Planetengetrieben. *Dissertation*. Bochum: Ruhr-Universität, 1989.
130. Karaivanov, D. Theoretical and experimental studies of influence of the structure of coupled two-carrier planetary gear trains on its basic parameters. *Dissertation*. Sofia: University of Chemical Technology and Metallurgy, 2000 (in Bulgarian).
131. Karaivanov, D. On the kinematic optimization of Wolfrom planetary gear trains. *Proceedings of the TMCE 2010*, edited by Z. Rusak, Ancona [Italy], Apr. 12–16, 2010, pp. 1757–1766. ISBN/EAN (CD-ROM) 987-90-5155-060-3.

132. Karaivanov, D. Structural analysis of compound planetary gear trains. *Balkan Journal of Mechanical Transmissions*. 2011, Vol. 1, No. 1, pp. 33–45. ISSN 2069-5497.
133. Karaivanov, D. and K. Arnaudow. Die zusammengesetzten Mehrsteg-Planetengetriebe und ihre Gesetzmäßigkeiten. Zbornik radova sa Naučno-stručnog skupa Istraživanje i razvoj mašinskih elementa i sistema JAHORINA – IRMES'2002, 1/2, Srpsko Sarajevo – Jahorina [Bosna & Hercegovina], Sep. 19–20, 2002, pp. 19–26.
134. Karaivanov, D. and S. Troha. Examining the possibilities for using coupled two-carrier planetary gears in two-speed mechanical transmissions. *Machinebuilding and Electrical Engineering*. 2006, No. 5–6, pp. 124–127. ISSN 0025-455X.
135. Karaivanov, D., S. Troha and R. Pavlova. Experimental study of the losses in a three-stage planetary gear train. *Proceedings of the 3rd International Conference "Power Transmissions'09"*, edited by A. Mihailidis, Kallithea [Greece], Oct. 1–2, 2009, pp. 527–532. ISBN 978-960-243-662-2.
136. Karaivanov, D., S. Troha and R. Pavlova. Investigation into self-looking planetary gear trains through the lever analogy. *Transactions of FAMENA*. 2012, Vol. 36, No. 1, pp. 13–24. ISSN 1333-1124.
137. Karaivanov, D., M. Velyanova and V. Bakov. Kinematic and power analysis of multi-stage planetary gearboxes through the torque method. *Machines, Technologies, Materials*. 2018, Vol. 12, No. 3, pp. 102–108. ISSN 1313-0226 (print) 1314-507X (online).
138. Klein, B. and Z. Li. Analytisches Verfahren zur parametrischen Optimierung von Zahnradgetrieben. *Antriebstechnik*. 1995, No. 3. pp. 57–78.
139. Kondrashov, Y. and others. Lastverteilung in Planetengetrienen. *Konstruktion*, 1974, No. 8, pp. 323–324.
140. Konzernnorm ZFN 201 Zahnradschäden. Friedrichshafen: Begriffsbestimmung Zahnradfabrik Friedrichshafen AG, 1990.
141. Kos, M. Bewertung der Ausgleichssysteme in Planetengetrieben mit dynamischen Kraftbeiwerten. *Konstruktion*. 1981, Vol. 33, No. 3, pp. 91–96.
142. Kos, M. Kraftverteilung über die Zahnbreite bei Planetengetrieben. Einfluß des Ausgleichssystems. *Antriebstechnik*. 1982, No. 11, pp. 553–557.
143. Kos, M. Über den Kräfteausgleich in Planetengetrieben mit Sonnenhohlrad. *Die Maschine*. 1982, No. 12, pp. 382–388.
144. Kos, M. Neue funktionsgerechte Ausgleichsysteme von Planetengetrieben. *Konstruktion*. 1983, Vol. 35, No. 11, pp. 427–431.
145. Kreynes, M. Efficiency and speed ratio of a gear mechanism. *Proceedings of the Seminar on Theory of Machines and Mechanisms*, Vol. 1, Acad. Of Sciences of USSR, Moscow [Russia], 1941, pp. 21–48 (in Russian).
146. Krüger, D. and others. Increasing the load capacity of gear teeth by asymmetric gear tooth design. *VDI-Berichte*. 2013, Vol. 2199, pp. 1327–1340. ISSN 0083-5560.
147. Kudryavtsev, V. *Planetary Gear Trains*. Moscow, Leningrad: Mashinostroenie, 1966 (in Russian).
148. Kudryavtsev, V., Y. Derzhavets and E. Glukharev. *Arrangements and Calculation of Gearboxes*. Leningrad: Mashinostroenie, 1971 (in Russian).
149. Kudryavtsev, V., Y. Kirdyashev and others. Planetary Gear Trains. Manual. Leningrad: Mashinostroenie, 1977 (in Russian).
150. Kurth, F. Efficiency determination and synthesis of complex-compound planetary gear transmissions. *Dissertation*. Munich: Technical University, 2012.
151. Kutzbach, K. Mehrgliedrige Radgetriebe und ihre Gesetze. *Maschinenbau*. 1927, No. 22, pp. 1080.
152. Lamparski, C. Einfache Berechnungsgleichungen für Lastüberhöhungen in Leichtbauplanetengetrieben. *Dissertation*. Bochum: Ruhr-Universität, 1995.
153. Langedon, H.R. Table of gear combinations for planetary gear ratios. *Machine Design*. 1960, No. 9, pp. 161–170.
154. Lee, H.-K. and others. A research for the planetary gear noise development in FF 6th speed automatic transmission. *Proceedings of the ASME International Design Engineering Technical Conferences & Computers and Information in Engineering Conference IDETC/CIE 2011*, Washington, DC [USA], Aug. 28–31, 2011, DETC2011-47129.

155. Leistner, F., G. Lörsch and O. Wilhelm. Umlaufrädergetriebe. 3. Auflage. Berlin: VEB Verlag Technik, 1987.
156. Linke, H., U. Trempler and F. Baumann. Analysis of the stress of toothings on planetary gears. *Proceedings of the 2nd International Conference "Power Transmissions'06"*, Novi Sad [Serbia & Montenegro], Apr. 25–26, 2006, pp. 143–152.
157. Linke, H., F. Hantschack and U. Trempler. Präzisierte Berechnung der Tragfähigkeit von Innenverzahnungen für Planetengetriebe. *Proceedings of the 3rd International Conference "Power Transmissions'09"*, edited by A. Mihailidis, Kallithea [Greece], Oct. 1–2, 2009, pp. 215–222. ISBN 978-960-243-662-2.
158. Linke, H. and others. Stirnradverzahnung – Berechnung, Werkstoffe, Fertigung. 2. Auflage. München/Wien: Carl Hanser Verlag, 2010.
159. Liu, H. and others. Analysis and test of dynamic load sharing behavior in compound planetary gears. *VDI-Berichte*. 2013, Vol. 2199, No. 2, pp. 1055–1066. ISSN 0083-5560.
160. Liu, G. and others. Mesh stiffness of cylindrical gear including tooth surface. Friction based on FEA method. *VDI-Berichte*. 2017, Vol. 2294, No. 1, pp. 807–814. ISSN 0083-5580.
161. Looman, J. *Zahnradgetriebe – Grundlagen, Konstruktion, Anwendung in Fahrzeugen*. 3. Auflage, Berlin: Springer-Verlag, 1996.
162. *MAAG-Taschenbuch*. 2. Auflage. Zürich: MAAG-Zahnräder AG, 1985.
163. Marinov, C., O. Alipiev and R. Dolchinkov. Synthese von Zykloidengetrieben mit angenäherten Hypo-Epizykloideneingriff. Tagung "Zahnradgetribe". Teil 1, Dresden, 1989.
164. Mihailidis, A. and C. Pupaza. Simulation driven design of internal gears. Multicriteria optimization of internal gears. *VDI–Berichte 2108-2*, 2010, pp. 725–740. ISSN 0083-5560.
165. Mihailidis, A. and others. Finite element method analysis of planetary gear systems considering backlash and manufacturing errors. *VDI-Berichte*. 2015, Vol. 2255, No. 1, pp. 79–88. ISSN 0083-5560.
166. Müller, H.W. *Epicyclic Drive Trains*. Detroit, MI: Wayne State University Press, 1982. ISBN 0-8143-1663-8.
167. Müller, H.W. *Die Umlaufgetriebe – Auslegung und vielseitige Anwendungen*. 2. Auflage. Berlin: Springer-Verlag, 1998.
168. Müller, H.W. Zum Problem des Lastausgleiches in Planetengetrieben. *Konstruktion*. 1975, Vol. 27, No. 3, pp. 106–109.
169. Müller, H.W. and W.F. Schäfer. Geometrische Voraussetzungen bei innenverzahnten Getriebestufen kleinster Zähnezahldifferenzen. *Forschung im Ingenieurwesen*. 1970, Vol. 36, No. 5, VDI-Verlag.
170. Nakagawa, M. and others. Development of novel test stand for three-axis driving planetary gear train control and analysis of novel PGT for noise reduction. *VDI-Berichte*. 2017, Vol. 2294, No. 1, pp. 331–340. ISSN 0083-5580.
171. Nenov, P., B. Kaloyanov and E. Angelova. Geometrical blocking contours as an instrument for high-technology design of geometry of involute external and internal gear drives. *Mechanisms and Machine Science*. 2013, 13, Power Transmissions, *Proceedings of the 4th International Conference*, edited by G. Dobre, Sinaia [Romania], June 20–23, 2012, pp. 379–388. ISSN 2211-0984.
172. Neufond, J. and others. The iterative spectral method for computing the planetary gear dynamic response. *VDI-Berichte*. 2017, Vol. 2294, No. 1, pp. 209–218. ISSN 0083-5580.
173. Neumann, R. Offene Umlaufrädergetriebe und ihre Anwendung. *Maschinenbautechnik*. 1973, No. 12, pp. 533–538.
174. Neumann, R. Technische Anwendungen des Umlaufräderprinzips. *Maschinenbautechnik*. 1976, No. 2, pp. 50–57.
175. Niemann, G. and H. Winter. *Maschinenelemente. Band 2. Zahnradgetriebe – Grundlagen, Stirnradgetriebe*. 2. Auflage. Berlin: Springer-Verlag, 1995.
176. Novikov, A.S., A.G. Paikin, V.L. Dorofeyev, V.M. Ananiev and A.L. Kapelevich. Application of gears with asymmetric teeth in turboprop engine gearbox. *Proceedings of the 10th International ASME Power Transmission and Gearing Conference*, Las Vegas, NV [USA], Sept. 4–7, 2007. DETC2007-34823.

177. Ogorkiewicz, R. *Tanks. 100 Years of Evolution*. London: Osprey Publ., 2015. ISBN 978-1-4728-2981-8.
178. ÖNORM M6700 Zahnradschäden, Bezeichnungen, Merkmale, Ursachen, 1968.
179. Orlić, Ž. and G. Orlić. *Planetni prijenosi*. Rijeka: Zigo, 2006.
180. Pandev, G. Gear ratio determination of a high-ratio gear reducer. *Proceedings of the 3rd International Conference "Power Transmissions'09"*, edited by A. Mihailidis, Kallithea [Greece], Oct. 1–2, 2009, pp. 155–158. ISBN 978-960-243-662-2.
181. Papies, J. Load distribution in planetary gears considering effects of manufacturing tolerances with a statistical approach. *VDI-Berichte*. 2017, Vol. 2294, No. 1, pp. 783–793. ISSN 0083-5580.
182. Park, Y. and others. Characteristics analysis of wind tubine gearbox considering fatigue loads caused by wind fluctuation. *Proceedings of the ASME International Design Engineering Technical Conferences & Computers and Information in Engineering Conference IDETC/CIE 2011*, Washington, DC [USA], Aug. 28–31, 2011. DETC2011-47316.
183. Pedrero, J.I., M. Pleguezuelos and M. Muños. Simplified calculation method for the efficiency of involute spur gears. *Proceedings of the ASME International Power Transmissions and Gearing Conference (PTG)*, San Diego, CA [USA], Aug. 30–Sep. 2, 2009. ISBN (CD-ROM) 978-0-7918-3856-3.
184. Pedrero, J.I., M. Pleguezuelos and M.B. Sanches. Load sharing model for spur gears with tip relief. An investigation on the influence of profile modifications on the mesh stiffness. VDI-Berichte. 2017, Vol. 2294, No. 1, pp. 795–806. ISSN 0083-5580.
185. Peng, C., C. Li and B. Chen. Development of a test rig for transmission performances of precision planetary reducer. *VDI-Berichte*. 2017, Vol. 2294, No. 2, pp. 1127–1132. ISSN 0083-5580.
186. Petrov, I. Driving axle for statorless electric motor. BG Patent No 10088, B 62 d, f.
187. Ping, S.T., Y.M. Shao and L.M. Wang. The influence of mesh misalignment on planetary gear set with time-varying mesh stiffness. *VDI-Berichte*. 2017, Vol. 2294, No. 1, pp. 219–230. ISSN 0083-5580.
188. Pleguezuelos, M., J.I. Pedrero and M.B. Sanches. Analitycal model of the efficiency of spur gears: Study of the influence of the design parameters. *Proceedings of the ASME International Design Engineering Technical Conferences & Computers and Information in Engineering Conference IDETC/CIE 2011*, Washington, DC [USA], Aug. 28–31, 2011, DETC2011-47662.
189. Popov, V. Überdeckungsgrad und Tragfähigkeit von Exzenter-Getrieben mit Evolventenverzahnung. *Dissertation*. Magdeburg: Otto-von-Guericke-Universität, 1996.
190. Predki, W., F. Jarchow and J. Kettler. Calculation method for the determination of the oil sump temperature of industrial planetary gears. *VDI-Berichte*. 2005, Vol. 1665, pp. 507–522. ISSN 0083-5560.
191. Prueter, P.L., R.G. Parker and F. Cunliffe. A study of gear root strains in a multi-stage planetary wind turbine gear train using a three dimensional finite element/contact mechanics model and experiments. *Proceedings of the ASME International Design Engineering Technical Conferences & Computers and Information in Engineering Conference IDETC/CIE 2011*, Washington, DC [USA], Aug. 28–31, 2011, DETC2011-47451.
192. Radzevich, S.P. *Theory of Gearing: Kinematics, Geometry and Synthesis*. 2nd edition. Boca Raton, FL: CRC Press, 2018, 898 p. ISBN 9781138585553.
193. Rahnejat, H. *Multi-Body Dynamics: Vehicles, Machines and Mechanisms*. Milton: Wiley-Blackwell, 1998, 370 p. ISBN 978-1-860-58122-9.
194. Rajasri, I. *Synthesis and Analysis of Epicyclic Gear Trains: Graph Theory*. Beau Bassin: LAP LAMBERT Academic Publishing, 2016, 252 p. ISBN 978-3659855405.
195. Reshetov, D. About one drawback of planetary mechanisms with a large gear ratio. *Vestnik Mashinostroeniya*. 1957, No. 5, pp. 14–15 (in Russian).
196. Reshetov, D. *Rational Mechanisms Design*. Moscow: Mashinostroenie, 1972 (in Russian).
197. Richter, W. Auslegung von Innerverzahnungen und Planetengetrieben. *Konstruktion*. 1962, No. 12, pp. 489–497.
198. Rockwell, P.D. Profile shift in external parallel-axis cylindrical involute gears. Gear Technology. Nov./Dec. 2001, pp. 18–25. ISSN 0743-6858.
199. Röhle, E. Theoretische und experimentelle Untersuchungen zum Lastausgleich in Planetengetrieben. *Dissertation*. Dresden: Tecnnische Universität, 1977.

200. Rudenko, V. Planetary and Harmonic Transmissions. Atlas. Moscow: Mashinostroenie, 1980 (in Russian).
201. Rudnick, L.R. *Lubricant Additives: Chemistry and Application*. 3rd edition. Boca Raton, FL: CRC Press, 2017, 693 p. ISBN 9781498731720.
202. Sadykova, A. and L. Shuster. Structural synthesis of planetary change-gears using computer programs. Ufa: Ufa State Aviation Technical University, 2009 (in Russian).
203. Scheffler, M. and F. Kurth. *Fördertechnik – Grundlagen der Fördertechnik Einführung, Elemente der mechanischen Ausrüstung, Antriebe, Grundlagen des Stahlbaus*. Berlin: VEB Verlag Technik, 1987.
204. Schlecht, B., T. Schulze and K. Riedel. Tooth contact analysis under load on double helical-geared planetary gearboxes. *VDI-Berichte*. 2017, Vol. 2294, No. 1, pp. 83–96. ISSN 0083-5580.
205. Schnetz, K. Optimierung zusammengesetzter Planetengetriebe. *Fortschrittsberichte der VDI-Zeitschriften*, Reihe 1, VDI-Verlag, Düsseldorf, 1971, No. 30.
206. Schoo, A. Verzahnungsverlustleistungen in Planetengetrieben. *Dissertation*. Bochum: Ruhr-Universität, 1985.
207. Schubert, M. Einfluß der Befestigungsart und Radkranzdicke auf die Zahnfußspannung von Innenstirnrädern. *Dissertation*. München: Technische Universität, 1984.
208. SDP-SI. *Elements of metric gear technology*. www.sdp-si.com/PDFS/Elements-of-Metric-Gear-Technology.pdf.
209. Seeliger, K. Das einfache Planetengetriebe. *Antriebstechnik*. 1964, pp. 216–221.
210. Sholomov, N. Determination of the stress in tooth root of the gears with a thin rim. *Vestnik Mashinostroeniya*. 1984, No. 4, pp. 63–68 (in Russian).
211. Shyamsananth, M., A. Manish and S.P. Ganesan. Influence of rim-thickness on planet bending stresses and load sharing. *Proceedings of the ASME International Design Engineering Technical Conferences & Computers and Information in Engineering Conference IDETC/CIE 2011*, Washington, DC [USA], Aug. 28–31, 2011, DETC2011-47643.
212. Skvortsova, N. Internal involute meshing for the case where the difference in number of gear teeth is equal to one. Collection Strength Calculation in Machinery. MSTU, Vol. 11, Mashgiz, Moscow, 1950, pp. 143–166 (in Russian).
213. Slavov, V., I. Slavova and I. Angelov. Vibroacoustics of gear reducer and its parts. *Proceedings of the 3rd International Conference on Manufacturing Engineering (ICMEN)*, Kallithea [Greece], Oct. 1–3, 2008, pp. 561–570. ISBN 978-960-243-649-3.
214. Slavov, V., I. Slavova and I. Angelov. Vibration-acoustic and computer aided design of gear train hood. *Bulgarian Journal for Engineering Design*. TU-Sofia. 2009, Vol. 2, No. 2, pp. 162–166. ISSN 1313-7530 (in Bulgarian).
215. St. SEV 5744-86 Cylindrical gears. Load capacity calculations (in Russian).
216. Stahl, K. Vorlesung Maschinenelemente II. Technische Universität München, 2013.
217. Stangl, M. Methodik zur kinematischen und kinetischen Berechnung mehrwelliger Planeten-Koppelgetriebe. *Dissertation*. München: Tecnnische Universität, 2007.
218. Stefanović-Marinović, J. and M. Milovančević. The optimization possibilities at the planetary gear trains, *Journal of Mechanics Engineering and Automation* (David Publishing). 2012, Vol. 2, No. 6, pp. 365–373. ISSN 2159-5275 (print) 2159-5283 (online).
219. Stefanović-Marinović, J., M. Petković and I. Stanimirović. Application of the ELECTRE method to planetary gear train optimization, *Journal of Mechanical Science and Technology*. 2014, Vol. 29, No. 2, pp. 647–654. www.springerlink.com/content/1738-494x. doi:10.1007/s12206-015-0124-z. ISSN 1738-494X (print) 1976-3824 (online).
220. Stefanović-Marinović, J., S. Troha and M. Milovančević. An application of multicriteria optimization to the two-carrier two-speed planetary gear trains. *FACTA Universitatis, Series: Mechanical Engineering*. 2017, Vol. 15, No. 1, pp. 85–95. ISSN 0354-2025 (print) 2335-0164 (online).
221. Stocco, L.J. and R. Gloeckner. Smashing the efficiency barrier – a practical comparison of planetary and orbitless gear-heads. *VDI-Berichte*. 2017, Vol. 2294, No. 1, pp. 319–330. ISSN 0083-5580.
222. Stoeckicht, W. Planetengetriebe. *Konstruktion*. 1953, Vol. 5, No. 4, pp. 108–109.

223. Stoyanov, S., V. Dobrev and A. Dobreva. Investigating dynamic behavior of planetary gear trains through the systematic approach. *VDI-Berichte*. 2017, Vol. 2294, No. 1, pp. 197–208. ISSN 0083-5580.
224. Strauch, H. Die Fundamentalgesetze der Planetengetriebe. *Antriebstechnik*. 1967, pp. 265–270.
225. Suzuki, A., T. Aoyama and T. Shimizu. Influence of bearing clearance on load sharing in planetary gears. *Proceedings of the ASME International Design Engineering Technical Conferences & Computers and Information in Engineering Conference IDETC/CIE 2011*, Washington, DC [USA], Aug. 28–31, 2011, DTEC2011-47622.
226. Talbot, D.C., A. Kahraman and A. Singh. An experimental investigation of the efficiency of planetary gear sets. *Proceedings of the ASME International Design Engineering Technical Conferences & Computers and Information in Engineering Conference IDETC/CIE 2011*, Washington, DC [USA], Aug. 28–31, 2011, DETC2011-47126.
227. Tenberge, P., D. Kupka and T. Panero. Influential criteria on the optimization of a gearbox, with application to an automatic transmission. *VDI-Berichte*. 2017, Vol. 2294, No. 1, pp. 253–265. ISSN 0083-5580.
228. Tenberge, P and M. Weibring. High ratio gearbox with very low bearing load. New single-stage planetary gear arrangement with stepped planets. *VDI-Berichte*. 2015, Vol. 2255, No. 1, pp. 3–14. ISSN 0083-5560.
229. Terplán, Z. *Dimensionierungsfragen der Zahnrad-Planetengetriebe*. Budapest: Akademiai Kiado, 1974.
230. Troha, S. Analysis of a planetary change gear train's variants. *Dissertation*. Rijeka: Engineering Faculty, University of Rijeka, 2011 (in Croatian).
231. Troha, S., P. Petrov and D. Karaivanov. Regarding the optimization of coupled two-carrier planetary gears with two coupled and four external shafts. *Machinebuilding and Electrical Engineering*. 2009, Vol. 58, No. 1, pp. 49–55. ISSN 0025-455X.
232. Troha, S., R. Žigulić and D. Karaivanov. Kinematic operating modes of two-speed two-carrier planetary gear trains with four external shafts. *Transactions of FAMENA*. 2014, Vol. 38, No. 1, pp. 63–76. ISSN 1333-1124.
233. Tsai, S.-J., S.-Y. Chang and G.-L. Huang. An approach for analysis of load sharing in planetary gear drives with a floating sun gear. *Proceedings of the ASME International Design Engineering Technical Conferences & Computers and Information in Engineering Conference IDETC/CIE 2011*, Washington, DC [USA], Aug. 28–31, 2011. DETC2011-047600.
234. Tsai, S.-J., L.-C. Chang and C.-H. Huang. Design of cycloid planetary gear drives with tooth number difference of two. *VDI-Berichte*. 2017, Vol. 2294, No. 1, pp. 61–73. ISSN 0083–5580.
235. Tsyplakov, Y. *Bi-planetary Mechanisms*. Moscow: Mashinostroenie, 1988 (in Russian).
236. Tkachenko, V. *Planetary Mechanisms*. Kharkiv: Kharkiv Aviation Institute Publ., 2003 (in Russian).
237. VDI-Richtlinie 2157 Planetengetriebe – Begriffe, Symbole, Berechnungsgrundlagen, 2012.
238. VDI-Richtlinie 2737 Berechnung der Zahnfußtragfähigkeit von Innenverzahnungen mit Zahnkranzeinfluß, 2005.
239. Volkov, D. and A. Kraynev. *Planetary, Harmonic and Combined Transmissions of Construction and Road Machinery*. Moscow: Mashinostroenie, 1968 (in Russian).
240. Volkov, D. and A. Kraynev. *Transmissions of Construction and Road Machinery*. Moscow: Mashinostroenie, 1974 (in Russian).
241. Vonderschmidt, R.W. Zahnkräfte in geradverzahnten Planetengetrieben (Lastverteilung). *Dissertation*. Bochum: Ruhr-Universität, 1982.
242. Vriesen, J.W. Topographical tooth modifications in planetary gears for reduction of the noise excitation without load capacity loss. *VDI-Berichte*. 2017, Vol. 2294, No. 1, pp. 185–196. ISSN 0083-5580.
243. Vulgakov, E. *Theory of Involute Gear Wheels*. Moscow: Mashinostroenie, 1995 (in Russian).
244. Vulić, A. and J. Stefanović-Marinović. Design parameters for planetary gear transmissions optimization. *Proceedings of the 2nd International Conference "Power Transmissions'06"*, Novi Sad [Serbia & Montenegro], Apr. 25–26, 2006, pp. 137–142.

245. Wagner, H.T. Versuche zur Lastaufteilung und zum Breitentragen in geradverzahnten Planetenradgetrieben. *Dissertation*. Bochum: Ruhr-Universität, 1984.
246. Waldron K.J., G.L. Kinzel and S.K. Agrawal. *Kinematics, Dynamics, and Design of Machinery*, 3rd edition. Milton: Wiley-Blackwell, 2016, 718 p. ISBN 978-1-118-93328-2.
247. Wasfy, T.M. and M.L. Stark. Multibody dynamics model for predicting the vibration response and transient tooth loads for planetary gear systems. *Proceedings of the ASME International Design Engineering Technical Conferences & Computers and Information in Engineering Conference IDETC/CIE 2011*, Washington, DC [USA], Aug. 28–31, 2011, DETC2011-48814.
248. Willis, R. *Principles of Mechanism*. London: John W. Parker, 1841.
249. Winkelmann, L. Lastverteilung in Planetengetrieben. *Dissertation*. Bochum: Ruhr-Universität, 1987.
250. Winter, H., T. Hösel and W. Schmidt. Tragfähigkeit von Innenverzahnungen. Teil I: Untersuchungen zur Grübchentragfähigkeit. *Antriebstechnik*. 1988, No. 11, pp. 58–61.
251. Winter, H. and K. Michaelis. Untersuchungen zum Wärmehaushalt von Getrieben. *Antriebstechnik*. 1981, No. 1, pp. 70–74.
252. Winter, H. and B. Podlesnik. Zahnfedersteifigkeit von Stirnradpaaren. Teile 1, 2 und 3. *Antriebstechnik*. 1983, No. 3 and 5; 1984, No. 11.
253. Winter, H. and others. Tragfähigkeit von Innenverzahnungen. Teil II: Untersuchungen zur Zahnfußtragfähigkeit. *Antriebstechnik*. 1992, No. 3, pp. 94–98.
254. Wolf, A. *Die Grundgesetze der Umlaufgetriebe*. Braunschweig: Friedr. Vieweg und Sohn, 1958.
255. Wolfrom, U. Der Wirkungsgrad von Planetenrädergetrieben. *Werkstattechnik*. 1912, H. 12, pp. 615–617.
256. Xiang, C.-I. and others. Research on natural vibration characteristics and sensitivity for rotational and translation vibration of two-stage planetary gears. *VDI-Berichte*. 2013, Vol. 2199, No. 2, pp. 1079–1090. ISSN 0083-5560.
257. Yastrebov, V. 3K planetary gear trains with one-rim planet. *Vestnik Mashinostroeniya*. 1960, No. 3, pp. 17–20 (in Russian).
258. Yastrebov, V. Investigation of planetary gear trains with two internal meshes of one-rim planets. *Dissertation*. Uralsk: Uralsk Polytechnic University, 1971 (in Russian).
259. Zablonskiy, K. and I. Gorobets. *Planetary Gear Trains*. Kiev: Tekhnika, 1972 (in Russian).

Index

E

F

G

H

I

K

L

M

T

V

W